The Butterflies and Moths (Lepidoptera) of Kentucky: *An Annotated Checklist*

The Butterflies and Moths (Lepidoptera) of Kentucky: *An Annotated Checklist*

For my close friend and wonderful colleague Varley Wiedeman with many thanks and warmest regards.

Charlie
12-7-1999

Charles V. Covell Jr.

Charles van Orden Covell, Jr., Ph.D.
Professor of Biology
University of Louisville
Louisville, Kentucky 40292 USA

Kentucky State Nature Preserves Commission
Scientific and Technical Series Number 6
1999

Published in 1999 by: Kentucky State Nature Preserves Commission
801 Schenkel Lane
Frankfort, KY 40601–1403

It is the mission of the Kentucky State Nature Preserves Commission to protect Kentucky's natural heritage by: (1) identifying, acquiring, and managing natural areas that represent the best known occurrences of rare native species, natural communities, and significant natural features in a statewide nature preserves system; (2) working with others to protect biological diversity; and (3) educating Kentuckians as to the value and purpose of nature preserves and biodiversity.

An Equal Opportunity Employer M/F/D

Printed with state funds

Citation: Covell, C.V., Jr. 1999. The butterflies and moths (Lepidoptera) of Kentucky: an annotated checklist. Kentucky State Nature Preserves Commission Scientific and Technical Series 6:1–220.

ISSN 1526-4394
ISBN 0-9673646-5-5
Library of Congress Catalog Card Number 99-64169

DEDICATION

I dedicate this book to all the Kentucky Lepidopterists, past and present, whose contributions have been recorded in this work. In particular, I dedicate it to Loran D. Gibson, with whom I have spent many happy hours at moth lights throughout Kentucky, and whose individual new state records have been legion. I further dedicate it to the memory of the late James R. Merritt, whose enthusiastic pioneering study of Kentucky butterflies inspired me to undertake this project; and also to the memory of the late Carl C. Cornett, who gave significant collecting and curating assistance during his lifetime, and whose financial support for Kentucky entomology continues to this day.

PREFACE

The value of any state's published faunal or floral checklist is dependent on several factors. These include how well the state's natural resources have been studied, and how much of an effort the author made to accurately research study material, museum collections, and the published literature. Further, the extent of the annotations adds to the usefulness of the publication and transforms it from a mere list to a catalogue or atlas.

If the above criteria are any indication, I believe this publication by Charles Covell more than meets the test of a valuable state checklist. In fact this is probably the most complete such checklist published for any U.S. state.

Dr. Covell has tirelessly and persistently worked on this list for the Kentucky butterfly and moth fauna since his 1964 arrival in the state. His founding of the Society of Kentucky Lepidopterists and his encouragement of its membership, acknowledged in the checklist, were major factors in the thorough exploration of Kentucky's fauna. Moreover, Charlie has sought out the services of most national experts on butterflies and moths to identify or confirm the identification of the species found in Kentucky. Thus both the completeness and accuracy of the names in this list are exceptionally high.

The extent of any state list of Lepidoptera is determined by a combination of the actual extent of the state's fauna and the sampling efforts made to discover it (Powell 1995). For example, no amount of sampling effort would ever cause the species richness of a relatively depauperate state such as Washington (which has no published state checklist) to be greater than an exceptionally rich state such as Arizona or Texas, neither of which has an adequate published state checklist (Opler 1995).

Among the few states with published checklists the extent of the Kentucky fauna reported herein matches up well with Maine (slightly less than 3,000 in Brower 1974, 1983, 1984), but is exceeded by that of Florida (Kimball 1965). Forbes (1923, 1948, 1954, 1960) contained 2,439 species from New York (Borror et al. 1989).

This checklist should serve as an example to those who would contemplate the effort needed to publish such a monumental work. It is really one of the major achievements of an outstanding career in science education, research, and public outreach.

Paul A. Opler, Ph.D.
Department of Bioagricultural Sciences
Colorado State University
Fort Collins, Colorado 80523

ACKNOWLEDGMENTS

Financial support for this work has been provided through grants to enable field, lab, and clerical work over the years. I am indebted to former deans of the College of Arts and Sciences at the University of Louisville – Richard L. Barber, A. J. Slavin, and Lois S. Cronholm – for such support, and also to the Arts and Science Research Committees over the years for numerous small grants that have added up to well over $10,000. I am also grateful for additional support to vice presidents of the University of Louisville Graduate Programs and Research over the years: John A. Dillon, Jr., X.J. Mussachia, and Patrick W. Flanagan. Support has also been provided by the Kentucky Academy of Science through the Marcia Athey Fund and by the Southern Regional Education Board for two travel grants for research visits to the National Museum of Natural History in Washington, D.C. The East Kentucky Power Cooperatives, Incorporated, through Mr. Jeff Hohman, supported biodiversity studies in transmission corridors in the Daniel Boone National Forest (1995–96), providing many county records we were able to include here. The Society of Kentucky Lepidopterists has also provided several small grants for clerical assistance in developing this manuscript. To all these organizations and agencies I extend my sincere gratitude.

The late Carl C. Cornett assisted me with the University of Louisville Collection from 1970 until his death in 1982 by curating material, recording data, designing and making equipment, and collecting moths with his black-light at a variety of localities, mostly state parks. He was a major leader in the Kentucky Society of Natural History and "Mr. Kentucky State Parks" with respect to our Lepidoptera survey. However, I am most indebted to Mr. Cornett for establishing the Carl Cornett Endowment which has supported the development of this and other entomological projects at the University of Louisville since 1982.

It has been appropriate to the focus of this study that the Kentucky State Nature Preserves Commission be involved. I am extremely grateful to Director Donald S. Dott, Jr., for publishing this work, to former Directors Richard R. Hannan and Robert McCance, Jr., for their support, and to staff members Joyce Bender, Ronald Cicerello, Brainard L. Palmer-Ball, Jr., and especially Ellis L. Laudermilk, for their encouragement and editorial assistance.

A few people have made major contributions of information to this study, including first and/or only state records of species. The late Annette F. Braun hosted several visits which I made to her home in Ohio in 1972–74, where she provided data on Kentucky microlepidoptera she had collected or reared from the earliest years of this century. We had several memorable taping sessions together, in which scientific information was interwoven with fascinating tales of her field trips with her sister, Lucy, during times long gone by in eastern Kentucky.

The late Burt L. Monroe, Jr., collected butterflies and other insects in Jefferson County since his boyhood in the 1940's, and made an extensive study of the county butterfly fauna for an undergraduate thesis (Monroe 1952). He also supported this work in many ways during his 23-year tenure as Chair of the Biology Department at the University of Louisville. Burt's son, Mark S. Monroe, joined his father on many collecting trips during the past few years to record butterflies in most of Kentucky's counties, adding greatly to our distributional database.

James R. Merritt, late Professor and former Dean of the University of Louisville School of Law, has also contributed many records to the Jefferson County butterfly and moth fauna, including his published butterfly checklist (Merritt 1948). Much of his collection has been incorporated into the Lovell Insect Museum, University of Louisville.

The late Siegfried Scholz, who moved from Hamburg, Germany to Kentucky in 1969, made an intensive collection of moths at black-light in the Waverly Park area of Valley Station. His collection of over 20,000 specimens was purchased by the University of Louisville. His work represents the most intensive collecting in a single locality for moths ever carried out in the state. He contributed many unique records to this list, as well as data in depth for many common species.

The late Robert V. Gregg collected many moths in Goshen, Oldham County, in his latter years, and donated many hours to assisting me in curating the

University of Louisville Collection. Most of the records from Nelson County, and many from Harlan County, were the result of extensive collecting by the late Gregory Florence, who also assisted in developing the collections at both the University of Louisville and the University of Kentucky during his college years. He was joined in the field by his wife, Jevetta, during the latter part of his tragically short career.

While a student at the University of Louisville, Alan J. Brownell collected light and Malaise trap samples for Lepidoptera in Bernheim Arboretum and Research Forest, Bullitt County, and recorded over 600 moth species there. He added many Valley Station records as well, and many of his captures were new state records. He also carried out rearing studies of Lepidoptera feeding on *Celtis* spp. While the results of his study were never published, his records have been included here through his generosity. His Bernheim Forest survey was augmented by Amy L. Harlow in 1990.

Extensive collecting over the past 15 years has been carried out by Richard A. Henderson, who has concentrated on the fauna of Otter Creek Park, Meade County, with the kind cooperation of Park Director, John Rowe. Henderson has contributed many new state records, contributing specifically in the *Papaipema* fauna. He also co-founded the Society of Kentucky Lepidopterists with me.

Extensive collecting by Loran D. Gibson has made the most significant contributions of new and often sole state records, particularly of micro-lepidoptera. Some of his most intensive collecting has been at Big Bone Lick State Park in Boone County, Markland Dam in Gallatin County, and also his late "Mom's back porch" in Owsley County, three miles southeast of Booneville. His material is always beautifully spread and usually excellent in condition, winning praise from professional as well as amateur colleagues. He has recently been joined in the field by Donald J. Wright, who has produced results of like quality.

A former resident of Lexington, John S. Nordin, collected heavily in several localities, notably in Red River Gorge (Menifee County), Woodford County, along the Kentucky River, and in Fayette County. A number of new and sole state records bear his collection labels, and he donated many specimens to the University of Louisville Lovell Insect Museum.

William R. Black, Jr., of Paducah, has provided many records from McCracken County. He and his wife Nancy have hosted several memorable field trips made to the Ray Black farm and nearby localities by the Society of Kentucky Lepidopterists, and his three sons have also helped him collect. While Scoutmaster of Troop 1 Paducah, Boy Scouts of America, he encouraged scouts to participate in collecting and studying Lepidoptera. He collected the only known Kentucky specimen of a rare migrant from the tropics, *Thysania zenobia*, and discovered and documented breeding colonies of the gulf fritillary (*Agraulis vanillae*) and the great purple hairstreak (*Atlides halesus*). His enthusiasm for this work has been contagious.

Leroy C. Koehn has collected extensively in many parts of Kentucky, and has provided several state records, particularly from Butcher Hollow in Whitley County. His knack for finding new and productive collecting spots, and his tireless pursuit of collecting goals have resulted in significant contributions. His improved designs in bait and light traps have yielded many interesting records among moths as well. He has been one of the most faithful supporters of this project.

Specimens of Lepidoptera were made available to me at several other colleges and universities in Kentucky through the kindness of the following: W. Blaine Early III, Cumberland College; Paul H. Freytag, Department of Entomology, University of Kentucky; Hunter Hancock, Murray State University; Richard Kessler, Campbellsville University; Gerald L. DeMoss, Morehead State University; Mary McGlasson, Eastern Kentucky University; and Herbert E. Shadowen, Western Kentucky University. Indirectly, many of their students have contributed to our knowledge of the state's Lepidoptera fauna through specimens collected for classes and deposited at the various Kentucky colleges and universities.

Some of the amateur collectors and professional colleagues who have contributed data on either butterflies, moths, or both over the years include the following: James K. Adams, Roman Anderson, George Balogh, Doug Barnett, the late Brett Barrett, Mike Beiser, Eric Belcher, Jeanne Bennett, James

Bess, the late Robert Bosserman, Paul H. Brand, Laura S. Burford, John V. Calhoun, Richard Cassell, Thomas W. Carr, Keith Collins, Carl Cook, Alice S. Covell, Charles V. Covell III, Elizabeth Katherine Covell, Robert B. Covell, Dennis A. Currutt, Zack Dembo, C. Jack Dempwolf, Douglas J. Durbin, John Enz, James K. Ettman, Richard Fellows, Douglas C. Ferguson, David and Geri Flaim, Adam Flora, Jevetta B. Florence, Paul A. Florence, Jeremy Fraser, R.S. Funk, Lee Guidry, Amy L. Harlow, Lana Hays, Chris Herin, Leonard D. Hilley II, Christa Anderson Hoffman, Jeffrey D. Hooper, Jessica Hutti, John A. Hyatt, Leroy Isaacs, the late Tom Johnson, Ron King, John P. Knudsen, Herbert H. Kulman, Michael Lacki, Howard Ladd, Gene Leppert, Ronald Leuschner, Sean Lube, Vincent P. Lucas, John R. MacGregor, Logan Mahaffey, Michael S. Mahoney, Monica C. Mahoney, Lindsey Marean, Robert A. Mattingly, H. Elliott McClure, Michael L. McInnis, J.A. Merkle, Eric H. Metzler, Mark S. Monroe, Michael M. Moore, Barry S. Nichols, Gerald Nordin, Paul A. Opler, Brainard L. Palmer-Ball, Jr., Thomson Paris, Nathan Payne, Troy Payne, John Peacock, Paul Pfenninger, Michael Plagens, Daniel Potter, Ronald Rockwell, Kilian Roever, Patricia Liles Rudicil, Edward Sadler, Rudy Scheibner, Amos Showalter, John Shuey, Philip Sisto, Ray E. Stanford, the late Gerald Straley, Donald Tangren, Lee Townsend, Sr., Jim Vargo, Warren H. Wagner, Reed A. Watkins, Charles Watson, J.B. and Lark L. Wood, LaKeith Wood, Charles Wright, D.M. Wright, Kenneth V. Yeargan, and Roger A. Zebold.

Kentucky botanists have lent great assistance by guiding us to foodplant colonies, thus assisting us in finding certain Lepidoptera. I am indebted especially to the late Raymond Athey, William S. Davis, Max Medley, Harry K. Woodward, and David M. Wright for their identification of plants and botanical information that has led to fruitful habitats.

Many specialists and museum curators in various groups of Lepidoptera have identified specimens for this survey, or have sent Kentucky records of species they have treated in their research. The lepidopterists at the Department of Entomology, National Museum of Natural History, Washington, D.C., have for over 25 years assisted me immeasurably by identifying many specimens. These include John M. Burns, the late J.F. Gates Clarke, Don R. Davis, Donald W. Duckworth, Douglas C. Ferguson, the late William D. Field, Ronald W. Hodges, Michael Pogue, Robert W. Poole, Robert K. Robbins, Maria Alma Solis, and the late Edward L. Todd. Specialists at the Biosystematics Research Institute, Agriculture Canada in Ottawa who have provided similar assistance include: Klaus B. Bolte, P.T. Dang, David F. Hardwick, J. Donald Lafontaine, Bernard Landry, Jean-Francois Landry, W.C. McGuffin, Eugene G. Munroe, and Akira Mutuura. Jane Ruffin at The Academy of Natural Sciences in Philadelphia, and Allen Young and Susan Borkin, Milwaukee Public Museum, have also provided assistance. Other professional, semiprofessional, and experienced amateur lepidopterists who have been most helpful in this regard include H. David Baggett, Jacksonville, Florida; the late Auburn E. Brower, Augusta, Maine; Richard L. Brown, Mississippi Entomological Museum, Mississippi State University, Starkville, Mississippi; Everett D. Cashatt, Illinois State Museum, Springfield, Illinois; the late Harry K. Clench and John Rawlins, Carnegie Museum, Pittsburgh, Pennsylvania; Julian P. Donahue and Brian Harris, Natural History Museum of Los Angeles County, California; Thomas Eichlin, California Department of Food and Agriculture, Sacramento, California; John G. Franclemont, Cornell University, Ithaca, New York; George L. Godfrey, formerly of the Illinois Natural History Survey, Urbana, Illinois; J. Richard Heitzman, Independence, Missouri; the late Roger L. Heitzman, Washington, D.C.; John B. Heppner, Florida State Collection of Arthropods, Gainesville, Florida; Jean-Francois Landry, Centre for Land and Biological Resources Research, Agriculture Canada, Ottawa; Tim McCabe, New York State Museum, Albany, New York; Eric H. Metzler, Columbus, Ohio; William E. Miller, University of Minnesota, St. Paul, Minnesota; Herbert H. Neunzig, North Carolina State University, Raleigh, North Carolina; Daniel Potter, University of Kentucky, Lexington, Kentucky; Jerry A. Powell, University of California, Berkeley; Eric Quinter and Frederick H. Rindge, American Museum of Natural History, New York, New York; Dale F. Schweitzer, The Nature Conservancy, Boston, Massachusetts; John Shuey, Indiana Chapter, The Nature Conservancy, Indianapolis, Indiana; Fred W. Stehr, Michigan State University; and Barry Wright, Nova Scotia Museum, Halifax, Nova Scotia.

I wish to acknowledge special thanks to the late Burt L. Monroe, Jr., former Chairman of the Department of Biology, University of Louisville. During his tenure in that office he provided support for both survey activities and development of the insect collection, not to mention many butterfly records from around the Commonwealth.

I am greatly indebted to the following for assisting me in curating over 100,000 Lepidoptera specimens in the University of Louisville Collection, recording data, preparing the manuscript, and carrying out many related logistical chores: Forrest Bledsoe, Alan J. Brownell, Rozenna B. Carr, David Dove, W. Blaine Early III, Nancy Esarey, Mary Anne Evans, Linda Finch, the late Gregory Florence, Susan E. Heady, Martha Heeter, Dariush Hosseini, Leah Kohn, Gene Leppert, Carol McClure, Martha Neudecker, Barry Nichols, Troy Payne, Katherine Potter, Susan Reigler, Lillian Rossini Stevenson, Lisa Summers, Kelly Thompson, James P. Tuttle, Brenda Sue Vance, and Kathy Zinn.

Special thanks go to Rozenna B. Carr, Nancy J. Esarey and Stephen Roszell for many hours on early drafts of the manuscipt. I am particularly grateful to my good friend and colleague Varley E. Wiedeman for many hours of work preparing this manuscript.

I am grateful to James R. Merritt, William E. Miller, and the late Burt L. Monroe, Jr., for reviewing portions of this manuscript in its early stages, and Ellis L. Laudermilk, Eric Metzler and Paul A. Opler for reviewing the later versions.

Finally I wish to thank my family for their varied and unstinting assistance: My late father, the Reverend Charles V. Covell, and my mother, Katherine S. Covell, Hendersonville, North Carolina; my wife's parents, the late Captain Robert M. Barnes (United States Navy) and the late Elizabeth G. Barnes; my children Chuck, Robert, and Elizabeth Katherine; and most of all, my wife Betty, whose enabling support and devoted partnership these many years have provided the logistical basis for the completion of this project.

TABLE OF CONTENTS

INTRODUCTION

This work is an annotated list of all the described, and a small number of undescribed, species of Lepidoptera (moths and butterflies) known to have been collected or otherwise authoritatively recorded in the Commonwealth of Kentucky. It is a result of a third of a century of work by me and many colleagues. The main purpose of the work is to establish a scientific record of these species for the Bluegrass State, giving biogeographic data to assist scientists in knowing where and when these species occur within its borders. With the present emphasis on biodiversity studies and habitat censusing, this book provides a baseline of information for many habitat types and specific localities in Kentucky. A special effort has been made to document the occurrence of Lepidoptera in parks and nature reserves so that this knowledge may assist in the planning and monitoring of the ecology of these reserves, particularly if the gypsy moth (*Lymantria dispar*) causes widespread defoliation and prompts drastic control measures in Kentucky forests.

When this study began, emphasis among lepidopterists was mostly upon collecting and identification. Now the spectrum has broadened as butterfly "zoos" (e.g., Butterfly World in Florida, the Day Butterfly Center at Callaway Gardens, Georgia, and the Cincinnati Zoo) and the nationwide interest in butterfly gardening have popularized attracting, watching, and photographing living butterflies. In 1990 the Kentucky General Assembly designated the viceroy (*Limenitis archippus*) as Kentucky's state butterfly, attesting to this new interest. A permit is now required to collect in state parks, and Lepidoptera are increasingly appreciated by naturalists and interpretive guides at parks and nature centers.

Thus the present work will assist such agencies as the Kentucky State Nature Preserves Commission, the Kentucky Department of Parks, the Kentucky Department of Fish and Wildlife Resources, and organizations like The Nature Conservancy, the Kentucky Audubon Council, and the Kentucky Society of Natural History, to help preserve critical habitats essential for survival of Kentucky's rich flora and fauna, of which Lepidoptera comprise a major portion. While responsible collecting is not at present considered detrimental to any Kentucky species of Lepidoptera, and none are listed by the United States Fish and Wildlife Service as Endangered or Threatened, we do face loss of critical habitats. By preserving unique prairie, mountain hollow, and wetland habitats, the survival of our present Lepidoptera fauna, with few exceptions, can be assured.

While many Lepidoptera species, mostly moths, occurring in Kentucky have not yet been discovered (and so are not included in this book), I believe it represents the majority – about 85%. Records for many species included here represent notable extensions of their recorded ranges. Over a dozen unnamed species, new to science, are in the process of being described and named. Additions to this list will be added by means of supplements when sufficient information is available to warrant their publication. I would like to hear from anyone with information that would increase our understanding and knowledge of Kentucky's diverse and fascinating lepidopteran fauna.

THE HISTORY OF KENTUCKY LEPIDOPTEROLOGY

While this work is the first to bring together all known records of both butterflies and moths for Kentucky, a number of publications exist that summarize or include Kentucky records for species of Lepidoptera. Some, particularly those of V.T. Chambers and A.F. Braun, describe new species based on specimens collected as adults or reared from immature stages collected in Kentucky. Thus Kentucky serves as the type locality for a number of taxa, primarily in the families of small moths, generally known as "microlepidoptera." Following are some high points in the history of lepidopterological activities in the Commonwealth.

The early history of Kentucky lepidopterology is even more obscure than that of ornithology as traced by Mengel (1965). Among the first educated naturalists in Kentucky were John James Audubon (1770–1851) and Constantine S. Rafinesque (1783–1840). The former resided for some time in Louisville and Henderson, and the latter was a professor of natural history and other subjects at what is now Transylvania University in Lexington, Kentucky. While both no doubt observed butterflies and moths while in the Bluegrass State, I have found no record of Lepidoptera collections or written references to such. Rafinesque is noted mainly for his work with plants and fishes, but did publish articles on such insect groups as aphids, seventeen-year cicadas, and ants (Osborn 1937). Audubon included insects in his bird paintings, but these were often added by assistant artists, such as the gifted Maria Martin of Charleston, South Carolina (Ford 1952). Ford's book records an amusing incident involving the two when Rafinesque paid Audubon a visit in Henderson. A large beetle, probably *Dynastes tityus*, was mentioned, but no Lepidoptera.

A noted British lepidopterist named Edward Doubleday visited Kentucky briefly in 1837. He collected in various American localities in 1837–38 with a companion, Robert Foster (Calhoun 1989). The two disembarked from a riverboat at Cincinnati in September, 1837. Doubleday (1838) wrote "My first excursion was to the shore of the Ohio. I crossed [to the Kentucky side] on the ferry to where Mr. Bullock once lived; then I turned off by a little brook into the woods. The trees were large; one hundred feet high, and eighteen feet in circumference, was the size of the majority. There was no brush-wood; but in the open spaces there grew a great abundance of *Veronica proaelta*; I am not sure that is the right specific name: on its flowers were swarms of *D. archippus* and some few *Cynthia huntera*... Almost all the insects I took at Cincinnati were captured in this one walk." He doubtless visited and made collections in part of Kenton or Campbell County, and noted the butterflies now named *Danaus plexippus* and *Vanessa virginiensis*. Later he stopped at Louisville, where he noted that "many butterflies were also out." The next recorded activity of lepidopterists in Kentucky was not until 1873. Lepidoptera specimens are still in the Museum of Comparative Zoology, Harvard University, with the simple printed labels "Kentucky Sanborn." These were collected during a visit to Kentucky that summer by Alpheus Spring Packard who was studying cave fauna for a book later published on the subject. With him on this expedition was Francis G. Sanborn, a consulting naturalist from Andover, Massachusetts, whom Packard (1876) indicated as assisting N.S. Shaler, Director of the Kentucky Geological Survey. Material at Harvard constitutes the earliest substantial collection of medium-sized moths from Kentucky about which I know. The Geometridae from this collection provided information utilized by Packard (1876) in his monograph of that family.

Kentucky's first notable "native" lepidopterist was Vactor Toucey Chambers (1831–1883), a lawyer that resided in Covington. In 1871 he began what soon became a flood of published life histories and descriptions of new species in several families of microlepidoptera ("Tineina"). A large number of these new species resulted from collections made in and around Covington, Kenton County, particularly in Linden Grove Cemetery. Type locality designations in his publications are scanty and sometimes absent altogether. However, pin data from specimens in the Museum of Comparative Zoology have recently been reviewed by Miller and Hodges (1990), who have clarified some of the confusion regarding Chambers' type localities. While many species names attributed to Chambers have now been synonymized, many others still stand as valid recognized species, comprising a substantial portion of the American microlepidoptera fauna.

The second major contributor of new species of microlepidoptera to use Kentucky material extensively was Annette Frances Braun (1884–1978) of Cincinnati, Ohio. The earliest of her specimens that I have seen bear labels for the year 1900. She published many works between her revision of the genus *Lithocolletis* (now *Phyllonorycter*, Gracillariidae) in 1908, and her revision of the North American Tischeriidae in 1972, 64 years later! While Kentucky provided few type localities for her new species, she did rear many "micros" from Kentucky localities. She also provided much valuable information directly to me on species she had collected or observed while tramping the forests of eastern Kentucky with her botanist sister, Emma Lucy Braun, over the span of many years. The bulk of the Braun Collection is in The Academy of Natural Sciences Collection in Philadelphia, Pennsylvania (ANSP), but some is scattered in other collections, including 200 representative microlepidoptera she donated to the University of Louisville Collection in 1972. Lepidoptera data from Braun's herbarium of mined leaves collection, also at the ANSP, were also a source of valuable records.

Dr. Harrison Garman (1858–1944) came to Kentucky from Illinois in 1889 as the first entomologist at the Kentucky Agricultural Experiment Station at the University of Kentucky. He served in that capacity until 1929, and left a numbered listing of data for his collections of insects from around the state, plus those submitted to him for identification. Entries in the code book (Garman 1889–1944) are numbered and dated, beginning on July 16, 1889, and ending on May 23, 1944; so the code book spans his retirement period as well as his 40 years of official service. During the present study, many specimens of Lepidoptera in the University of Kentucky Collection that bear Garman number labels corresponding to these entries (but no other data labels) were identified. Data from many of these records are included here. Full data labels have been affixed to these specimens, and they have been returned to the University of Kentucky Collection. Much valuable information covering this early period of Kentucky entomology has been garnered from these records. Garman's code book entries indicate that his concern was chiefly with economically important insects of all kinds: their identification, life cycles, and types and extent of damage to food and fiber throughout Kentucky. General collecting trips, light and "oil" trap collections are also recorded there.

Other scientists also added Lepidoptera to the University of Kentucky Collection during this period, notably Mary L. Didlake, H.H. Jewett, Walter A. Price, and L.H. Ray. Lee H. Townsend, Sr., contributed significantly from the mid-1930's until his recent retirement. Add to that the many incidental additions to the Experiment Station Collection by researchers, students, and the public over nearly a century, and we have a major source of data housed there, at present under the curation of Dr. Mike Sharkey. The previous curator, Dr. Paul H. Freytag, added many Lepidoptera to the collection, much of it collected by means of Malaise traps. Butterflies collected by this method are tabulated in Covell and Freytag (1979).

In 1934 H. Elliott McClure made light trap collections in Hart County in the vicinity of Horse Cave. He kindly provided jars of unspread specimens from these collections to the University of Louisville Collection in 1990 for this survey, and moth data from these collections are included here.

The first annotated list of any portion of the Kentucky Lepidoptera fauna was Hattie Warner's *Kentucky Butterflies* (Warner 1894). A student at Kentucky State College (as the University of Kentucky was then called), she based her list of 60 species on specimens in the Experiment Station Collection. The following year she republished the list, this time through the university, adding one species (*Erynnis juvenalis*) for a total of 61 (Warner 1895). Frank Merrill Wheat published a list as a Bachelor of Arts thesis (Wheat "1908" [1909]) which included 65 species. Wheat ("1908" [1909]) did not mention Warner's (1894, 1895) papers, and information included in their works leaves uncertainty as to whether some of the same specimens were used by both of them. While both leave out dates and refer to occurrence of species (common to rare), Wheat ("1908" [1909]) was more vague, only occasionally mentioning a specific locality (usually just "Lexington"). Warner (1894, 1895) mentioned a number of localities around Kentucky as sources, including East Cairo, Ballard County, for her one record of *Speyeria diana*. Neither author mentioned Dr. Garman in acknowledgments, and Warner (1894) implied that there was little interest in butterfly collecting apparent from the material present at the Experiment Station.

The first county faunal study of butterflies is an unpublished undergraduate thesis by Weniger (1946), who listed 42 species taken in various parts of Jefferson County, divided into five habitat types. He seems to have been accurate in identifying most of the butterflies listed; but his misidentification of *Speyeria cybele* as *S. aphrodite* casts doubt on the reliability of his single record of *Phyciodes gorgone* (although that species has subsequently been recorded from elsewhere in the state).

It was the Jefferson County butterfly list of the late James R. Merritt, however, that was the first county butterfly survey to be published (Merritt 1948). Merritt reported 64 species, combining his own field studies of 1947 and 1948 with the information reported by Weniger (1946). He added considerably to the number of species Weniger (1946) had reported, but failed to find four species: *Agraulis vanillae*, *Speyeria aphrodite*, *Charidryas gorgone*, and *Calycopis cecrops*.

In 1952 the late Burt L. Monroe, Jr., of Anchorage, Jefferson County, submitted to the University of Louisville Biology Department, an undergraduate senior thesis entitled *Butterflies of the Louisville Region of Kentucky* (Monroe 1952). This unpublished study encompassed Bullitt, Jefferson, Meade, and Oldham counties. Monroe built upon the records of Weniger (1946) and Merritt (1948), adding his own plus those from other collectors such as Paxton Gibbs, Jerry Smith, Robert Steilberg and Dr. David A. Young. Monroe (1952) listed seven additions to the area record: *Urbanus proteus*, *Lerema accius*, *Hesperia metea*, *Amblyscirtes vialis*, *Erynnis baptisiae*, *Glaucopsyche lygdamus*, and *Polygonia faunus smythi*. In addition, he included the record of Kentucky's first wild-caught hybrid *Limenitis archippus* x *L. arthemis astyanax*. This latter record, from the Falls of the Ohio, was the only portion of the work to be formally published (Monroe 1953).

During the late 1940's and 1950's there were others active in lepidopterology in Kentucky. In 1952 James M. Unseld, Jr., of Gravel Switch, Marion County, assumed editorship of the monthly mimeographed newsletter of the Moth and Butterfly Club (*Club Notes*), and began writing short commentaries on Kentucky butterflies. Additional information was contributed by another member, Howard O. Ladd of Elizabethtown, Hardin County. Unfortunately, I have not located a complete set of this newsletter, so some of their contributions could not be included in the present work. Ladd (1952) reported seeing a small Kentucky collection of a person he did not name, in which were Kentucky specimens of "*Terias mexicana* and *Terias flava*," as well as *Zerene cesonia* and *Agraulis vanillae*. Since *Eurema mexicana* has been found as a stray in Illinois (Opler and Krizek 1984), it is possible that this record is valid, though the lack of precise data is unfortunate. The "*Terias flava*" is merely the yellow morph of *E. nicippe,* a species sometimes "abundant" in parts of Kentucky (Holland 1931). Ladd (1953) also reported capture of *Kricogonia lyside*, presumably at Elizabethtown. He mentions other interesting captures too, particularly *Nathalis iole*. Ladd's (1954) *Butterflies of Kentucky and Their Prevalence* listed 48 species (excluding skippers), and indicated "C" for "casual," "R" for "rare," and "P" for "plentiful at times." The only questionable identification was *Speyeria atlantis*, which I believe must have been misdetermined, even though he indicated it as "VR" (very rare). Unseld also made observations in August 1953 (J.M. Unseld, Jr., *pers. comm.*) that "The weather has been rather warm and the commonest specimen is *Pieris protodice*, and *Enodia creola* is now out and can be found in the cool forest." Unseld and Ladd apparently had no contact with Merritt, Monroe, and other Louisville area collectors during this period, and apparently never joined the Lepidopterists' Society, which is the major organization uniting those in all phases of lepidopterology. Mr. Unseld left Kentucky for the Korean War in 1955, and did not resume collecting afterwards. He recently donated much of what remained of his collection to the University of Louisville.

James A. Shields made a collection mostly from Sullivan, Union County, stored in home-made Riker mounts. This collection was given to Murray State University, and was stored in a closet there until Dr. Hunter Hancock donated it to the University of Louisville Collection. Fortunately, typed collection data labels had been included with the specimens, and some (i.e., several *Agraulis vanillae)* were in good enough condition to pin and retain. His material was mostly collected in 1954.

Also active during this period were Carl Cook of Crailhope, Green County, and John A. Bishop, M.D., of Jeffersontown, Jefferson County. They were charter members of the Lepidopterists' Society when it was formed in 1947. Cook (1948) reported the first Kentucky capture of *Erora laetus* along with records of several other butterfly species. During that general period, Cook collected many moths and butterflies in that area, some of which have been deposited in major museum collections such as the National Museum of Natural History. Data from some of his captures are included in this work. I visited Dr. Bishop in the 1960's and learned that he had long been inactive as a collector, and that his small collection had been mostly destroyed by pests.

The University of Louisville's insect collection apparently began in 1936 with the arrival of the late Dr. David A. Young, a specialist in leafhoppers. He left some Lepidoptera after moving to the North Carolina State University, in Raleigh, where he spent the remainder of his career. The late Theodore Spilman, a University of Louisville biology student, curated the collection briefly in the 1950's, and later became a coleopterist at the National Museum of Natural History, specializing in the beetle family Tenebrionidae. Next came another leafhopper specialist, Dr. Paul Christian, who was an entomologist in the Biology Department from 1958 to 1962, when he departed to head the Biology Department at Bethel College in Minnesota. Between 1962 and 1964, the collection, housed in 40 Cornell drawer cabinets, was moved among several buildings without proper supervision, resulting in some damage to specimens. Also, some Lepidoptera that had been donated to the university by J.R. Merritt apparently disappeared. There were about 1,500 specimens of Lepidoptera in the collection in August, 1964, when I assumed the position of entomologist at the University of Louisville. The beginnings of my own survey efforts began in July 1959, when I was living in Norfolk, Virginia, and made my first visit to the Bluegrass State by car over U.S. 460, enroute to Fort Knox for annual summer camp with the United States Army Reserve. On the night of July 29, I entered Kentucky in Pike County, and stopped to collect some moths attracted to lights at a drive-in restaurant along the highway in Pikeville. That same day I had made contact in Blacksburg, Virginia, with Entomology Department Head Dr. James M. Grayson, who influenced me to attend graduate school in entomology (1960–64). That in turn led to my coming in 1964 to the University of Louisville.

After my arrival at the university, I began building the collection through my own collecting activities, and through donations and occasional purchases. The Lovell Insect Collection (named for the late faculty member Harvey B. Lovell) grew from about 28,000 specimens in 1964 to 70,000 in 1968, of which about 7,000 were Lepidoptera (Covell 1968). A recent count shows the collection to contain nearly 250,000 specimens, with Lepidoptera comprising approximately fifty percent.

A county-by-county survey through the state was initiated to collect butterflies, and in some localities, moths. On July 13, 1974, my elder son Chuck and I collected in Allen County, the last to be visited in my program of recording at least five butterfly species in each of Kentucky's 120 counties. Some localities, particularly those close to Louisville, were visited regularly, but many were represented by only a single stop.

In 1970 Mr. Carl Cornett, a retired personnel manager for Reynolds Aluminum Company and amateur naturalist, joined me as a volunteer assistant in the Lepidoptera survey. He visited many localities in Kentucky, especially state resort parks, and put out black-lights to collect moths for us. Later, he employed light traps. He brought in a vast amount of material from which we were able to identify many new state records, plus provide baseline date for a number of the parks and reserves. His substantial contributions continued after his death in 1982 in that he bequeathed an endowment for the study of Kentucky insects at the University of Louisville.

On November 23, 1974, a small group of interested friends convened to found the Society of Kentucky Lepidopterists, whose purpose is to promote knowledge and fellowship among lepidopterists, and to provide records of Kentucky moths and butterflies for the survey. Since that year we have had many group collecting trips ranging from Big Black Mountain in the east to Hickman in the west. Members of this group have brought in many new state records of both butterflies and moths, and we are still getting identified some of the material collected in years past. The Society now numbers about 170 members, and new additions to the membership are welcome and can be obtained by contacting the author.

A number of people from 1974 to the present have contributed greatly to the records of Kentucky Lepidoptera, and their names will appear often in the following species accounts. Several also served as student assistants in the Lovell Insect Collection. The contributions of these valued colleagues are detailed more fully in Acknowledgments.

In 1965 a card file was begun in which were recorded localities, capture dates, collector's names, and other pertinent information on all Lepidoptera collected by amateur and professional lepidopterists. Literature references to species were also recorded. In order to stimulate a response by anyone with knowledge of Kentucky butterflies, a preliminary checklist of Kentucky butterflies was published (Covell 1974), with a brief summary of previously published lists and notes. Covell (1974) listed 123 butterfly species, and indicated the total known Lepidoptera fauna as "more than 1,350 species."

Short papers dealing with Kentucky Lepidoptera during the 1960's into the 1980's include the following: publication of several moth records (Donahue 1967); a short account of 30 butterfly species sampled in the Cumberland River floodplain prior to filling of Lake Barkley (Covell 1969); first report of *Amblyscirtes belli* in Kentucky (Henderson 1976); a description of a new all-black hindwing aberration of *Catocala micronympha* (Covell 1978); the addition of six new state butterfly records (Covell et al. 1979); an analysis of butterfly collections in Malaise traps (Covell and Freytag 1979); the first Kentucky record of *Poanes yehl* (Covell 1981); papers dealing with the biology of *Celastrina nigra* (then called *ebenina*) with some Kentucky information (Wagner and Showalter 1976, Wagner and Mellichamp 1978); and numerous Kentucky references and illustrated moths in Covell (1984).

Additional records of Kentucky Lepidoptera appeared in the annual *Field Season Summary* in publications of the Lepidopterists' Society beginning with notes from Cook (1948). Some Kentucky data have been published there in most years since then. A second source has been issues of *Kentucky Lepidopterist*, a quarterly newsletter published by the Society of Kentucky Lepidopterists from 1975 to the present. A third has been the *Kentucky Cooperative Economic Insect Report* and *Kentucky Cooperative Plant Pest Report* series from the Kentucky Agricultural Experiment Station during the 1970's, edited by Douglas Barnett. And finally, I have conducted 15 Xerces Society July Fourth Butterfly Counts in the vicinity of the Horner Wildlife Sanctuary, Oldham County (conducted between 1975–1998). B. L., Jr., and M. S. Monroe began a count in 1990 for the Red River Gorge Geological Area, Menifee, Powell, and Wolfe counties (now sponsored by the North American Butterfly Association).

We have supplied data on Kentucky Lepidoptera to writers and researchers for many years. Thus new state records, particularly of moths, have appeared in a number of books and articles, including fascicles of *The Moths of America North of Mexico*. Acknowledgments of such contributed information appear in Rindge (1971), Munroe (1976), Ferguson (1978, 1985a), Hodges (1978, 1986), Miller (1979, 1983a, 1985), Lafontaine (1987), Neunzig (1990), and Opler and Krizek (1984).

This book treats 2,388 taxa in 65 families of Lepidoptera. The work of adding to the state records of Lepidoptera continues as you read these pages. Future records may extend the number of known Kentucky species by as many as 300–400.

THE PHYSIOGRAPHY OF KENTUCKY

Some information about Kentucky as a habitat for moths and butterflies is in order. Mengel (1965) summarized Kentucky features from the standpoint of an ornithologist, and Burr and Warren (1986) provided a fine description, with an emphasis on aquatic habitats. The information below is modified from these sources.

The 105,149 square km (40,598 square mi) of the Bluegrass State are bounded irregularly by the Ohio River on the north, the crest of Pine Mountain and other mountains on the east, a nearly straight border with Tennessee on the south, and a short span of the Mississippi River on the west. It extends from approximately 36° 30' to 39° 09' North latitude, and from 81° 58' to 89° 35' West longitude. The state is roughly divided into three provinces, east to west (Fig.1): I. The Coastal Plain (subdivided as Mississippi Alluvial Plain, Eastern Gulf Coastal Plain, and Tennessee River Plain); II. The Interior Low Plateaus (subdivided as Highland Rim, Shawnee Hills,

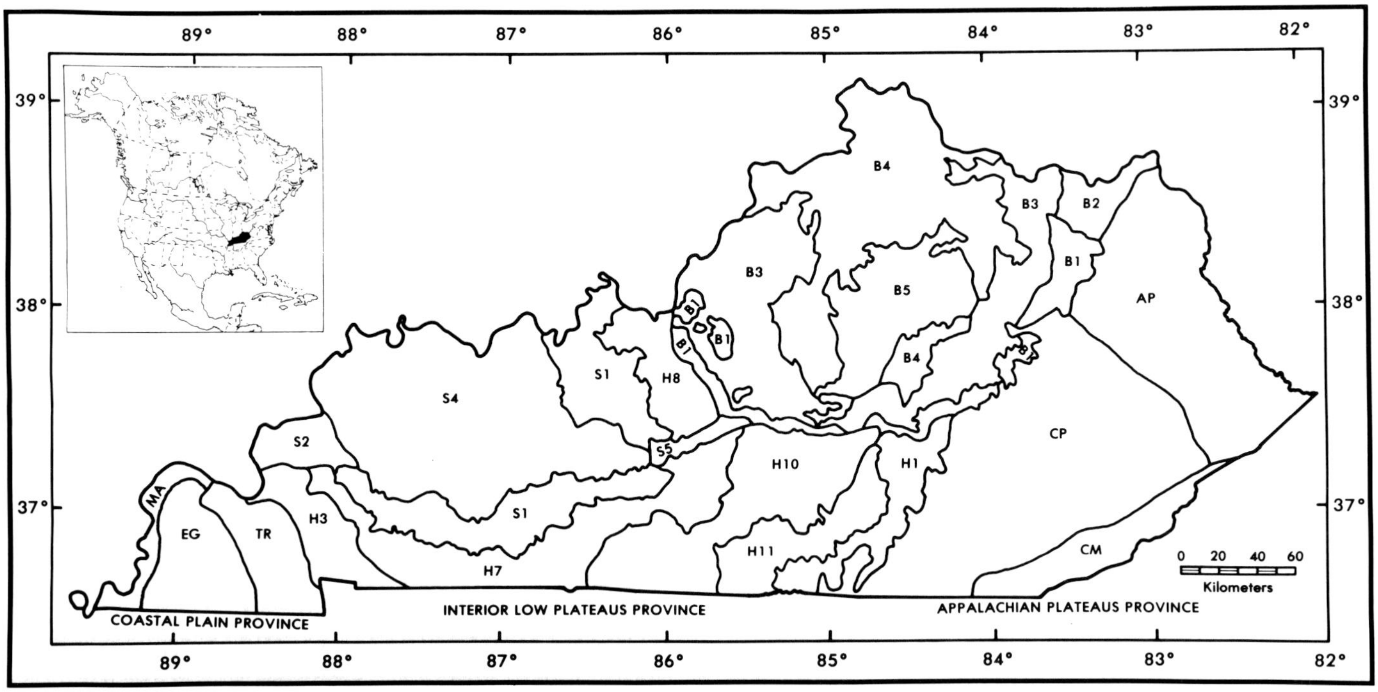

Fig. 1. Physiographic provinces, sections, and subsections of Kentucky (from Burr and Warren 1986).

COASTAL PLAIN PROVINCE
- MA Mississippi Alluvial Plain
- EG Eastern Gulf Coastal Plain
- TR Tennessee River Plain

INTERIOR LOW PLATEAUS PROVINCE*

Highland Rim Section
- H1 Eastern Highland Rim Subsection
- H3 Western Highland Rim Subsection
- H7 Pennyroyal Plain Subsection
- H8 Elizabethtown Plain Subsection
- H10 Greensburg Upland Subsection
- H11 Cumberland Enclave Subsection

Shawnee Hills Section
- S1 Mammoth Cave Plateau Subsection
- S2 Marion Subsection
- S4 Ohio River Hills and Lowlands Subsection
- S5 Brush Creek Hills Subsection

Blue Grass Section
- B1 Knobstone Escarpment and Knobs Subsection
- B2 Northeastern Blue Grass Subsection
- B3 Outer Blue Grass Subsection
- B4 Eden Shale Belt Subsection
- B5 Inner Blue Grass Subsection

APPALACHIAN PLATEAUS PROVINCE
- AP Unglaciated Allegheny Plateau Section
- CP Cumberland Plateau Section
- CM Cumberland Mountains Section

* Coding for this province after Quarterman and Powell (1978)

and Blue Grass sections); and III. The Appalachian Plateaus (including the Unglaciated Allegheny Plateau, Cumberland Mountains, and Cumberland Plateau sections). Each of these provinces and subdivisions is described in greater detail in Burr and Warren (1986).

The highest point in the state is the crest of Big Black Mountain in Harlan County at approximately 1,262 m (4,139 ft), abutting the Virginia state line. While I have collected butterflies from all 120 counties and many unique habitats such as Big Black Mountain, only some of the three major provinces and their subdivisions have been sampled for moths. There are specialized habitats yet to be sampled, and there is ample opportunity for extensive further exploration where relict plant and animal communities still exist with the promise of exciting discoveries. I hope this book will stimulate students of Kentucky wildlife to discover the rich faunal corners of the state not yet studied.

CONSERVATION

Insects, in general, are renewable resources. Biotic potential is usually high, with one female usually laying 100 to 1,000 eggs. Multiple generations per year add to the ability of species to rebuild populations reduced by adverse environmental conditions (heat, drought, cold, food shortages). Populations fluctuate over the years – some species are more common one year – others more abundant another. Long-term dynamics of populations are difficult to trace for this reason. It may be that permanent environmental conditions are affecting species in ways we do not comprehend. The virtual disappearance of the regal fritillary (*Speyeria idalia*) in Kentucky is perhaps the best case in point right now. On the other hand, the arrival of the meadow fritillary (*Boloria bellona*) in about 1970, and its spread to the westernmost county by 1993, is equally puzzling.

Lepidoptera form an important component in natural habitats of many types. Those in Kentucky vary from the mountainous eastern region to the coastal plain in the Jackson Purchase of western Kentucky. A number of unique hollows, prairie remnants, wetlands and other sites are already preserved as state parks, nature preserves, national and state forests, and local parks and reserves. The biological components of some are known, but the depth of that knowledge is variable. The survey of insect life in almost all, however, has yet to be completed. With thousands of insects (many of them as yet unnamed), it will be a monumental task to carry out the census work needed to learn which species are naturally present in these parks and preserves.

A major ecological catastrophe that is very likely to occur, but as of this writing has not yet befallen Kentucky's forests, is an outbreak of the gypsy moth (*Lymantria dispar*). This species is devastating forests to the north and east of our state. Dr. John Peacock (*pers. comm.*), United States Forest Service (retired), who has worked in gypsy moth control, believes it is only a matter of time before major forest defoliation from this pest occurs here. Whether the moth larvae denude trees to deprive native species of foodplant, or treatment with pesticides kills target and nontarget insects alike in huge quantities, a deleterious effect on Kentucky's native moth and butterfly species can be expected. Thus survey activities to learn what species are present in our parks and reserves should be increased as soon as possible. Only then will we have a database enabling proper evaluation of the effects of control efforts and direct impacts caused by the gypsy moth.

The conservation of Lepidoptera and the collection of butterflies and moths can be compatible. First, I strongly believe that making collections of insects is important in training and motivating people to go into the field of entomology. Second, only by studying the smaller moths by means of modern techniques can reliable identifications be made. Responsible collecting seldom reduces a population of butterflies or moths any more than a natural predator or environmental extreme such as drought or cold would. Collecting butterflies with a net is challenging and rarely affects a significant portion of the population then present. With forests full of butterflies and moths, responsible collecting makes no appreciable dent. The main cause of species extinction or rarity comes when critical habitats are destroyed by human activities which alter that habitat, physically or chemically. Thus conservation of Lepidoptera and other wildlife is best accomplished by habitat preservation.

I favor responsible, well-informed collecting as a means to provide experience for amateurs, students, and professionals, and specimens for research and education. Collecting should be carried out with respect for our natural heritage, within the limits of the law, correct respect for landowners (ask permission), and courtesy to other people. A good set of guidelines for collecting has been adopted and published by the Lepidopterists' Society and can be found in its publications.

A recent development that threatens to disrupt butterfly ecology in Kentucky as well as other parts of North America is the practice of "releasing butterflies" at weddings and other festive occasions. Butterflies are raised in large numbers by several dealers in Florida and California for interstate sale to special event companies for this purpose. The results can mean: (1) the addition of butterflies not native to the area where they are released; (2) butterflies from other areas "polluting" the local gene pool; and (3) disappointments when the butterflies are too dehydrated, hot, or cold to fly when released. I documented one case in June 1996 in which I discovered queens (*Danaus gilippus berenice*), which had been released at the Louisville Zoo, flying and mating there ten days later. Since the queen is not a part of the Kentucky fauna, I was suspicious. I later found that the Animal and Plant Health Inspection Service, United States Department of Agriculture (APHIS), had issued a permit for them to be shipped from Florida to Kentucky and released, along with the monarch and painted lady. While APHIS agents are now more aware of the problem of introducing alien butterflies to an area, releases continue – a practice that has become controversial among ecologists.

This work should be viewed as a beginning, a baseline of information about Kentucky's Lepidoptera fauna. There is much localized survey work to be done everywhere in the Commonwealth, and several preliminary inventories have been accomplished in the last five years. In the future the surveys should be repeated to compare the present fauna with what is known from the past. Our natural history collections should be better housed and curated, and combined into a state museum (at one site, or as an umbrella covering collections at various sites) to assure that these collections will not be destroyed or lost. They are, in my opinion, important natural resource treasures. Finally, I urge the establishment of a Kentucky Biological Survey to maintain our collections, add to them, see that they are fruitfully used, and to assure that inventory and biomonitoring activities are supported in the years ahead.

HOW TO USE THIS BOOK

Species accounts are arranged by suborder, superfamily, family, subfamily, genus, and species, and generally follow the latest check list to North American Lepidoptera (Hodges et al. 1983). However, there are some modifications resulting from more recent publications that are cited in the text and listed in the bibliography. These especially concern higher classification, such as that of the family Nymphalidae, which has been altered to conform to that proposed by Harvey (1991). Consequently, some of the Hodges et al. (1983) numbers are out of their original positions in sequence. Not incorporated here are some major rearrangements in the microlepidoptera families summarized in Scoble (1992). These reflect recent changes in the higher classification proposed by several specialists. To include them here would make the numbering system of Hodges et al. (1983) hopelessly confusing.

Along with species reference numbers corresponding to those in Hodges et al. (1983), genus and species names are followed by the authors. If an author's name is in parentheses, the species is listed in a genus other than that in which the author originally named it. Common names follow for species that have them, and follow nomenclature in Cassie et al. (1995) and Covell (1984) for butterflies and moths, respectively. Below each species are collection data in detail if five or fewer different county records for the species are known. In some cases, collection data were included for six or seven counties. Year and collector's name are given for unique records only, unless that information has special value (as with very old records). If there are more than five counties recorded, information is summarized under the headings: PARKS AND RESERVES, COUNTIES, and DATES. Abbreviations were used for parks and nature reserves (Table 1). Two reserves, Covered Bridge and Horine Boy Scout camps no longer exist. The former is now a housing development called Covered Bridge Crossings, and the latter is a small part of Jefferson County Memorial Forest. Big Black Mountain (BBM) is not a reserve, but is an ecologically special habitat which includes the highest point in Kentucky (ca. 1,262 m), and species of plants and animals (including insects) found nowhere else in the state. Road abbreviations used in collection site localities are as follows: FS = United States Forest Service Road; SR = State Route; and US = United States Route. Counties are abbreviated to correspond to the legend and map in Figure 2.

The dates given are those for the earliest and latest adult records in a year, with months in between indicated if the species is reported from all weeks within those months. Thus "March 25–June 3" would indicate records in every week between those extremes, including April and May. If a gap of more than one week occurs in our records, that is so indicated (e.g., "May 1–12, 31–June 4" indicates 18 days without records between May 12 and 31).

A REMARKS section is added for some species, giving additional information, such as taxonomic notes, range extension comments, and life history or ecological details. However, if a simple occurrence comment such as "Rare" or "Common" is added, it is merely appended to the date information. If no statement regarding commonness of occurrence is included, it was omitted because I believe that evidence to make a judgment was insufficient. Such statements are based on evidence available at the time of publication, and may not be true in every year or at every locality where the species is found.

Indications of "rare" indicate that the species is believed to be genuinely present in extremely few places and in very small numbers. "Uncommon" means that while widespread in the state, numbers of individuals seen are usually small and one cannot count on finding it in appropriate habitat and season. "Frequent" means that one finds the species often, has a relatively fifty-fifty chance of finding it in appropriate habitat and season, and while occasionally numbers may be substantial, they are usually low. "Common" refers to the fact that a moderate to heavy number of individuals can be found when sought in appropriate place and season. "Very common" indicates generally wide overall distribution and high numbers in season. Finally, "abundant" is used when the species occurs regularly in extremely high numbers – 100 or more indi-

Table 1. Abbreviations for Kentucky parks, reserves, and other important habitat sites.

AJJ	A.J. Jolly County Park, Campbell County.
ASC	All Saints Episcopal Conference Center, Rough River Lake, Grayson County.
AUD	John James Audubon State Park, Henderson County.
BAN	Barren River Lake State Resort Park, Barren County.
BBL	Big Bone Lick State Park, Boone County.
BBM	Big Black Mountain (summit), Harlan County.
BCN	Boone County Cliffs State Nature Preserve, Boone County.
BEC	Beargrass Creek State Nature Preserve, Jefferson County.
BIP	Breaks Interstate Park, Pike County.
BLA	Blackacre State Nature Preserve, Jefferson County.
BLD	Ballard Wildlife Management Area, Ballard County.
BMF	Bernheim Arboretum and Research Forest, Bullitt County.
BOW	Boone Woods Park, Boone County.
BUC	Buckhorn Lake State Resort Park, Perry County.
CAC	Carter Caves State Resort Park, Carter County.
CAH	Camp Horine Boy Scout Reservation, Jefferson County (now part of JCF).
CAV	Cave Run Recreation Area, Bath County.
CBB	Covered Bridge Boy Scout Reservation, Oldham County (now a subdivision called Covered Bridge Crossings).
CBF	Cumberland Falls State Resort Park, Whitley County.
CCR	Camp Cedar Ridge, Jefferson County.
CER	Camp Ernst, Boone County.
CEW	Camp Earl Wallace, Wayne County.
CMI	Camp Michael Boy Scout Reservation, Boone County.
COB	Columbus-Belmont State Park, Hickman County.
DBNF	Daniel Boone National Forest, several counties.
DHW	Dale Hollow Lake State Resort Park, Clinton County.
FAO	Falls of the Ohio, Jefferson County (bedrock and shore below cliffs).
FKN	Fort Knox Military Reservation, Bullitt, Hardin, and Meade counties.
GBO	Greenbo Lake State Resort Park, Greenup County.
GBS	General Burnside Island State Park, Pulaski County.
GBT	General Butler State Resort Park, Carroll County.
HHW	Higginson Henry Wildlife Management Area, Union County.
HNK	Hancock Biological Station, Calloway County.
HWS	Horner Bird and Wildlife Sanctuary, Oldham County.
IRO	Iroquois Park, Jefferson County.
JCF	Jefferson County Memorial Forest, Jefferson County.
JWS	Jenny Wiley State Resort Park, Floyd County.
KDV	Kentucky Dam Village State Resort Park, Marshall County.
KGC	Kingdom Come State Park, Harlan and Letcher counties.
KIN	Kincaid Lake State Park, Pendleton County.
KLK	Kenlake State Resort Park, Marshall County.
KRC	Koomer Ridge Campground, Daniel Boone National Forest, Wolfe County.
LBK	Lake Barkley State Resort Park, Trigg County.
LBL	Land Between the Lakes, Lyon and Trigg counties.
LCP	Lake Cumberland State Resort Park, Russell County.
LMP	Lake Malone State Park, Muhlenberg County.
LVJ	Levi Jackson Wilderness Road State Park, Laurel County.
LWO	Lilly Cornett Woods, Letcher County.
MCV	Mammoth Cave National Park, Barren, Edmonson, and Hart counties.
MUR	Murphy's Pond, Hickman County.
NBP	Natural Bridge State Resort Park, Powell County.
OCP	Otter Creek Park, Meade County.
OKH	My Old Kentucky Home State Park, Nelson County.
PCP	Pine Creek Barrens Nature Preserve, Bullitt County.
PMS	Pine Mountain Settlement School, Harlan County.

PNM Pine Mountain State Resort Park, Bell County.
PYR Pennyrile Forest State Resort Park, Caldwell and Christian counties
RBF Robinson Forest, Breathitt County.
RGR Rough River Dam State Resort Park, Grayson County.
RNR Reelfoot National Wildlife Refuge, Fulton County.
RRG Red River Gorge Geological Area, Menifee, Powell and Wolfe counties.
SGS Shanituck Girl Scout Reservation, Bullitt County.
SWR Frank Sauerheber Unit, Sloughs Wildlife Management Area, Henderson County.
TIO Tioga Falls Park, Hardin County.
WKM West Kentucky Wildlife Management Area, McCracken County.
WVP Waverly Park, Jefferson County.
YEL Yellowbank Wildlife Management Area, Breckinridge County.
YHF Yahoo Falls, Big South Fork National River and Recreation Area, McCreary County.

viduals at a time in appropriate habitat and season. Sometimes the reader will find adjectives to modify these statements of abundance, and should remember that numbers often fluctuate widely from place to place and year to year.

In cases of records taken from the literature, citations are made to the bibliography at the back of this book. I would be grateful to receive information on citations from the literature, as well as additional records of Lepidoptera species, that have been otherwise omitted herein.

While there are no figures by which to identify the Lepidoptera listed, I recommend the following references as companions when using this book. They also provide other useful information such as overall ranges and foodplant information for the species they cover. For butterflies, Opler and Malikul (1992) should be the most helpful, with excellent plates also in Opler and Krizek (1984), Iftner et al. (1992), and Shull (1987). For moths, Covell (1984) and Rings et al. (1992) will provide illustrations of most of the larger and middle-sized moths included here. Holland (1903) also shows some species not illustrated in the first two works. References in the above works will lead you to other, more specialized coverages of some groups. However, many genera and even families are still badly in need of taxonomic revision.

There is some taxonomic terminology that may be useful to the reader. In many cases with microlepidoptera, the designation "Kentucky is the type locality" appears. This means that the author of the species name (usually Vactor T. Chambers) indicated that the specimen or specimens on which the name is based came from Kentucky (no specific locality in the state). *Holotype* is the term for the one specimen designated as basis for a species-level name, while *paratype* is the term for specimens other than the holotype that were used in formulating an original description and basing the name of the new species. *Syntypes* are two or more specimens used by the author when no holotype was originally designated, from which a revising scientist usually selects a *lectotype* – one of the syntypes to serve the purpose of a holotype.

Finally, institutional abbreviations where Kentucky Lepidoptera specimens are housed were used for several taxa and include the following: AMNH, American Museum of Natural History, New York, New York; ANSP, The Academy of Natural Sciences, Philadelphia, Pennsylvania; CMNHS, Museum of Natural History and Science, Cincinnati, Ohio; LACM, Natural History Museum of Los Angeles County, Los Angeles, California; MCZ, Museum of Comparative Zoology, Harvard University, Cambridge, Massachusetts; NHM, British Museum of Natural History, London, England; NMNH, National Museum of Natural History, Washington, District of Columbia; UK, University of Kentucky Collection, Department of Entomology, University of Kentucky, Lexington, Kentucky; and ULK, Lovell Insect Museum, University of Louisville, Louisville, Kentucky.

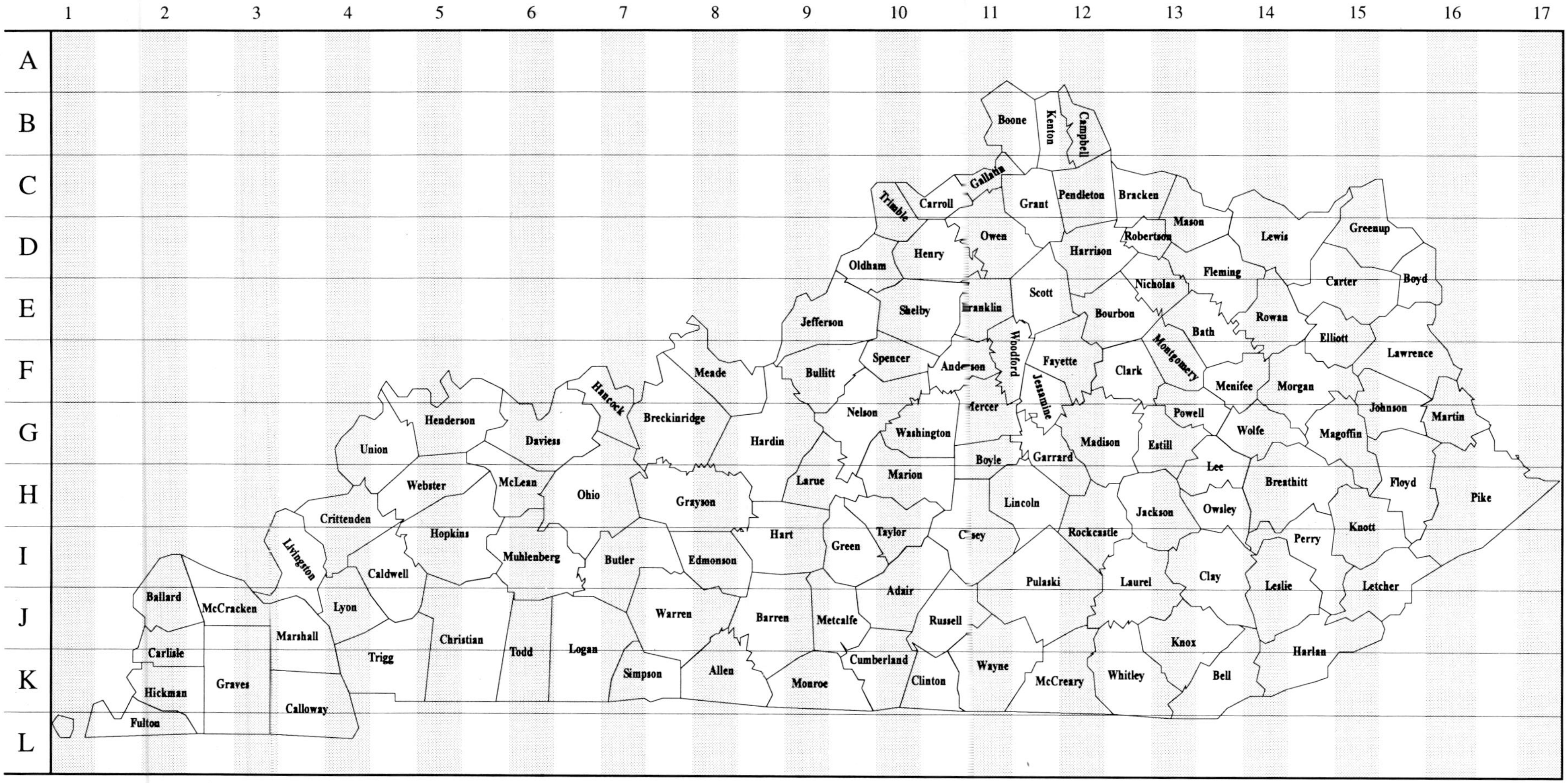

Adair, ADA, J10
Allen, ALL, K8
Anderson, AND, F11
Ballard, BAL, J2
Barren, BAR, J9
Bath, BAT, E13
Bell, BEL, K13
Boone, BOO, B11
Bourbon, BOU, E12
Boyd, BOD, D16
Boyle, BOL, G11
Bracken, BRA, C13
Breathitt, BRE, H14
Breckinridge, BRK, G8
Bullitt, BUL, F9
Butler, BUT, I7
Caldwell, CAD, I4
Calloway, CAL, K4
Campbell, CAM, B12
Carlisle, CAE, K2
Carroll, CRL, C10
Carter, CAR, E15
Casey, CAS, I11
Christian, CHR, J5
Clark, CLK, F12
Clay, CLA, I13
Clinton, CLI, K10
Crittenden, CRI, H4
Cumberland, CUM, K10
Daviess, DAV, G6
Edmonson, EDM, I8
Elliott, ELL, E15
Estill, EST, G13
Fayette, FAY, F12
Fleming, FLE, D13
Floyd, FLO, H16
Franklin, FRA, E11
Fulton, FUL, L2
Gallatin, GAL, C11
Garrard, GAR, G12
Grant, GRA, C11
Graves, GRV, K3
Grayson, GRY, H8
Green, GRE, I9
Greenup, GUP, D15
Hancock, HAN, G7
Hardin, HRD, G9
Harlan, HLN, J15
Harrison, HSN, D12
Hart, HRT, I9
Henderson, HND, G5
Henry, HEN, D10
Hickman, HIC, K2
Hopkins, HOP, I5
Jackson, JAC, H13
Jefferson, JEF, E9
Jessamine, JES, F12
Johnson, JOH, F15
Kenton, KEN, B12
Knott, KNO, H15
Knox, KNX, J13
Larue, LAR, H9
Laurel, LAU, I13
Lawrence, LAW, F16
Lee, LEE, H13
Leslie, LES, I14
Letcher, LET, I15
Lewis, LEW, D14
Lincoln, LIN, H11
Livingston, LIV, I3
Logan, LOG, J7
Lyon, LYO, J4
Madison, MAD, G12
Magoffin, MAG, G15
Marion, MAR, H10
Marshall, MRL, J3
Martin, MTN, G16
Mason, MAS, D13
McCracken, MCC, J3
McCreary, MCY, K12
McLean, MCL, H6
Meade, MEA, F8
Menifee, MEN, F14
Mercer, MER, G11
Metcalfe, MET, J9
Monroe, MON, K9
Montgomery, MTG, F13
Morgan, MOR, F14
Muhlenberg, MUH, I6
Nelson, NEL, G10
Nicholas, NIC, E13
Ohio, OHI, H7
Oldham, OLD, D10
Owen, OWE, D11
Owsley, OWS, H13
Pendleton, PEN, C12
Perry, PER, I14
Pike, PIK, H16
Powell, POW, G13
Pulaski, PUL, I12
Robertson, ROB, D13
Rockcastle, ROC, H12
Rowan, ROW, E14
Russell, RUS, J11
Scott, SCO, E12
Shelby, SHE, E10
Simpson, SIM, K7
Spencer, SPE, F10
Taylor, TAY, I10
Todd, TOD, K6
Trigg, TRI, K4
Trimble, TRM, C10
Union, UNI, G4
Warren, WAR, J8
Washington, WAS, G10
Wayne, WAY, K11
Webster, WEB, H5
Whitley, WHI, K12
Wolfe, WOL, G14
Woodford, WOO, F11

Fig. 2. The counties of Kentucky. Three-letter abbreviations used in species accounts are presented after the county name.

LIST OF RECORDED LEPIDOPTERA OF KENTUCKY

ORDER **LEPIDOPTERA**

SUBORDER **ZEUGLOPTERA**

Superfamily **MICROPTERIGOIDEA**

Family **MICROPTERIGIDAE**

Mandibulate Moths

Genus ***Epimartyria*** Wlsm.

1 ***E. auricrinella*** Wlsm.
Letcher Co., Cumberland Valley, May 24, and Colliers Creek, June 8; Powell Co., May 23, 25; all collected by A. F. Braun during daytime. Rare.

SUBORDER **DACNONYPHA**

Superfamily **ERIOCRANIOIDEA**

Family **ERIOCRANIIDAE**

Eriocraniid Moths

Genus ***Dyseriocrania*** Spuler

3 ***D. griseocapitella*** (Wlsm.)
Boone Co., BBL, several specimens at light, Apr. 25; Bullitt Co., BMF, in Malaise trap, May 4–8; Laurel Co., DBNF, Holly Bay Campground, Apr. 19, and ridge top SW of Ned Branch, FS 4176, Apr. 24; McCreary Co., Blue Heron Campground, Apr. 27; Powell Co., RRG, Tunnel Ridge, at light, Apr. 28. Locally common.

SUBORDER **EXOPORIA**

Superfamily **HEPIALOIDEA**

Family **HEPIALIDAE**

Ghost Moths or Swifts

Genus ***Sthenopis*** Pack.

22 ***S. auratus*** Grt.
Laurel Co., Bald Rock, June 11, 1978, R. Loechel; Menifee Co., along Kelley Branch of East Fork of Indian Creek, FS 9B, in light trap, June 15, 1985, J. S. Nordin. Difficult to find.

SUBORDER **MONOTRYSIA**

Superfamily **NEPTICULOIDEA**

Family **NEPTICULIDAE**

Nepticulid Moths

Genus ***Obrussa*** Braun

37 ***O. ochrefasciella*** (Cham.)
Kentucky is the type locality (Chambers 1873), where it was collected in June. Herbert H. Kulman (*pers. comm.*) has also recorded this species from Kentucky.

Genus ***Ectoedemia*** Bsk.

43 ***E. platanella*** (Clem.)
Listed by Chambers from Kentucky, which is also the type locality of synonym *maximella* (Chambers 1873). Jefferson Co., May 26–30, 1979, C. V. Covell, Jr.

44 ***E. clemensella*** (Cham.)
Kentucky is the type locality (Chambers 1873). Listed in the original description as being from "Kentucky; Pennsylvania," but the lectotype is from Kentucky (Miller and Hodges 1990).

45 ***E. similella*** (Braun)
Listed from Kentucky by Braun *in* Forbes (1923).

48 ***E. rubifoliella*** (Clem.)
Clay Co., Red Bird River, on *Rubus* (blotch mine), emerged July 19 and 29, A. F. Braun.

51 ***E. quadrinotata*** (Braun)
Listed from Kentucky by Braun *in* Forbes (1923).

Genus ***Fomoria*** Beirne

60 ***F. pteliaeella*** (Cham.)
Kentucky is the type locality (Chambers 1880b).

Genus ***Stigmella*** Schr.

62 ***S. crataegifoliella*** (Clem.)
Listed from Kentucky by Braun *in* Forbes (1923).

70 ***S. prunifoliella*** (Clem.)
Kentucky is the type locality for synonym *serotinaeella* (Chambers 1873), which he described from miners in wild cherry. Magoffin Co., on *Prunus*, emerged June 7, A. F. Braun.

72 ***S. intermedia*** (Braun)
Listed from Kentucky by Braun *in* Forbes (1923).

73 ***S. rhoifoliella*** (Braun)
Listed from Kentucky by Braun *in* Forbes (1923).

81 ***S. apicialbella*** (Cham.)
Kentucky is the type locality (Chambers 1873), where it was collected in June.

82 ***S. fuscotibiella*** (Clem.)
Kentucky is the type locality of synonym *ciliae-*

fuscella (Chambers 1873), who wrote "Taken at lamp, Aug. 23rd." That information, plus a Kentucky label, is on a syntype in the MCZ (Miller and Hodges 1990); but the label does not give a locality.

84 ***S. populetorum*** (F. & B.)
Listed from Kentucky by Braun *in* Forbes (1923).

87 ***S. saginella*** (Clem.)
Kentucky is the type locality for synonyms *quercicastanella* Chambers (reported mining in chestnut oak in the latter part of July), and *fuscocapitella* Chambers ("Captured in Kentucky in June.") (Chambers 1873).

88 ***S. latifasciella*** (Cham.)
Kentucky is the type locality (Chambers 1878b). The holotype female in the MCZ was collected "on chestnut-trees, August, Chambers."

89 ***S. nigriverticella*** (Cham.)
Kentucky is the type locality (Chambers 1875a), where the holotype female was collected on May 12.

90 ***S. castaneaefoliella*** (Cham.)
Kentucky is the type locality (Chambers 1875a). Powell Co., emerged Aug. l, A. F. Braun.

91 ***S. flavipedella*** (Braun)
Listed from Kentucky by Braun *in* Forbes (1923).

92 ***S. corylifoliella*** (Clem.)
Kentucky is the type locality for synonym *minimella* (Chambers 1873), who collected it at light in Aug. Campbell Co., Claryville, emerged May 16–25; Fleming Co., emerged July 27; both A. F. Braun, as synonym *opulifoliella* Braun.

93 ***S. ostryaefoliella*** (Clem.)
Listed from Kentucky by Braun *in* Forbes (1923).

95 ***S. juglandifoliella*** (Clem.)
A specimen of this species in the MCZ is labelled "Kentucky, Chambers." Chambers (1878c) described and named this as his own new species, but it had already been named by Clemens in 1861 (Miller and Hodges 1990).

96 ***S. unifasciella*** (Cham.)
Kentucky is the type locality (Chambers 1875a).

98 ***S. tiliella*** (Braun)
Listed from Kentucky by Braun *in* Forbes (1923).

99 ***S. quercipulchella*** (Cham.)
Kentucky is the type locality (Chambers 1878b). Also listed from Kentucky by Forbes (1923) under *terminella* Braun, which is now a synonym of *quercipulchella*.

106 ***S. amelanchierella*** (Clem.)
Listed from Kentucky by Braun *in* Forbes (1923).

111 ***S. resplendensella*** (Cham.)
Kentucky is the type locality (Chambers 1875a). The lectotype is labelled "Kentucky, 23 May" (Miller and Hodges 1990).

Genus ***Microcalyptris*** Braun

112 ***M. thoracealbella*** (Cham.)
Kentucky is the type locality; Chambers (1873) wrote "Captured in June in Kentucky."

Family **OPOSTEGIDAE**

Opostegid Moths

Genus ***Peudopostega*** Kozlov

122 ***P. quadristrigella*** (Cham.)
The type locality is "Camp Bee Spring, Edmonson Co., Ky." where it was collected in the early part of July (Chambers 1875a).

123 ***P. nonstrigella*** (Cham.)
Kentucky is the type locality (Chambers 1881).

Family **TISCHERIIDAE**

Tischeriid Moths

Genus ***Tischeria*** Zell.

126 ***T. citrinipennella*** Clem.
Kentucky is the type locality of synonym *quercivorella* (Chambers 1875a). Carter Co.; Edmonson Co., MCV; Laurel Co.; Letcher Co.; Madison Co.; dates unknown; all A. F. Braun.

129 ***T. badiiella*** Cham.
Kentucky is the type locality (Chambers 1875a). Campbell Co. (questionably); Carter Co.; Edmonson Co. MCV; Fleming Co.; Letcher Co., Big Black Mountain; Pulaski Co.; Rowan Co.; dates unknown; all A. F. Braun.

134 ***T. simulata*** Braun
Fleming Co., Fox Mountain; Logan Co., Russellville, a paratype locality; Rowan Co., Morehead; dates unknown; all A. F. Braun.

135 ***T. purinosella*** Cham.
Kentucky is the type locality (Chambers 1875a).

138 ***T. clemensella*** Cham.
Although Chambers (1875a) implied and Braun (1972) stated the type locality to be Kentucky, Miller and Hodges (1990) found the type was labelled "Tex." Larue Co., Upton, date unknown, A. F. Braun.

139 ***T. fuscomarginella*** Cham.
Kentucky is the type locality (Chambers 1875a). Carter Co., CAC; McCreary Co., Natural Arch; Whitley Co., CBF; dates unknown; all A. F. Braun.

140 ***T. castaneaeella*** Cham.
Kentucky is the implied type locality (Chambers 1875a). Carter Co., CAC; Fleming Co., Fox Mountain; Powell Co., NBP; Rowan Co., Morehead; dates unknown; all A. F. Braun.

143 ***T. zelleriella*** Clem.
Kentucky is the type locality for synonym *latipennella* (Chambers 1878b). Carter Co., Olive Hill; Edmonson Co., MCV; McCreary Co.; Rowan Co., Morehead; dates unknown; all A. F. Braun.

144 ***T. quercitella*** Clem.
Kentucky is the type locality of synonym *tinctoriella* (Chambers 1875a). Fayette Co., Lexington, reared on *Quercus palustris*, emerged Aug. 4, 1939 (identified as *tinctoriella*); Lewis Co.; Powell Co.; no dates; A. F. Braun.

145 ***T. malifoliella*** Clem.
Rowan Co., date unknown, A. F. Braun; Rowan Co., Morehead, reared on *Malus*, emerged Aug. 16, 1963, T. N. Freeman.

147 ***T. roseticola*** F. & B.
Braun (*pers. comm.*) was confident that Chambers had reared this species from Kentucky.

148 ***T. agrimoniella*** Braun
Edmonson Co., MCV; Wolfe Co., DBNF; dates unknown; both A. F. Braun.

149 ***T. aenea*** F. & B.
Lewis Co.; Rowan Co., dates unknown; A. F. Braun. Rowan Co., Morehead, reared on *Rubus*, emerged Sept. 14 and 19, 1962, T. N. Freeman.

154 ***T. amelanchieris*** Braun
Whitley Co., CBF; date unknown; A. F. Braun.

156 ***T. solidagonifoliella*** Clem.
Fleming Co.; Letcher Co., Pine Mountain; Rowan Co., Morehead; dates unknown; all A. F. Braun.

157 ***T. astericola*** Braun
Bell Co., Log Mountain; Letcher Co., Big Black Mountain; Lewis Co., dates unknown; all A. F. Braun. Oldham Co., CBB, reared from an unidentified leaf, emerged Aug. 5, 1974, C. V. Covell, Jr.

159 ***T. heliopsisella*** Cham.
Kentucky is the type locality (Chambers 1875a).

160 ***T. ambrosiaeella*** Cham.
Kentucky is the type locality (Chambers 1875a).

161 ***T. helianthi*** F. & B.
Kenton Co., Kenton; Powell Co., NBP; dates unknown; A. F. Braun.

Superfamily **INCURVARIOIDEA**

Family **INCURVARIIDAE**

Incurvariid Moths

Genus ***Paraclemensia*** Busck

181 ***P. acerifoliella*** (Fitch)
Maple Leafcutter Moth
Harlan Co., BBM, May; Owsley Co., 4.8 km (3 mi) NE of Booneville, May 10; Trimble Co., Milton, mining in *Acer saccharum*, Aug. 29.

Family **PRODOXIDAE**

Yucca Moths

Genus ***Lampronia*** Steph.

175 ***L. russatella*** (Clem.)
Harlan Co., BBM; Letcher Co., Cumberland Valley; Menifee Co., RRG, May 17, 18; Owsley Co., 4.8 km (3 mi) NE of Booneville, May 29; Powell Co., May.

Genus ***Tegeticula*** Zell.

198 ***T. yuccasella*** (Riley)
Yucca Moth
Bullitt Co., BMF, June 5, 10; Jefferson Co., June 18; Rowan Co., Morehead, June 20. Adults in flowers of *Yucca filamentosa*. Locally common.

Genus ***Prodoxus*** Riley

200 ***P. quinquepunctella*** (Cham.)
Specimen labelled "Ky." in NMNH (Davis 1967); Fayette Co., Lexington, at black-light, June 11; Gallatin Co., Markland Dam, June 15; Owsley Co., 4.8 km (3 mi) NE of Booneville, one male, June 14; Whitley Co., Goldbug, June 5.

Family **ADELIDAE**

Fairy Moths

Genus ***Adela*** Latreille

227 ***A. caeruleella*** Wlk.
PARKS AND RESERVES: BCN, BMF, RRG, YHF. **COUNTIES:** BOO, BUL, JEF, LET, MCY, MEN, OWS. **DATES:** Apr. 27–June 8. **REMARKS:** Kentucky is the type locality of synonym *bella* (Chambers 1873). This species has been found on flowers during the day and at lights at night. Uncommon.

228 ***A. ridingsella*** Clem.
Kentucky is the type locality for synonym *corruscifasciella* (Chambers 1873). Boone Co., BCN, June 15; Letcher Co., Big Black Mountain, June 8, and Cumberland Valley, June 26; Meade Co., OCP, at lights, June 7, 15; Owsley Co., 4.8

km (3 mi) NE of Booneville, May 27; Powell Co., RRG, Tunnel Ridge, May 10, 18. Uncommon and hard to find.

Family **HELIOZELIDAE**

Shield Bearer Moths

Genus ***Heliozela*** H.-S.

230 ***H. aesella*** Cham.
The holotype was from "Kentucky, near Covington, Apr. 24" (Chambers 1877a). Carter Co., CAC, May 2, 1936, A. F. Braun.

Genus ***Antispila*** Hbn.

236 ***A. isabella*** Clem.
Rowan Co., Morehead, reared from *Vitis*, emerged May 28, 31, June 3, 5, 1963, T. N. Freeman.

237 ***A. viticordifoliella*** Clem.
Kentucky is the type locality. Miller and Hodges (1990) report 2 syntypes in the MCZ from Kentucky. Fayette Co., reared from grape, emerged July 23, 1901, H. Garman; Rowan Co., Clack Mtn., mining in *Vitis aestivalis*, July 2, 1951, A. F. Braun.

240 ***A. ampelopsifoliella*** Cham.
Kentucky is the type locality (Chambers 1874a). Chambers (1874a) wrote "known only in the larval state," but three Kentucky adults are in the MCZ (Miller and Hodges 1990).

243 ***A. hydrangaeella*** Cham.
Kentucky is the type locality (Chambers 1874a). Carter Co., mining in *Hydrangea arborescens*, Oct. 19; Letcher Co., Big Black Mountain, mining in *Hydrangea arborescens* Oct. 21; both A. F. Braun.

Genus ***Coptodisca*** Wlsm.

244 ***C. diospyriella*** (Cham.)
Kentucky is the type locality (Chambers 1874a). Edmonson Co., Camp Bee Spring, mining in persimmon, Aug. 24; Rowan Co., Morehead, on persimmon, emerged May 26; both A. F. Braun.

247 ***C. lucifluella*** (Clem.)
Letcher Co., Big Black Mountain, on hickory, emerged July 5, A. F. Braun.

249 ***C. magnella*** Braun
Bell Co., Fern Lake, emerged July 29; Edmonson Co., MCV, mining in *Vaccinium arboreum*, Aug. 17; Fleming Co., mining in *Vaccinium vacillans*, July 12; Powell Co., emerged from *Vaccinium* on Aug. 7; Whitley Co., CBF, mining in *Vaccinium*; all A. F. Braun.

254 ***C. splendoriferella*** (Clem.)
Chambers (1873) refers to this species as follows: "At Linden Grove Cemetery . . . it occurs by the million, mining in cherry." This site is in Covington, Kenton Co.

SUBORDER **DITRYSIA**

Superfamily **TINEOIDEA**

Family **TINEIDAE**

Clothes Moths and relatives

Genus ***Nemapogon*** Schr.

261 ***N. acapnopenella*** (Clem.)
A specimen of this species in the NMNH is labelled "Ky." and "Aug. Busck Coll."

263 ***N. auropulvella*** (Cham.)
The type locality is "Kentucky, July" based on syntypes in the MCZ and NMNH (Miller and Hodges 1990). Letcher Co., Big Black Mountain, June 8, A. F. Braun.

266 ***N. granella*** (L.)
European Grain Moth or Wolf Moth
Kentucky is the type locality for synonyms *costotristrigella* Chambers ("August and September") and *fuscomaculella* (Chambers 1873). Powell Co., NBP, Aug. 21, A. F. Braun.

Genus ***Doleromorpha*** Braun

276 ***D. porphyria*** Braun
Bell Co., PNM, July 7, 8; Clark Co., Red Bird River, July 8; Fleming Co., July 12; Letcher Co., Big Black Mountain, July 19, Colliers Creek, July 7, and Pine Mountain, June 24, July 1, 7; all A. F. Braun.

Genus ***Eudarcia*** Clem.

277 ***E. eunitariaeella*** Cham.
Kentucky is considered to be the type locality. Chambers (1873) states that it occurs in Kentucky and the Gulf States." In 1877 he sent Kentucky specimens of this species to the NHM under the name *Tinea coemetariiella* [*sic*] (Chambers 1876b), a synonym (Miller and Hodges 1990).

278 ***E. simulatricella*** Clem.
Letcher Co., Cumberland Valley, July 1, 26 and Pine Mountain, July; Rowan Co., Clack Mtn., May; all A. F. Braun.

Genus ***Diachorisia*** Clem.

279 ***D. velatella*** Clem.
Kentucky is the type locality for synonyms *fuscopulvella* (Chambers 1872) and *maculabella* (Chambers 1873). Fayette Co., Lexington, in black-light trap, June 6, 7; Letcher Co., Big Black Mountain, June 8, Cumberland Valley, June 26, July 1.

Genus *Leucomele* Dietz

282 ***L. miriamella*** Dietz
Jefferson Co., Valley Station, July 14, 1975, A. J. Brownell.

Genus *Oenoe* Cham.

283 ***O. hybromella*** Cham.
Kentucky is the type locality (Chambers 1874a).

Genus *Homosetia* Clem.

284 ***H. argentinotella*** (Cham.)
Kentucky is the type locality (Chambers 1876b). A syntype in the MCZ is labelled "Kentucky June" (Miller and Hodges 1990).

285 ***H. argentistrigella*** (Cham.)
Kentucky is the type locality (Chambers 1873). Chambers (1873) called it "a very handsome species."

286 ***H. auricristatella*** (Cham.)
Kentucky is the type locality (Chambers 1873).

287 ***H. bifasciella*** (Cham.)
Boone Co., CER, Aug. 17, 1979, L. D. Gibson.

290 ***H. cristatella*** (Cham.)
Kentucky is the type locality ("June, Chambers") (Chambers 1875b). Letcher Co., Big Black Mountain, July 8, A. F. Braun.

291 ***H. fasciella*** (Cham.)
Kentucky is the type locality (Chambers 1873). Letcher Co., Pine Mountain, June 11, A. F. Braun.

292 ***H. fuscocristatella*** (Cham.)
Kentucky is the type locality (Chambers 1873). Miller and Hodges (1990) indicate "syntypes presumed lost."

293 ***H. marginimaculella*** (Cham.)
Chambers (1875d) did not state a type locality, but the species described in the article were mostly from Canada. Forbes (1923) states "I have seen the type and believe it belongs in *Diachorisia*. It is certainly no *Tinea*. It came from Kentucky."

294 ***H. miscecristatella*** (Cham.)
Kentucky is the type locality (Chambers 1873). Letcher Co., Pine Mountain, July 1; McCreary Co., Pine Knot, June 14; Whitley Co., CBF, June 25; all A. F. Braun.

295 ***H. tricingulatella*** (Clem.)
Letcher Co., Line Fork, July 2; Letcher Co., Pine Mountain, June 10; both A. F. Braun.

Genus *Isocorypha* Dietz

299 ***I. mediostriatella*** (Clem.)
Type locality information for synonym *auristrigella* (Chambers 1873) is "Ky. in June." Hardin Co., Glendale, in Malaise trap, Aug. 6–14; McCreary Co., near Whitley City, Aug.

Genus *Hybroma* Clem.

300 ***H. servulella*** Clem.
Kentucky is the type locality for synonym *aurosuffusella* (Chambers 1873). Clay Co., Nolan's Branch, July 3, A. F. Braun.

Genus *Homostinea* Dietz

301 ***H. curviliniella*** Dietz
Boone Co., Florence, Aug. 13, 1979, L. D. Gibson. Miller and Hodges (1990) also cite a Kentucky specimen in the LACM.

Genus *Dryadaula* Meyr.

307 ***D. visaliella*** (Cham.)
The type series was collected resting on an oak tree in June at Visalia, Kentucky (Chambers 1873). Letcher Co., Pine Mountain, June 11; McCreary Co., Saltpeter Cave, June 22; Powell Co., June 26; Todd Co., Clifty Creek, June 15; all A. F. Braun. **NOTE:** The species name was misspelled in Hodges et al. (1983) as "*vesaliella*."

Genus *Scardiella* G. Robinson

NOTE: Names in this genus and the next reflect changes made by Robinson (1986).

308 ***S. approximatella*** (Dietz)
Letcher Co., Cumberland Valley, July 1; Menifee Co., RRG, in black-light trap, June 28.

Genus *Montescardia* Amsel

310 ***M. fuscofasciella*** (Cham.)
Kentucky is the type locality (Chambers 1875b). Harlan Co., PMS, July 8, 1976, C. V. Covell, Jr.

Genus *Scardia* Tr.

311 ***S. anatomella*** (Grt.)
PARKS AND RESERVES: BBM, BMF, CAV, OCP, PNM, RRG. **COUNTIES:** BAT, BEL, BUL, HLN, JEF, MEA, MEN, OLD. **DATES:** Apr. 26, May 1, 2, 15, 31, July 8, Aug. 28, Sept. 17, 23. Locally common.

Genus *Diataga* Wlsm.

315 ***D. leptosceles*** Wlsm.
Christian Co., PYR, June 20, 1980, R. A. Henderson.

Genus *Xylesthia* Clem.

317 ***X. pruniramiella*** Clem.
Kentucky is the type locality for synonym *clemensella* (Chambers 1873). Jefferson Co., Louisville, June 6, 24, July 14; Menifee Co., RRG, July 4, 5.

Two specimens in the NMNH are labelled "Ky." and "Aug. Busck collector."

Genus *Kearfottia* Fern.

319 ***K. albifasciella*** Fern.
PARKS AND RESERVES: BMF, KGC, PNM, RRG. **COUNTIES:** BEL, BUL, CLI, HLN, KEN, LET, MEN, OWS, WHI. **DATES:** June 10–30, July 1–18, Aug. 9. Common.

Family ACROLOPHIDAE

Acrolophid Moths

Genus *Amydria* Clem.

329 ***A. brevipennella*** Dietz
Bullitt Co., BMF, June 19; Harlan Co., BBM, July 14.

333 ***A. dyarella*** Dietz
Bell Co., PNM, July 18, 1975, A. J. Brownell.

334 ***A. effrentella*** Clem.
Fayette Co., Lexington, Aug. 28–31, Sept. 1–5, 15–29; Harlan Co., BBM, July 14.

335 ***A. margoriella*** Dietz
Hardin Co., Glendale, in Malaise trap, July 1–8, 1975, P. H. Freytag.

Genus *Acrolophus* Poey

340 ***A. arcanella*** (Clem.)
Donahue (1967) published some of the records for this species as well as for *plumifrontella*, *mortipenella*, and *popeanella*. Fayette Co., Lexington, June 19, 27–30, July 5, 17; Trigg Co., LBL, July 10. Common.

347 ***A. cressoni*** (Wlsm.)
Trigg Co., LBL, July 10, 1965, C. V. Covell, Jr.

366 ***A. mortipennella*** (Grt.)
Fayette Co., Lexington, June 5, 9, 20.

367 ***A. morus*** (Grt.)
Carroll Co., GBT, Oct. 11; Carter Co., near Cascade Caves, Oct. 19; Jefferson Co., Valley Station, Aug. 14, Oct. 30, Nov. 1–8; Menifee Co., RRG, Oct. 12–13; Rowan Co., 3.2 km (2 mi) W of SR 519 on SR 1274. Common.

372 ***A. plumifrontella*** (Clem.)
PARKS AND RESERVES: BMF, CBF, HWS, KRC, LBL. **COUNTIES:** BUL, FAY, JEF, JOH, MCC, MEA, NEL, OLD, TRI, WAR, WHI, WOL. **DATES:** June 7–July 1, Aug. 14. Common.

373 ***A. popeanella*** (Clem.)
PARKS AND RESERVES: BMF, HWS, LBL, OCP, RGR. **COUNTIES:** BRK, BUL, FAY, GRY, HRD, JEF, MCC, MEA, MER, OLD, TRI. **DATES:** June 6–15, 23–30, July, Aug. 1–21. Common.

374 ***A. propinquus*** (Wlsm.)
Letcher Co., Pine Mountain, June 10, A. F. Braun.

383 ***A. texanella*** (Cham.)
Bell Co., PNM, July 18; Letcher Co., Cumberland Valley, July 6; Trigg Co., LBL, July 13, 15.

Genus *Tinea* L. **NOTE: Return to Tineidae**

392 ***T. apicimaculella*** Cham.
Kentucky may be the type locality (A. F. Braun, *pers. comm.*). There is also an old specimen labelled "Ky." and "Aug. Busck Coll." in the NMNH. Boone Co., Florence, Aug. 13; Bullitt Co., BMF, in Malaise trap, Apr. 13–17; Fayette Co., Lexington, Oct. 6–13; Jefferson Co., Louisville, June 11, 15. Common.

393 ***T. behrensella*** Cham.
Bullitt Co., BMF, Apr. 24–27, in Malaise trap; Fayette Co., Lexington, Oct. 6–13. These specimens are in the NMNH, along with one labelled "Ky." and "Aug. Busck Coll."

394 ***T. carnariella*** Clem.
Barren Co., 8.0 km (5 mi) W of Cave City, May 24, 1980, L. D. Gibson.

396 ***T. croceoverticella*** Cham.
Kentucky is the type locality. Also listed by Forbes (1923).

400 ***T. mandarinella*** Dietz
Kentucky is the type locality for *bimaculella* (Chambers 1873), a junior homonym of *bimaculella* Thunberg (Miller and Hodges 1990). Chambers wrote: "Kentucky in August and September. Taken flying." Letcher Co., Pine Mountain, July 6, A. F. Braun.

401 ***T. misceella*** Cham.
Kentucky is the type locality (Chambers 1873).

405 ***T. pellionella*** L.
Casemaking Clothes Moth
This pest was reported as follows in the *Kentucky Cooperative Plant Pest Report* infesting wool or crawling on ceilings in homes: Clark and Jefferson counties (Nov., 1977); Bourbon and Fayette counties (Dec., 1978); Mason and Mercer counties (Oct., 1979). Old Lexington records from the UK Collection are dated Apr. 9, 22, 30, Aug., Sept. 18. Larvae were present on baseboards of a house in Jefferson Co., Louisville, Mar. 7, 1985. Widespread throughout Kentucky.

406 ***T. straminiella*** Cham.
Type locality information in Chambers (1873) indicates "Kentucky in June."

407 *T. thoracestrigella* Cham.
Forbes (1923) says "definitely known only from Texas, but is doubtfully credited to Kentucky in Dyar's list." Chambers (1876b) did not designate a type locality, and the syntypes are presumed lost (Miller and Hodges 1990); inclusion here is provisional.

Genus ***Niditinea*** Petersen

411 ***N. fuscella*** (L.)
Kentucky is the type locality of synonym *griseella* (Chambers 1873). The change in species name is from Robinson and Nielsen (1983). Letcher Co., Pine Mountain, June 16; Menifee Co., RRG, Aug. 25–26.

Genus ***Monopis*** Hbn.

415 ***M. crocicapitella*** (Clem.)
A specimen of this species in the NMNH is labelled "Ky." and "Aug. Busck Coll."

416 ***M. dorsistrigella*** (Clem.)
Jefferson Co., Valley Station, June 13, 1975, A. J. Brownell.

417 ***M. marginistrigella*** (Cham.)
Kentucky is the type locality (Chambers 1873). Jefferson Co., Valley Station, June 6; Woodford Co., 1.6 km (1 mi) W of SR 68 on Ky. River, June 8.

418.1 ***M.*** species near ***monachella*** (Hbn.)
D.R. Davis (*pers. comm.*) considers this to be a named, introduced moth, the correct name of which is still uncertain. Oldham Co., Goshen, Sept. 21, 30, Oct. 6, 17, 30; Owsley Co., 4.8 km (3 mi) NE of Booneville, May 23, June 14, Aug. 27.

Genus ***Eccritothrix*** Bradley and Hinton

422 ***E. trimaculella*** (Cham.)
Kentucky is the type locality (Chambers 1873).

Genus ***Tineola*** H.-S.

426 ***T. bisselliella*** (Hummel)
Webbing Clothes Moth
Fayette Co., Lexington, Oct. 15, 1906, "emerged in lab May 3, 1907, lived 2 weeks"; Jefferson Co., Louisville, found in home, June 5, 1986. Certainly much more common than these records indicate.

Family **PSYCHIDAE**

Bagworm Moths

Genus ***Solenobia*** Zell.

435 ***S. walshella*** Clem.
Carter Co., CAC, Apr. 11, C. V. Covell, Jr.

Genus ***Zamopsyche*** Dyar

440 ***Z. commentalis*** Dyar
Oldham Co., Goshen, at lights, 1 male, Sept. 17, 1995, R. V. Gregg.

Genus ***Astala*** Davis

443 ***A. confederata*** (G. & R.)
Bullitt Co., BMF, in Malaise trap, June 22–30; Fayette Co., Lexington, June 22, 26; Jefferson Co., Louisville, July 4; Oldham Co., HWS, June 16.

Genus ***Thyridopteryx*** Steph.

457 ***T. ephemeraeformis*** (Haw.)
Evergreen Bagworm Moth
PARKS AND RESERVES: BMF, GBT, HWS, SWR, WVP. **COUNTIES:** BUL, CRL, FAY, FUL, HND, JEF, MAD, MCC, OLD. **DATES:** Sept. 8–Oct. 20 (adults). **REMARKS:** A common and destructive pest throughout Kentucky. It feeds on many plants, but prefers red cedar (Covell 1984). See Cox and Potter (1986) regarding aerial dispersal of larvae by "ballooning."

Family **LYONETIIDAE**

Lyonetiid Moths

Genus ***Philonome*** Cham.

462 ***P. clemensella*** Cham.
Kentucky is the type locality (Chambers 1874a). A series of 7 syntypes, one of which is in the NMNH labelled "*Philonome clemensella* Chambers Ky." in Chambers' handwriting. McCreary Co., Natural Bridge, June 28; Mercer Co., Shaker's Landing, July 4, 5.

Genus ***Bedellia*** Staint.

466 ***B. somnulentella*** (Zell.)
One specimen in the NMNH is labelled "Ky. Aug. Busck Collector." Fayette Co., Lexington, in Malaise trap, July 11–14, P. H. Freytag.

Genus ***Lyonetia*** Hbn.

472 ***L. speculella*** Clem.
Kentucky is type locality of synonym *apicistrigella* (Chambers 1875a). A syntype labelled "Kentucky, August, Chambers" is in the MCZ (Miller and Hodges 1990).

Genus ***Proleucoptera*** Bsk.

474 ***P. smilaciella*** (Bsk.)
Lewis Co., emerged July 5; McCreary Co., on *Smilax*, emerged June 30, July 5; Powell Co., emerged Apr. 10 and May; Rowan Co., Morehead, emerged

from *Smilax* Aug. 22, 1963; all collected by A. F. Braun.

Genus ***Paraleucoptera*** Heinr.

475 ***P. albella*** (Cham.)
Listed from Kentucky by Forbes (1923). Kentucky is the type locality (Chambers 1871a).

Family **BUCCULATRICIDAE**

Ribbed Cocoon Maker Moths

Genus ***Bucculatrix*** Zell.

487 ***B. magnella*** Cham.
Six syntypes in the MCZ are from Kentucky (Miller and Hodges 1990). Forbes (1923) states the range of this species as "Connecticut to Kentucky," but Braun (1963) does not mention Kentucky in her treatment of the species.

488 ***B. needhami*** Braun
Barren Co., Hiseville, reared from galls on stems of *Helianthus angustifolius*, June 3, 12, A. F. Braun.

515 ***B. agnella*** Clem.
Kentucky is the type locality for *capitealbella* (Chambers 1873), a synonym of this species. There are also 3 specimens in the NMNH labelled "Ky. Aug. Busck Collector."

518 ***B. ambrosiaefoliella*** Cham.
Kentucky is the type locality (Chambers 1875a). Two syntypes in the MCZ are labelled "Kentucky, bred from leaves of *Ambrosia trifida*" (Miller and Hodges 1990). Powell Co., Oct. 24, A. F. Braun.

522 ***B. angustata*** F. & B.
"Kentucky" indicated in range by Braun (1963). A specimen in the NMNH is labelled "Ky. Aug. Busck Collector."

526 ***B. polymniae*** Braun
Holotype male from Rowan Co., Clack Mtn., collected Aug. 26, 1946, reared on leaves of *Polymnia uvedalia*, emerged Sept. 8, 1946, A. F. Braun.

529 ***B. sexnotata*** Braun
Type locality is Powell Co., NBP, one male and one female, reared on *Aster divaricatus*, collected as larvae Sept. 12, 1924, adults emerged Apr. 18, 27, 1925; all A. F. Braun.

549 ***B. packardella*** Cham.
"Kentucky, near Cincinnati, Ohio" is the type locality, according to the original description.

554 ***B. trifasciella*** Clem.
Kentucky is the type locality of synonym *obscurofasciella* (Chambers 1873).

555 ***B. quinquenotella*** Cham.
Kentucky is the type locality (Chambers 1875a). Clay Co., Red Bird River, July 13, 1933, A. F. Braun.

559 ***B. coronatella*** Clem.
Rowan Co., Morehead, Triplett Creek, larvae on *Betula nigra* (no adults reared), July 9, A. F. Braun.

560 ***B. canadensisella*** Cham.
Birch Skeletonizer Moth
Harlan Co., BBM, date unknown, A. F. Braun.

563 ***B. luteella*** Cham.
Kentucky is the type locality. Miller and Hodges (1990) report 7 syntypes in the MCZ labelled "Ky., Mar." Edmonson Co., MCV, wings only, Sept. 9, A. F. Braun.

567 ***B. callistricha*** Braun
Laurel Co., E of London, 9 paratypes, May 8–19, 1935, A. F. Braun.

571 ***B. locuples*** Meyr.
Fleming Co., Crain Creek, Apr. 25–30, May 1–7, and Plummer's Landing, Apr. 16–30; Rowan Co., Morehead, May 2–12; all A. F. Braun.

574 ***B. anaticula*** Braun
Rowan Co., Clack Mtn., mines observed in early July and late August on leaves of *Ceanothus americanus* (Braun 1963).

Superfamily **GRACILLARIOIDEA**

Family **DOUGLASIIDAE**

Douglasiid Moths

Genus ***Tinagma*** Zell.

2487 ***T. obscurofasciella*** (Cham.)
Bullitt Co., BMF, in Malaise trap, Apr. 4–8, 1976, A. J. Brownell.

Family **GRACILLARIIDAE**

Leaf Blotch Miner Moths

Genus ***Caloptilia*** Hbn.

592 ***C. azaleella*** (Brants)
Fayette Co., Lexington, in Malaise trap, Aug. 22–25, P. H. Freytag.

594 ***C. belfrageella*** (Cham.)
Meade Co., OCP, May 3, 1980, L. D. Gibson.

595 ***C. bimaculatella*** (Ely)
Menifee Co., RRG, July 19–20; Owsley Co., 4.8 km (3 mi) NE of Booneville, May 10.

596 ***C. blandella*** (Clem.)
Kentucky is the type locality for synonym *juglandivorella* (Chambers 1873). Boone Co., BBL, Aug. 25; Fayette Co., Lexington, July 6–7; Jefferson Co., Valley Station, July 11; McCracken Co., Ray Black farm, 9.7 km (6 mi) SW of Paducah, June 22; Owsley Co., 4.8 km (3 mi) NE of Booneville, May 10, July 3, Aug. 22.

600 ***C. cornusella*** (Ely)
Menifee Co., RRG, FS 9B, Indian East Fork, Kelley Branch, 229 m (750 ft) elevation, Apr. 26, May 17, 18. D.R. Davis (*pers. comm.*) determined this as "species near *cornusella*."

607 ***C. glutinella*** (Ely)
Bell Co., Clear Creek, emerged July 27; Letcher Co., Cumberland Valley, emerged Aug.; both A. F. Braun.

608 ***C. hypericella*** (Braun)
Collected by A. F. Braun on *Hypericum cistifolium*, "Ohio River Bank, Gallatin Co., Ky., Apr. 23, 1939" (A. F. Braun herbarium of mined leaves, ANSP).

610 ***C. juglandiella*** (Cham.)
Kentucky is the type locality. Chambers (1872) wrote "Kentucky. Rather common." It was mining in *Juglans nigra*.

614 ***C. murtfeldtella*** (Bsk.)
Boone Co., BBL, June 25, 1980, L. D. Gibson.

615 ***C. negundella*** (Cham.)
Chambers (1876a) states that he never "met with any larvae of this genus [*sic*]" in Kentucky; but L.D. Gibson found this species in Boone Co., BBL, June 25, 1980.

620 ***C. packardella*** (Cham.)
Chambers sent a Kentucky specimen of this species to the NHM in 1877 (Miller and Hodges 1990). Campbell Co., Apr. 26; Franklin Co., Apr. 20; Madison Co., Berea, emerged from *Acer saccharum*, Apr. 8, 24; Meade Co., OCP, Apr. 11 and June 7.

630 ***C. rhoifoliella*** (Cham.)
Edmonson Co., MCV, mining in *Rhus copallina*, Aug. 24, 1941 (A. F. Braun, in her herbarium of mined leaves, ANSP).

638 ***C. speciosella*** (Braun)
Laurel Co., London, emerged from *Acer rubrum*, Apr. 8, 1938, A. F. Braun.

639 ***C. stigmatella*** (F.)
While Chambers (1872) did not state a type locality for his synonym *purpuriella*, he did send Kentucky material representing this name to the NHM in 1877 (Miller and Hodges 1990).

641 ***C. superbifrontella*** (Clem.)
Chambers probably had this species from Kentucky (Chambers 1875b). Letcher Co., Joe Day Branch, mining in *Hamamelis virginiana*, July 19; Menifee Co., RRG, in black-light trap, May 10, 17–18.

642 ***C. umbratella*** (Braun)
Todd Co., Clifty Creek, emerged in July, A. F. Braun.

644 ***C. violacella*** (Clem.)
Chambers (1872) refers to synonym *desmodifoliella* Clemens as follows: "Kentucky. Common." He notes it as feeding on leaves of *Desmodium*.

Genus *Micrurapteryx* Spuler

647 ***M. salicifoliella*** (Cham.)
Kentucky is the type locality. Chambers (1872) wrote "Common in Kentucky" in his original description. Several specimens in the NMNH were collected in "Ky." by August Busck. Rowan Co., Triplett Creek, mining in willow, Oct. 18, A. F. Braun.

Genus *Parectopa* Clem.

650 ***P. bumeliella*** Braun
Braun's type series is from Edmonson Co., MCV, emerged June 13, 28, 29, July 4, Sept. 12, mining in *Bumelia lyciodes*, Oct. 10, A. F. Braun.

651 ***P. geraniella*** Braun
A paratype is from Clay Co., Red Bird River, emerged July 7, 19. Carter Co., CAC, emerged June 23, 29; Edmonson Co., MCV, emerged July 14, Aug. 28; Perry Co., Clover Fork, emerged July 19, 25, and Lynn Fork, emerged July 25.

654 ***P. occulta*** Braun
The type series is from Powell Co., emerged July 12. Menifee Co., Wolfpen Creek, reared on *Vicea*, emerged Sept. 4; Rowan Co., Clack Mtn., emerged Sept. 3, Oct. 13, 30; Whitley Co., Buck Branch, emerged May 11, 15, all A. F. Braun.

655 ***P. pennsylvaniella*** (Engel)
A series in the NMNH bears labels "Ky." and "Aug. Bsk. Collector." Menifee Co., RRG, Apr. 26, 1981, L. D. Gibson.

656 ***P. plantaginisella*** (Cham.)
Kentucky is the type locality. Chambers (1872) indicated it as common and mining in leaves of *Plantago virginica* in Sept., Oct., and Nov.

657 ***P. robiniella*** Clem.
COUNTIES: BEL, BOO, FAY, LET, MCY, ROW. **DATES:** Apr. 19, July 1–17, Aug. 13, 15–28. Foodplants recorded by A. F. Braun are *Desmodium* and *Lespedeza hirta*.

Genus *Neurobathra* Ely

663 ***N. strigifinitella*** (Clem.)
Kentucky is the type locality for synonym *duodecemlineella* (Chambers 1872). He wrote "Kentucky; not common." Fayette Co., Lexington, June 6–7; Harlan Co., BBM, mining in *Castanea dentata*, July 22; McCreary Co., Natural Bridge, June 22.

Genus *Parornix* Spuler

673 ***P. geminatella*** (Pack.)
Kentucky is type locality for *prunivorella* (Chambers 1873), a synonym. Boone Co., Woolper Creek, Apr. 3, 1935, A. F. Braun, who recorded it mining in apple and wild cherry.

675 ***P. inusitatumella*** (Cham.)
The type locality is "near Covington, Kentucky" (Chambers 1873).

678 ***P. obliterella*** (Dietz)
Harlan Co., BBM, 1,220 m (4000 ft) elevation, emerged Sept. 13; Lewis Co., on *Betula lenta*, emerged Aug. 28; Powell Co., emerged June 14, 15, 18; all A. F. Braun.

681 ***P. quadripunctella*** (Clem.)
McCreary Co., mining in service berry, Sept., 1934; Powell Co., reared on *Amelanchier*, emerged Aug. l; Whitley Co., CBF, emerged Apr. 15; all A. F. Braun.

Genus *Acrocercops* Wallgr.

689 ***A. albinatella*** (Cham.)
Kentucky is the type locality (Chambers 1872).

692 ***A. astericola*** (F. & B.)
Menifee Co., RRG, Apr. 26; Powell Co., RRG, found mining in *Aster divaricatus*, Oct. 21, 1954, A. F. Braun.

694 ***A. quinquistrigella*** (Cham.)
Listed from Kentucky by Forbes (1923).

697 ***A. strigosus*** Braun
Carter Co., CAC, emerged June 16, 18, 21; Edmonson Co., mining in *Quercus alba*, July 4; Powell Co. (holotype); Whitley Co., CBF, mining in *Quercus alba*, June 23; all are A. F. Braun records.

Genus *Leucospilapteryx* Spuler

698 ***L. venustella*** (Clem.)
Listed from Kentucky by Forbes (1923). Kentucky is the type locality for synonym *eupatoriella* (Chambers 1872). He wrote "The larva may be found in the leaves of *Eupatorium ageratoides* from July to October, but it is rather rare."

Genus *Leucanthiza* Clem.

700 ***L. amphicarpeaefoliella*** Clem.
The holotype of synonym *saundersella* (Chambers 1871b) is "A single specimen taken in Kentucky in Aug." Letcher Co., Joe Day Branch, mining in *Amphicarpa monoica*, July 19, 1937, A. F. Braun.

701 ***L. dircella*** Braun
Carter Co., CAC, mining in *Dirca palustris*, Oct. 1, 1953; Whitley Co., CBF, mining in *Dirca palustris*, June 25, 1936; both A. F. Braun.

Genus *Marmara* Clem.

707 ***M. elotella*** (Bsk.)
Jefferson Co., Valley Station, May 27, 1975, A. J. Brownell.

708 ***M. fasciella*** (Cham.)
Kentucky is the type locality (Chambers 1875a), as well as that of synonym *quinquenotella*: "Kentucky June 10th to 15th" (Chambers 1877a).

718 ***M. smilacisella*** (Cham.)
Kentucky is the type locality (Chambers 1875a). Letcher Co., Pine Mountain, emerged Oct. 7; Lewis Co., Vanceburg, emerged Apr. 26–May 1; Powell Co., emerged Oct. 28, and NBP, Sept. 10, 17; all A. F. Braun.

Genus *Cremastobombycia* Braun

720 ***C. ambrosiella*** (Cham.)
"Southern Kentucky" is the type locality (Chambers 1871a). He described this species from "just dead" larvae in mines in *Smilax glabra* found in July with their dead parasitoids. Rowan Co., Morehead, emerged from *Ambrosia trifida*, Aug. 9, Dec. 18, T. N. Freeman.

722 ***C. ignota*** (F. & B.)
Kentucky is the type locality of synonym *helianthisella* (Chambers 1874c).

723 ***C. solidaginis*** (F. & B.)
Rowan Co., Morehead, emerged from *Solidago* Aug. 9, 1963, T. N. Freeman.

Genus *Phyllonorycter* Hbn.

725 ***P. aberrans*** (Braun)
Edmonson Co., MCV, emerged July 18; McCreary Co., emerged on *Desmodium*, Apr. 9; Monroe Co., Mud Lick, emerged on *Desmodium*, Apr. 11; Perry Co., Lynn Fork, emerged July 29; Whitley Co., CBF, emerged July 12; all A. F. Braun.

727 ***P. albanotella*** (Cham.)
Kentucky is the type locality (Chambers 1875a).

733 ***P. argentifimbriella*** (Clem.)
Kentucky is the type locality of synonym *fusco-*

costella (Chambers 1875a). Jefferson Co., Tom Wallace Lake, reared from *Quercus*, Sept. 24, 1975.

734 ***P. argentinotella*** (Clem.)
Chambers (1871b) indicated this species was common in Pennsylvania and Kentucky.

737 ***P. auronitens*** (F. & B.)
A. F. Braun reared this from alder as follows: Fleming Co., emerged Apr. 15, May 1, 8, 17, Crain Creek, emerged Apr. 20, and Plummer's Landing, emerged July 20; Montgomery Co., Jeffersonville, emerged Apr. 6; Rowan Co., emerged Apr. 3 and 13, and Triplett Creek, Oct. 18, 1941; Whitley Co., CBF, emerged Apr. 21.

738 ***P. basistrigella*** (Clem.)
Rowan Co., on *Quercus alba*, Apr. 2, A. F. Braun.

741 ***P. caryaealbella*** (Cham.)
Listed from Kentucky by Braun *in* Forbes (1923).

742 ***P. celtifoliella*** (Cham.)
Kentucky is the type locality (Chambers 1871b). Oldham Co., HWS, reared from *Celtis occidentalis*, emerged June 20, Aug. 28, Sept. 12, 23, 1975, A. J. Brownell.

743 ***P. celtisella*** (Cham.)
Kentucky is the type locality (Chambers 1871b). Bullitt Co., BMF, reared on *Celtis occidentalis*, emerged July 27, 30, Aug. 3, 11, A. J. Brownell.

744 ***P. clemensella*** (Cham.)
Kentucky is the type locality (Chambers 1871b). He wrote "Kentucky. Common."

746 ***P. crataegella*** (Clem.)
Fayette Co., Lexington, in Malaise trap, Aug. 22–25; Letcher Co., Big Black Mountain, emerged July 21, A. F. Braun.

748 ***P. deceptusella*** (Cham.)
Kentucky is the type locality (Chambers 1879).

750 ***P. diversella*** (Braun)
Edmonson Co., MCV, emerged Apr. 29, May 1; Madison Co., Berea, Apr.; Powell Co., emerged Apr. 23, May 5, 22, July 23; all A. F. Braun.

752 ***P. fitchella*** (Clem.)
Edmonson Co., MCV, emerged Apr. 3, 9, 12, 14, 16; Fleming Co., Fox Mountain, on *Quercus stellata*, emerged Mar. 31, Apr. 2, 3, 5; Letcher Co., Pine Mountain, June 22; all A. F. Braun.

755 ***P. hagenii*** (F. & B.)
Kentucky is the type locality for synonym *necopinusella* (Chambers 1878b).

761 ***P. kearfottella*** (Braun)
Letcher Co., Pine Mountain, mining in *Castanea dentata*, no date; Powell Co., larvae on chestnut, Aug. 15, 1908, A. F. Braun.

764 ***P. lucetiella*** (Clem.)
Chambers (1871b) reported this species from Pennsylvania and Kentucky, mining in the underside of *Tilia americana* leaves. McCreary Co., Yahoo Valley, mining, Oct. 11; Rowan Co., Clack Mtn., mining in *Tilia neglecta*, Oct. 3; all A. F. Braun.

765 ***P. lucidicostella*** (Clem.)
Chambers (1871b) wrote of this species: "Imago in Apr., May, July, and Aug. Abundant. Pennsylvania, Kentucky, and Alabama."

766 ***P. lysimachiaeella*** (Cham.)
Kentucky is the type locality (Chambers 1875a).

767 ***P. malimalifoliella*** (Braun)
Braun (1908) includes data on Kentucky specimens of this species reared on quince by Chambers.

770 ***P. martiella*** (Braun)
Rowan Co., Morehead, reared on *Quercus* sp., emerged Aug. 7, 1963, T. N. Freeman, who identified it as "near *martiella*."

773 ***P. morrisella*** (Fitch)
Chambers sent a Kentucky specimen of synonym *amphicarpeaeella* (Chambers 1877b) to the NHM in January, 1878 (Miller and Hodges 1990).

775 ***P. obscuricostella*** (Clem.)
Kentucky is the type locality for synonym *virginiella* (Chambers 1871b), which he reared from hornbeam. Carter Co., CAC, emerged Apr. 15, A. F. Braun.

781 ***P. ostryaefoliella*** (Clem.)
Powell Co., June 8, A. F. Braun.

783 ***P. populiella*** (Cham.)
Localities given in the original description are "Ohio and Kentucky" (Chambers 1878b).

785 ***P. quercialbella*** (Fitch)
Kentucky is the type locality of synonym *quercipulchrella* (Chambers 1878d).

790 ***P. robiniella*** (Clem.)
Boone Co., Florence, Aug. 13; Fayette Co., Lexington, Malaise trap, Aug. 19–25; Jefferson Co., Valley Station, July 14; McCreary Co., Natural Arch, mining in *Robinia* sp., June 15; Whitley Co., CBF, on *Robinia*, emerged June 19.

791 ***P. salicifoliella*** (Cham.)
Kentucky is the type locality. Chambers (1872) called it "common in Kentucky."

794 ***P. scudderella*** (F. & B.)
Kenton Co., Doe Run Lake County Park, reared from *Salix sericea*, collected 15 Oct 1997, two females emerged 5 April 1998, J.A. Merkle

795 ***P. sexnotella*** (Cham.)
Kentucky is the type locality (Chambers 1880a).

796 ***P. symphoricarpaeella*** (Cham.)
Kentucky is the type locality (Chambers 1875a). McCracken Co., emerged July 11, 1938, A. F. Braun.

797 ***P. tiliacella*** (Cham.)
Kentucky is the type locality. Chambers (1871b) wrote "Kentucky. Rare."

800 ***P. tritaenianella*** (Cham.)
Kentucky is the type locality: "A single specimen, captured in Apr. in Kentucky" (Chambers 1871b).

Genus ***Cameraria*** Chapman

803 ***C. aceriella*** (Clem.)
Chambers (1871b) noted that this species is "common in Kentucky." McCreary Co., mining in *Acer rubrum*, Aug. 30, 1941, A. F. Braun.

804 ***C. aesculisella*** (Cham.)
Chambers (1871b) indicated "Kentucky, Pennsylvania. Abundant" and that it mines the upper surface of leaves of the buckeye (*Aesculus glabra*).

805 ***C. affinis*** (F. & B.)
Rowan Co., Clack Mtn., emerged Mar. 29, mining in *Triosteum angustifolium*, Oct. 5, A. F. Braun.

809 ***C. bethunella*** (Cham.)
Rowan Co., Clack Mtn., emerged May 11, A. F. Braun.

811 ***C. caryaefoliella*** (Clem.)
Chambers (1871b) noted this species as follows: "Kentucky: Rare." Rowan Co., Clack Mtn., mining in hickory, Sept. 3, 1945, A. F. Braun.

812 ***C. castaneaeella*** (Cham.)
Kentucky is the type locality (Chambers 1875a). Rowan Co., Morehead, emerged Apr. 26, A. F. Braun.

815 ***C. cincinnatiella*** (Cham.)
Boone Co., near BBL, mining in *Quercus macrocarpa*, Sept. 27, 1941; Edmonson Co., MCV, emerged July 8; both A. F. Braun.

816 ***C. conglomeratella*** (Zell.)
Kentucky is the type locality for synonym *bicolorella* (Chambers 1878b).

817 ***C. corylisella*** (Cham.)
Chambers sent a Kentucky specimen of this species to the NHM in 1877 (Miller and Hodges 1990). Edmonson Co., MCV, emerged July 8; Powell Co., emerged May 28; both A. F. Braun.

819 ***C. fasciella*** (Wlsm.)
Kentucky is the type locality for synonym *unifasciella* (Chambers 1875a).

822 ***C. guttifinitella*** (Clem.)
Bell Co., PNM, July 18, 1975, A. J. Brownell.

823 ***C. hamadryadella*** (Clem.)
Solitary Oak Leafminer Moth
Fayette Co., mining in *Quercus macrocarpa* leaves, emerged Aug. 28, H. Garman; Rowan Co., Clack Mtn., emerged July 17, A. F. Braun. "Scattered moderate infestations in Rockcastle Co. in May," (*Kentucky Cooperative Economic Insect Report* 1976).

824 ***C. hamameliella*** (Bsk.)
Letcher Co., Big Black Mountain, mining in *Hamamelis virginiana*, July 19; Carter Co., Tygart's Creek, mining in *H. virginiana*, Oct. 16; McCreary Co., Natural Bridge, on *Hamamelis*, emerged May 13; Powell Co., emerged Aug. 2, 14, all A. F. Braun.

831 ***C. obstrictella*** (Clem.)
Kentucky is type locality of synonym *bifasciella* (Chambers 1878b). Carter Co., CAC, emerged Apr. 20; Letcher Co., Pine Mountain, emerged July 20; Perry Co., Lynn Fork, on *Quercus alba*, emerged July 19, 21; Rowan Co., Clack Mtn., on *Q. montana*, emerged in July; all A. F. Braun.

832 ***C. ostryarella*** (Cham.)
Kentucky is the type locality. Chambers (1871b) indicated that it was common. Powell Co., emerged Aug. 2, A. F. Braun.

838 ***C. tubiferella*** (Clem.)
Carter Co., mining in *Quercus alba*, Oct. 18, 1941, A. F. Braun.

839 ***C. ulmella*** (Cham.)
Kentucky is the type locality (Chambers 1871b).

Genus ***Chrysaster*** Kumata

842 ***C. ostensackenella*** (Fitch)
Reared on black locust by A. F. Braun, as follows: Bell Co., PNM, emerged July 5; Fayette Co., Lexington, emerged Aug. 20; Rowan Co., Morehead, emerged Aug. 7, 9.

Genus ***Porphyrosela*** Braun

843 ***P. desmodiella*** (Clem.)
Chambers (1871b) listed this species from Ken-

tucky: "Mines the under surface of leaves of different species of *Desmodium*. I have met with it only in August and September and rarely then. Pennsylvania and Kentucky."

Genus ***Phyllocnistis*** Zell.

844 ***P. ampelopsiella*** Cham.
Kentucky is the type locality. Chambers (1871b) indicated it was common.

846 ***P. insignis*** F. & B.
Kentucky is type locality for *erechtiisella* (Chambers 1878c), a synonym. Letcher Co., Cumberland Valley, June 26; Oldham Co., CBB, from pupa in ground ivy, Nov. 8.

848 ***P. liquidambarisella*** Cham.
Kentucky is the type locality. Chambers (1875a) wrote "It mines in the upper surface of the leaves of the sweet gum, *Liquidambar styraciflua*, in southern Kentucky throughout the summer." Laurel Co., near London, mining in *L. styraciflua*, June 24; Monroe Co., mining in *L. styraciflua*, Aug. 21, A. F. Braun.

849 ***P. liriodendronella*** Clem.
Powell Co., Sept. 17, A. F. Braun.

851 ***P. magnoliella*** Fbs.
Letcher Co., Big Black Mountain, July 19; Letcher Co., Pine Mountain, emerged June 29; both A. F. Braun.

852 ***P. populiella*** Cham.
Kentucky is the type locality (Chambers 1875a).

853 ***P. vitegenella*** Clem.
Clay Co., Red Bird River, on grape, emerged July 12; Fayette Co., Lexington, reared on grape, emerged Aug. 11.

854 ***P. vitifoliella*** Cham.
The type locality is either Kentucky or Wisconsin depending on the selection of a lectotype. Chambers (1871b) wrote "The larva mines the upper surface of wild grape leaves from May to Oct. — Kentucky, Wisconsin."

Superfamily **GELECHIOIDEA**

Family **OECOPHORIDAE**

Oecophorid Moths

Genus ***Agonopterix*** Hbn.

856 ***A. hyperella*** Ely
Edmonson Co., MCV, A. F. Braun.

859 ***A. curvilineella*** (Beutenmüller)
Bracken Co., SR 1159 and Locust Cr., Sept. 25; Bullitt Co., BMF, in Malaise trap, Apr. 6–9; Harlan Co., BBM, Aug. 8; Oldham Co., CBB, May 14.

862 ***A. clemensella*** (Cham.)
Boone Co., Florence, Mar. 11, 1977, L. D. Gibson.

864 ***A. atrodorsella*** (Clem.)
Meade Co., OCP, Apr. 10; Powell Co., emerged in June.

866 ***A. eupatoriiella*** (Cham.)
Kentucky is the type locality (Chambers 1878c). Bullitt Co., BMF, Apr. 9–13; Jefferson Co., Valley Station, Apr. 22.

867 ***A. pulvipennella*** (Clem.)
Boone Co., BBL, Mar. 14, Apr. 30, May 22, Aug. 25, and Florence, Apr 3; Grayson Co., ASC, Sept. 25; Jefferson Co., Valley Station, Apr. 1; Oldham Co., HWS, May 19; Owsley Co., near Booneville, July 5.

869 ***A. walsinghamella*** (Bsk.)
Rowan Co., Tackett Mountain, emerged in June, A. F. Braun.

880 ***A. flavicomella*** (Engel)
Bullitt Co., BMF, in Malaise trap, May 11–30, June 1–6; Menifee Co., RRG, May 26, 27, June 8; Oldham Co., Goshen, June 6, 7, 11; Rowan Co., Tackett Mountain, emerged in June.

881 ***A. senicionella*** (Bsk.)
Oldham Co., HWS, June 22, 1966, C. V. Covell, Jr.

882 ***A. robiniella*** (Pack.)
PARKS AND RESERVES: BBM, GBT, HWS, LBL, OCP, PYR. **COUNTIES:** CHR, CRL, FAY, HLN, JEF, MEA, OLD, TRI. **DATES:** Apr. 26, June 14–27, July 10, 14, Aug. 6, Sept. 12, Oct. 3, 10. Common.

884 ***A. thelmae*** Clarke
Carroll Co., GBT, Oct. 11; Fayette Co., Lexington, in Malaise trap, July 7–11; Jefferson Co., CAH, Aug. 7; Oldham Co., Goshen, July 6, Aug. 2, 28, Sept. 4, 5, 18. Common.

889 ***A. argillacea*** (Wlsm.)
Pendleton Co., overlook of Ohio River along SR 8, July 2, 1987, D. J. Wright.

Genus ***Semioscopis*** Hbn.

913 ***S. merriccella*** Dyar
Harlan Co., KGC, Apr. 20; Menifee Co., RRG, Mar. 12, Apr. 5; Owsley Co., 4.8 km (3 mi) NE of Booneville, Mar. 20, 29, Apr. 3, 22; Powell Co., Mar. 12.

915 ***S. megamicrella*** Dyar
Boone Co., BBL, Mar. 25, and BCN, Mar. 24, Apr. 3; Menifee Co., RRG, Mar. 12.

916 ***S. aurorella*** Dyar
Boone Co., BBL, Mar. 25; Carroll Co., GBT, Mar. 29; Meade Co., OCP, Mar. 16; Oldham Co., Goshen, Mar. 7, 11, 21.

Genus *Depressaria* Haw.

922 ***D. pastinacella*** (Dup.)
Parsnip Webworm Moth
Montgomery Co., US 460 at I-64 near Mt. Sterling, as larvae on flowerhead of *Zizia aurea*, collected May 24, 1987 by B. S. Nichols, pupated June 1, 1987, adult emerged "late June 1987."

Genus *Nites* Hodges

945 ***N. maculatella*** (Bsk.)
Letcher Co., Big Black Mountain, emerged in June, A. F. Braun.

Genus *Machimia* Clem.

951 ***M. tentoriferella*** Clem.
Gold-Striped Leaftier Moth
PARKS AND RESERVES: BIP, BMF, CBF, GBT, KLK. **COUNTIES:** BRA, BUL, CRL, HLN, JEF, MRL, NEL, PIK, WHI. **DATES:** Apr. 28, June 1, 2, 15, July 16, Sept. 18–25, Oct. 3–16. **REMARKS:** A. J. Brownell reared this species from *Celtis* sp. Common.

Genus *Psilocorsis* Clem.

955 ***P. quercicella*** Clem.
PARKS AND RESERVES: BMF, CER, OCP, RRG. **COUNTIES:** BOO, BUL, HIC, JEF, MEA, MEN, OWS, ROW. **DATES:** May 10, 22–31, June 1–6, 15, 16, July 11, Aug. 9–26, Sept. 11. **REMARKS:** A. F. Braun reared this from larvae collected at Morehead, and adults emerged in April. Common.

956 ***P. cryptolechiella*** (Cham.)
Kentucky is the type locality for synonym *faginella* (Chambers 1872). Rowan Co., Clack Mtn., emerged in Apr., A. F. Braun.

957 ***P. reflexella*** Clem.
PARKS AND RESERVES: BBL, BBM, BMF, CBB, KGC, LBL, NBP, OCP, RRG. **COUNTIES:** BOO, BUL, HLN, JEF, MEA, MEN, NEL, OLD, OWS, POW, SPE, TRI, WHI. **DATES:** Apr. 11, May 4–26, June 2, 30, July 8–15, 23, Oct. 13. Common.

Genus *Ethmia* Hbn.

982 ***E. macelhosiella*** Bsk.
Boone Co., BBL, Oct. 26; Fayette Co., Lexington, 1 female at black-light, Oct. 27; Oldham Co., Goshen, Oct. 20. Uncommon.

992 ***E. zelleriella*** (Cham.)
PARKS AND RESERVES: BBL, BBM, BIP, BMF, CBB, HWS, LCP, NBP, OCP, PNM, RRG. **COUNTIES:** BEL, BOO, BUL, HLN, JEF, MEA, MER, OLD, PIK, POW, ROC, RUS. **DATES:** Apr. 21–29, May 5–18, June 8, 22, July 8, 18, 19, Aug. 5, 17. Common.

999 ***E. longimaculella*** (Cham.)
A lectotype for this name was selected by Powell (1973), with data as follows: "Kentucky, June 4, V. T. Chambers." Thus Kentucky is the type locality.

1003 ***E. trifurcella*** (Cham.)
The type locality is Kentucky (Chambers 1873). Bullitt Co., BMF, June 19; Jefferson Co., emerged May 12–16; Jefferson Co., near Cedar Creek, mining in *Cynoglossum virginianum*, July 22; Owsley Co., 4.8 km (3 mi) NE of Booneville, May 29, June 14; Trigg Co., LBL, July 13.

Genus *Antaeotricha* Zell.

1011 ***A. schlaegeri*** (Zell.)
PARKS AND RESERVES: BBL, BBM, BMF, CAC, CBF, HWS, KRC, PNM, RBF. **COUNTIES:** BEL, BOO, BRE, BUL, CAR, FAY, HLN, HRD, JEF, LET, MCC, OLD, OWS, WHI, WOL. **DATES:** Apr. 13–21, May 1, 10–June 4, 15, July 11, 15, 27–Aug. 11, 26, Sept. 5. Common.

1014 ***A. leucillana*** (Zell.)
PARKS AND RESERVES: BAN, BBL, BUC, CAC, HNK, HWS, PMS, RRG. **COUNTIES:** BAR, BOO, BRE, CAL, CAR, FAY, FUL, HLN, JEF, MEN, MER, OLD, PER. **DATES:** Apr. 20, 28, May 1–17, June 15, 20, July 8, Aug. 8–15, Sept. 8, 22. Common.

1019 ***A. humilis*** (Zell.)
Letcher Co., Pine Mountain on red oak, emerged July 15, 25, A. F. Braun.

Genus *Setiostoma* Zell.

1026 ***S. xanthobasis*** Zell.
Harlan Co., KGC, July 11; McCreary Co., Blue Heron Campground, Apr. 27; Owsley Co., 4.8 km (3 mi) NE of Booneville, May 28; Powell Co., no date.

Genus *Menesta* Clem.

1030 ***M. tortriciformella*** Clem.
Kentucky is the type locality of synonym *coryliella* (Chambers 1875b). Bullitt Co., BMF, in Malaise trap, June 1–15; Letcher Co., Pine Mountain, June 16, 21, A. F. Braun.

1031 ***M. melanella*** Murt.
Harlan Co., Big Black Mountain, mining, Sept. 28;

Larue Co., Upton, emerged July 15, 18; Owsley Co., 4.8 km (3 mi) NE of Booneville, May 29; Powell Co., mining in *Quercus montana*, A. F. Braun; Whitley Co., CBF, mining in *Q. prinus*, July 24; all A. F. Braun.

Genus ***Gonioterma*** Wlsm.

1032 ***G. mistrella*** (Bsk.)
Barren Co., BAN, Oct. 3; Fayette Co., Lexington, in Malaise trap, June 23–30, July 7–11, Sept. 8–15; Rockcastle Co., Sept. 13.

Genus ***Inga*** Bsk.

1034 ***I. sparsiciliella*** (Clem.)
Laurel Co., FS 4176, July 28; Owsley Co., 4.8 km (3 mi) NE of Booneville, July 20; Russell Co., LCP, Aug. 14.

Genus ***Decantha*** Bsk.

1042 ***D. boreasella*** (Cham.)
PARKS AND RESERVES: BBL, OCP, RRG. **COUNTIES:** BOO, JEF, MCY, MEA, MEN, OWS, POW. **DATES:** Apr. 27, May 16, June 1–25. **REMARKS:** Kentucky is the type locality (Chambers 1873).

1043 ***D. stecia*** Hodges
Bell Co., PNM, July 19, 1975, A. J. Brownell.

Genus ***Callima*** Clem.

1046 ***C. argenticinctella*** Clem
PARKS AND RESERVES: AJJ, BEC, HWS. **COUNTIES:** BOO, CAM, FAY, FUL, HOP, JEF, OLD, OWS, ROW, WHI. **DATES:** June 5–28, July 1, 3, 11–16, Aug. 13, Sept 7. Common.

Genus ***Fabiola*** Bsk.

1050 ***F. shaleriella*** (Cham.)
The type locality is Edmonson Co., Camp Bee Spring (Chambers 1875a). Harlan Co., BBM, May; Letcher Co., Big Black Mountain, June; Morgan Co., near Ezel, May; Wolfe Co., May; all A. F. Braun.

1051 ***F. tecta*** Braun
The type locality is Saltpeter Cave near Natural Bridge, McCreary Co., June 22; also Whitley Co., CBF, June; Letcher Co., Cumberland Valley, July; all A. F. Braun.

1053 ***F. edithella*** (Bsk.)
A synonym is *amplicincta* Braun, which Braun reported from Bell Co., PNM, June 11, and Powell Co. (paratype of *amplicincta*), June 26, 1921.

Genus ***Polix*** Hodges

1058 ***P. coloradella*** (Wlsm.)
Bell Co., PNM, June 6, 1979, C. Cornett.

Genus ***Mathildana*** Clarke

1059 ***M. newmanella*** (Clem.)
Bullitt Co., BMF, in Malaise trap, June 1–6; Kenton Co., Taylor Mill, May 25, June 9.

Genus ***Martyringa*** Bsk.

1065 ***M. latipennis*** (Wlsm.)
Harlan Co., BBM and KGC, July 11; Oldham Co., Goshen, June 17, 29–30, July 1–20, Aug. 2–20; Powell Co., RRG, Tunnel Ridge, July 23; Rowan Co., Morehead, June 27–30. Locally common.

Genus ***Eido*** Cham.

1068 ***E. trimaculella*** (Fitch)
Type locality for synonym *albapalpella* (Chambers 1872) "Captured in June in Kentucky." Letcher Co., Colliers Creek, June; Powell Co., NBP, Aug.; both A. F. Braun.

Genus ***Cyphacma*** Meyr.

1073 ***C. tragiae*** Braun
Jefferson Co., Cedar Creek, mining in *Tragia macrocarpa*, collected July 22, Sept. 3, 1940, A. F. Braun.

NOTE: Numbers for the remaining three genera in Oecophoridae reflect their transfer from Blastobasidae by Adamski and Brown (1989).

Genus ***Sceptea*** Wlsm.

1135 ***S. aequepulvella*** (Cham.)
Chambers (1872) gave the type locality as "Kentucky" and later stated (Chambers 1875b) that he had "taken it frequently in Kentucky..." McCracken Co., Paducah, at lights, Aug. 2; Meade Co., OCP, July 25.

Genus ***Glyphidocera*** Wlsm.

1137 ***G. democratica*** Meyr.
Jefferson Co., Valley Station, June 20, 1974, A. J. Brownell.

1142 ***G. septentrionella*** Bsk.
Letcher Co., Big Black Mountain, July 7, A. F. Braun.

Genus ***Gerdana*** Bsk.

1144 ***G. caritella*** Bsk.
Campbell Co., AJJ, July 15, 1982, L. D. Gibson.

Family **ELACHISTIDAE**

Elachistid Moths

Genus ***Stephensia*** Staint.

1079 ***S. cunilae*** Braun
Fleming Co., Blue Bank, June 20 (Braun 1948).

Genus *Elachista* Tr.

1098 ***E. hiberna*** Braun
Menifee Co., RRG, larvae reported mining in leaves (Braun 1948).

1119 ***E. solitaria*** Braun
Powell Co., June 11, July 10 (Braun 1948).

1120 ***E. radiantella*** Braun
Fleming Co., July 25; Powell Co., June 10 (Braun 1948).

Genus *Cosmiotes* Clem.

1129 ***C. illectella*** Clem.
"Kentucky" (Braun 1948); Bullitt Co., BMF, Mar. 27–31, Apr. 1–27; Fayette Co., Lexington, in Malaise trap, 11 specimens, June 9–20, July 11–14, Aug. 19–29, Sept. 1–8; Jefferson Co., Louisville, Apr. 19, June 22.

Genus *Dicranoctetes* Braun

1132 ***D. brachyelytrifoliella*** (Clem.)
Carter Co., CAC, emerged Apr. 11 and May 11, 1937; Edmonson Co., MCV, emerged Sept. 28, 1940 and Apr. 16–May 28, 1941 (Braun 1948); Hart Co., near Mammoth Cave, mining in *Muhlenbergia*, Sept. 8; Hart Co., near Mammoth Cave, mining in *Uniola latifolia*, Sept. 8 (Braun 1948).

Family **BLASTOBASIDAE**

Blastobasid Moths

NOTE: Differences here from Hodges et al. (1983) reflect changes by Adamski and Brown (1989). These include movement of some genera to Oecophoridae.

Genus *Blastobasis* Zell.

1162 ***B. glandulella*** (Riley)
Acorn Moth
Boone Co., CER, Aug. 17; Oldham Co., HWS, Aug. 18; Wolfe Co., KRC, July 15.

Genus *Asaphocrita* Meyr.

1212 ***A. sciaphilella*** (Zell.)
Kentucky is the type locality for synonym *triangularisella* (Chambers 1875b).

Genus *Pigritia* Clem.

1232 ***P. laticapitella*** Clem.
Kentucky is the locality of a syntype of synonym *luteopulvella* (Chambers 1875c).

1246 ***P. murtfeldtella*** (Cham.)
Fourteen syntypes of this species in the MCZ are labelled as from Kentucky (Miller and Hodges 1990).

Family **COLEOPHORIDAE**

Casebearer Moths

Genus *Coleophora* Hbn.

1254 ***C. malivorella*** Riley
Kentucky is the type locality for synonyms *multipulvella* and *cinerella* (Chambers 1878b). A. F. Braun's records: Harlan Co., BBM, emerged in June; Lewis Co., Big Sulphur Creek, mining in *Prunus serotina*, May 19; Powell Co., NBP, May 16, "June."

1261 ***C. querciella*** Clem.
Harlan Co., BBM, mining in *Castanea dentata*, July 22; Letcher Co., Pine Mountain, June 21, A. F. Braun.

1269 ***C. vernoniaeella*** Cham.
Kentucky is the type locality (Chambers 1878b).

1271 ***C. pruniella*** Clem.
Kentucky is the type locality for synonyms *nigralineella* (Chambers 1876b) and *ochrella* (Chambers 1878b).

1273 ***C. kalmiella*** (McD.)
Carter Co.; Whitley Co., CBF; both A. F. Braun.

1276 ***C. gaylussaciella*** Heinr.
Rowan Co., Clack Mtn., adults emerged in June, A. F. Braun.

1278 ***C. viburniella*** Clem.
Blue Lick, county unrecorded (probably Lincoln or Nicholas), mining in *Viburnum rufidulum*, Sept. 22, emerged in May, A. F. Braun.

1291 ***C. laticornella*** Clem.
Miller and Hodges (1990) indicated that Chambers sent Kentucky specimens of synonym *caryaefoliella* Clemens to the NHM in 1877. Edmonson Co., MCV, mining in hickory, Aug. 19, 1941, A. F. Braun.

1292 ***C. corylifoliella*** Clem.
Powell Co., no date, A. F. Braun.

1295 ***C. ostryae*** Clem.
Larue Co., Knob Creek, mining in *Ostrya virginiana*, May 19, 1940, A. F. Braun.

1299 ***C. umbratica*** Braun
Rowan Co., near Morehead, mining in wild plum, Oct. 17, 1937, A. F. Braun.

1333 ***C. subapicis*** Braun
Rowan Co., Tackett Mountain, emerged July 26, A. F. Braun.

1337 ***C. duplicis*** Braun
Menifee Co., RRG, FS 9B, Indian East Fork, Kel-

ley Branch, 220 m (720 ft) elevation, Aug. 17–18, light trap; Whitley Co., CBF, Sept.

1349 ***C. lineapulvella*** Cham.
A specimen labelled "Kentucky" in NMNH may be Chambers' type, which he indicated was from "Kentucky. Taken at the lamp" (Chambers 1878a). Forbes (1923) also attributed *amaranthella* Braun, now a synonym, to Kentucky.

1358 ***C. quadrilineella*** Cham.
Kentucky is the type locality (Chambers 1878b).

1361 ***C. borea*** Braun
Menifee Co., RRG, FS 9B, Indian East Fork, Kelley Branch., 220 m (720 ft) elevation, Sept. 6–7, 1985, in light trap, J. S. Nordin.

1365 ***C. cratipennella*** Clem.
Kentucky is the type locality of synonyms *gigantella* (Chambers 1874a) and *shaleriella* (Chambers 1875a). Boone Co. BBL, May 31; Fulton Co., RNR, May 13; Jefferson Co., Anchorage, May 22.

1372 ***C. coenosipennella*** Clem.
Big Black Mountain, county and date unknown; Rowan Co., Lick Fork, Apr., and Morehead, May; all A. F. Braun.

1378 ***C. glaucicolella*** Wood
Menifee, RRG, FS 9B, Indian East Fork, Kelley Branch, 220 m (720 ft) elevation, May 10–11, 1986, in light trap, J. S. Nordin.

1387 ***C. mayrella*** (Hbn.)
Bullitt Co., BMF, June 10; Edmonson Co., Bee Spring Camp of the Kentucky Geological Survey near Mammoth Cave (Chambers 1878c); Jefferson Co., Anchorage, May 23; Kenton Co., Covington. Formerly known as *C. spissicornis* (Haworth) (Landry and Wright 1993).

1388 ***C. trifolii*** (Curt.)
Gallatin Co., Markland Dam, May 25, June 2, 7, L. D. Gibson and D. J. Wright.

1389 ***C. apicialbella*** Braun
Bullitt Co., BMF, June 19; Campbell Co., AJJ, reared from *Silene virginica*; Fayette Co., Lexington, in Malaise trap, June 16–20.

1392 ***C. aenusella*** Cham.
Kentucky is the type locality (Chambers 1874a), but the type material is lost or destroyed, and the species is known only from the original description (Landry and Wright 1993).

1396 ***C. indefinitella*** Oudejans
Kentucky is the type locality for synonym *zelleriella* (Chambers 1874a). Status of this species is unclear; a *nomen dubium* according to J.-F. Landry (*pers. comm.*).

Family **MOMPHIDAE**

Momphid Moths

Genus ***Mompha*** Hbn.

1426 ***M. argentimaculella*** (Murt.)
Lewis Co., larva on *Oenothera*, emerged Aug. 22, A. F. Braun.

1433 ***M. cephalonthiella*** (Cham.)
Kentucky is the type locality. Chambers (1871b) noted in his original description: "Common. Kentucky. Mines in Buttonbush, *Cephalanthus occidentalis*. Pupa and adult in Oct." The name has sometimes been misspelled "*cephalanthiella*."

1434 ***M. circumscriptella*** (Zell.)
Christian Co., PYR, June 27; Hickman Co., COB, along Mississippi River, at black-light, June 21.

1443 ***M. eloisella*** (Clem.)
Jefferson Co., Louisville, June 4, 13, 25; Menifee Co., RRG, in black-light trap, June 27, July 14; Owsley Co., 4.8 km (3 mi) NE of Booneville, May 22, July 3, 5.

1445 ***M. luciferella*** (Clem.)
Letcher Co., Pine Mountain, July 11, A. F. Braun.

1448 ***M. murtfeldtella*** (Cham.)
Forbes (1923) indicated Kentucky in the distribution of this species. Boone Co., BBL, May 31; Fayette Co., Lexington, in Malaise trap, June 9–12. Uncommon.

1456 ***M. terminella*** (Westwood)
Clay Co., Nolan's Branch, July 3, A. F. Braun.

Family **AGONOXENIDAE**

Agonoxenid Moths

Genus ***Blastodacna*** Wocke

1461 ***B. curvilineella*** (Cham.)
Kentucky is the type locality (Chambers 1872).

Family **COSMOPTERIGIDAE**

Cosmopterigid Moths

Genus ***Euclemensia*** Grt.

1467 ***E. bassettella*** (Clem.)
Boone Co., BBL, July 18 Aug. 3; Christian Co., PYR, Aug. 21; Laurel Co., DBNF, FS 4176, July 28; Meade Co., OCP, at light, Aug. 10; Owsley Co., 4.8 km (3 mi) NE Booneville, July 24, Aug. 22. Powell Co., RRG, July 23. Common.

Genus ***Cosmopterix*** Hbn.

1472 ***C. pulchrimella*** Cham.
Type locality is Covington, Kentucky (Chambers

1875b). Elliott Co., Little Sandy River, one male, May 3, 1936, A. F. Braun (Hodges 1962b); Meade Co., OCP, Apr. 11, 1981.

1475 ***C. clandestinella*** Bsk.
Fleming Co., Hillsboro, Aug. 21; McCreary Co., Natural Bridge, one male, Oct. 10, 1950, on *Panicum*, emerged May 10, 1951; Powell Co., one male and one female, recorded as rearing record B. 927, emerged Apr. 27, May 28, and June 9, 1917; all A. F. Braun (Hodges 1962b).

1482 ***C. lespedezae*** Wlsm.
PARKS AND RESERVES: CBF, MCV. **COUNTIES:** EDM, LAU, LET, LEW, MCY, ROW, WHI. **DATES:** June 24–27, July 5–16, Aug. 20, 27, Sept. 2–11.

1490 ***C. gemmiferella*** Clem.
Bullitt Co., BMF, Malaise trap, June 6–10; Christian Co., PYR, June 20, 26; Letcher Co., Joe Day Branch, mining in *Panicum dichotomum*, May 3; Letcher Co., Pine Mountain, one male, June 21; McCreary Co., Pine Knot, one male, June 14, A. F. Braun, cited *in* Hodges (1962b).

1493 ***C. clemensella*** Staint.
Letcher Co., Big Black Mountain, reared adult emerged June 14 and a specimen collected July 17; Letcher Co., Pine Mountain, June 11, 21; Lewis Co., reared on *Carex*, emerged May 17, 19; Menifee Co., RRG, emerged June 3, 6; all A. F. Braun.

Genus *Tanygona* Braun

1500 ***T. lignicolorella*** Braun
Bell Co., PNM, June 8, A. F. Braun.

Genus *Stagmatophora* H.-S.

1509 ***S. wyattella*** B. and Bsk.
Boone Co., BBL, Sept. 2, and CER, Aug. 17; Campbell Co., AJJ, July 23; Jefferson Co., Valley Station, Aug. 19.

Genus *Limnaecia* Staint.

1515 ***L. phragmitella*** Staint.
Shy Cosmet
Boone Co., BBL, June 28; Hardin Co., Glendale, in Malaise trap, Aug. 22–31, Sept. 1–3; Jefferson Co., Valley Station, June 14, 20.

Genus *Triclonella* Bsk.

1524 ***T. pergandeella*** Bsk.
PARKS AND RESERVES: NBP, PNM. **COUNTIES:** BEL, HLN, HRD, JEF, LET, OWS, POW. **DATES:** May 18–23, June 30, July 1, 5, 18–22, Aug. 6–14. Common.

1527 ***T. determinatella*** (Zell.)
Four males and one female from Kentucky collected by Chambers are cited by Hodges (1962b). Bullitt Co., PCP, May 28, 1989, L. D. Gibson.

Genus *Periploca* Braun

1563 ***P. gleditschiaeella*** (Cham.)
Kentucky is the type locality (Chambers 1876b). Hodges (1962a) cites a series of 15 males and 6 females "reared from thorns of *Robinia* (no further locality given), emerged February 1903, A. Busck."

Genus *Stilbosis* Clem.

1596 ***S. ostryaeella*** (Cham.)
Kentucky is the type locality (Chambers 1874a).

1609 ***S. tesquella*** Clem.
Jefferson Co., Valley Station, June 6; Letcher Co., Pine Mountain, July 1; Menifee Co., RRG, May 10–11.

Genus *Chrysopeleia* Cham.

1612 ***C. purpuriella*** Cham.
Kentucky is the type locality (Chambers 1874a). Jefferson Co., near Louisville, mining in *Quercus velutina*, Sept. 25, 1937; Rowan Co., Triplett Creek, mining in willow, Oct. 18, 1941; both A. F. Braun.

Genus *Walshia* Clem.

1613 ***W. particornella*** (Bsk.)
Jefferson Co., Anchorage, May 22, B. L. Monroe, Jr.

1615 ***W. miscecolorella*** (Cham.)
Sweetclover Root Borer Moth
Jefferson Co.,Valley Station, May 27, June 4, 6, July 14; Whitley Co., Goldbug, June 5.

1616 ***W. amorphella*** Clem.
Gallatin Co., emerged May 13, 23, A. F. Braun (Hodges 1961); Trigg Co., Buffalo Lick, mining in *Gaura biennis*, Aug. 26, 1941, A. F. Braun.

Genus *Perimede* Cham.

1623 ***P. erransella*** Cham.
The type locality is Covington, Kentucky (Chambers 1874a). Letcher Co., Big Black Mountain, June 7, A. F. Braun.

1631 ***P. ricina*** Hodges
One paratype male is labelled "near Ezel, Morgan Co., July 17, 1935, A. F. Braun."

1632 ***P. falcata*** Braun
One male in the LACM bears the label "Kentucky" and August Busck's name as collector.

Family **SCYTHRIDIDAE**

Scythridid Moths

Genus ***Scythris*** Hbn.

1652 ***S. basilaris*** (Zell.)
Harlan Co., BBM, June 29, 1932; Letcher Co., Pine Mountain, June 11, 1933; both A. F. Braun.

1655 ***S. eboracensis*** (Zell.)
Fayette Co., Lexington, in Malaise trap, Aug. 19–22; Jefferson Co., Louisville, May 30; Letcher Co., Pine Mountain, June; Pulaski Co., "Cumberland National Forest" (= DBNF), Apr.; Rowan Co., Clack Mtn., June 20.

1659 ***S. fuscicomella*** (Clem.)
Bullitt Co., BMF, in Malaise trap, June 6–10; Fayette Co., Lexington, in Malaise trap, Aug. 19–22; Jefferson Co., Louisville, May 30, Sept. 4.

1668 ***S. immaculatella*** (Cham.)
Landry (1991) resurrected the above name and made *pacifica* McD. its junior synonym. Letcher Co., Pine Mountain, June 10, 1933, A. F. Braun.

Genus ***Assymetrura*** J.-F. Landry

1662 ***A. impositella*** (Zell.)
Bullitt Co., BMF, in Malaise trap, May 26–31, June 6–10; Powell Co., larvae on *Aster*, May; both A. F. Braun.

Family **GELECHIIDAE**

Gelechiid Moths

Genus ***Isophrictis*** Meyr.

1702 ***I. similiella*** (Cham.)
Kentucky is the type locality ("Kentucky, at the light, in August").

Genus ***Monochroa*** Heinemann

1706 ***M. absconditella*** (Wlk.)
Kentucky is the type locality of synonym *palpianulella* (Chambers 1872), who indicated it was common. Miller and Hodges (1990) note 4 syntypes from Kentucky in the MCZ.

1708 ***M. disconotella*** (Cham.)
Kentucky is the type locality (Chambers 1878b). Lewis Co., Aug. 4, A. F. Braun.

Genus ***Chrysoesthia*** Hbn.

1718 ***C. lingulacella*** (Clem.)
"Kentucky," mining in leaves of *Chenopodium* (Forbes 1923).

Genus ***Theisoa*** Cham.

1722 ***T. constrictella*** (Zell.)
A male syntype of synonym *bifasciella* (Chambers 1874a) from Kentucky is in the MCZ (Miller and Hodges 1990). Boone Co., Florence, Aug. 13, 1979, L. D. Gibson.

1724 ***T. pallidochrella*** (Cham.)
Kentucky is the type locality. Chambers (1873) recorded it in June.

Genus ***Stereomita*** Braun

1725 ***S. andropogonis*** Braun
Letcher Co., Pine Mountain, Sept. 11, A. F. Braun.

Genus ***Aristotelia*** Hbn.

1744 ***A. lespedezae*** Braun
PARKS AND RESERVES: MCV. **COUNTIES:** BAR, EDM, HRT, LET, MCY, MEA, POW, ROW. **DATES:** May 26–30, June 5–20, July 2, Sept. 4, 11, Oct. 7. **REMARKS:** Paratypes from Powell Co. were reared from *Desmodium* and emerged June 20, 1922.

1748 ***A. monilella*** B. and Bsk.
Bell Co., PNM; McCreary Co., Greenwood, June 14–21; Whitley Co., CBF, June 24; all A. F. Braun.

1752 ***A. physaliella*** (Cham.)
Kentucky is the type locality. Chambers (1872) reported two specimens reared from *Physalis viscosa*, which emerged on Apr. 14.

1755 ***A. psoraleae*** Braun
Fleming Co., emerged May 31; McCreary Co., Stearns, emerged June 30, July 5, 7, 8; all A. F. Braun.

1761 ***A. roseosuffusella*** (Clem.)
COUNTIES: FAY, JEF, LET, MCC, MCY, ROW. **DATES:** June 6–10, 22, 26, July 1, Aug. 2, Sept. 2, 11, Oct. 2, 3. **REMARKS:** Reared from *Quercus palustris* by B. S. Nichols.

1762 ***A. rubidella*** (Clem.)
PARKS AND RESERVES: BBM, NBP, PNM. **COUNTIES:** BEL, FLO, HLN, LET, LEW, MCY, POW, ROW. **DATES:** May 18–23, June 5–18, July 13, Aug. 4, 26–27. **REMARKS:** Kentucky is the type locality of synonym *rubensella* (Chambers 1872). Common.

Genus *Glauce* Cham.

1766 ***G. pectenalaeella*** Cham.
PARKS AND RESERVES: CBF, NBP, PNM, RRG. **COUNTIES:** BEL, LET, MCY, MEN, POW, ROW, WHI. **DATES:** May 28, 31–June 1, 10–11, 20–24.

Genus *Evippe* Cham.

1770 ***E. leuconota*** (Zell.)
Two specimens of synonym *plutella* (Chambers 1874b) labelled "Kentucky" are in the MCZ (Miller and Hodges 1990).

1771 ***E. prunifoliella*** Cham.
Kentucky is the type locality (Chambers 1873), where larvae were reported feeding under tips of "Red Wild Plum (*Prunus americana*)."

Genus *Agnippe* Cham.

1772 ***A. biscolorella*** Cham.
Kentucky is the type locality (Chambers 1872). Found in "Apr."

1775 ***A. fuscopulvella*** Cham.
Kentucky is the type locality (Chambers 1872). Found in "Apr."

Genus *Recurvaria* Haw.

1781 ***R. consimilis*** Braun
Meade Co., OCP, emerged May; Rowan Co., Clack Mtn., emerged June 12, on *Ceanothus*, June 10, July 30, and near Morehead, mining in *Ceanothus americanus*, Oct. 17; all A. F. Braun.

Genus *Coleotechnites* Cham.

1804 ***C. cristatella*** (Cham.)
Chambers (1875b) gave the type locality as Kentucky "in June." Two Kentucky syntypes in the MCZ are reported by Miller and Hodges (1990).

1805 ***C. vagatioella*** (Cham.)
Kentucky is the type locality (Chambers 1872). A. F. Braun reported the following under the synonym, *dorsivittella* Zell.: Carter Co., CAC, emerged May 12; Rowan Co., Clack Mtn., emerged Apr. 2 and 7.

1823 ***C. nigritus*** Hodges
Pulaski Co., "Cumberland National Forest" (= DBNF), emerged May 30, A. F. Braun.

1824 ***C. obliquistrigella*** (Cham.)
Kentucky is the type locality (Chambers 1872). Miller and Hodges (1990) report two syntypes from Kentucky in the MCZ.

1828 ***C. quercivorella*** (Cham.)
Kentucky is the type locality (Chambers 1872); the holotype female is in the MCZ (Miller and Hodges 1990). Rowan Co., Morehead, May 2, A. F. Braun.

1833 ***C. variella*** (Cham.)
Kentucky is the type locality (Chambers 1872).

Genus *Sinoe* Cham.

1834 ***S. robiniella*** (Fitch)
Three syntypes from Kentucky of synonym *fuscopalidella* (Chambers 1875b) are in the MCZ (Miller and Hodges 1990). Bell Co., PNM, emerged Aug. 4; Fayette Co., Lexington, in Malaise trap, June 16–20; Jefferson Co., Valley Station, July 14. Common.

Genus *Exoteleia* Wallgr.

1840 ***E. pinifoliella*** (Cham.)
Kentucky is the type locality (Chambers 1880a). Bell Co., PNM, June 24, July 19; Bullitt Co., BMF, June 19; Powell Co., NBP, June 8.

Genus *Trypanisma* Clem.

1841 ***T. prudens*** Clem.
Kentucky is the type locality for synonym *quinqueannulella* (Chambers 1872).

Genus *Taygete* Cham.

1842 ***T. attributella*** (Wlk.)
PARKS AND RESERVES: BBM, RRG. **COUNTIES:** HLN, LET, MCY, MEN, POW, ROW. **DATES:** June 5–28, July 1, 9. **REMARKS:** Kentucky is the type locality for synonym *difficilisella* (Chambers 1872). Hodges (1986) indicates this species should be placed in the Family Oecophoridae.

1846 ***T. saundersella*** (Cham.)
Kentucky is the type locality. Chambers (1876b) indicated "Kentucky in July" in the original description, and 2 syntypes are in the MCZ (Miller and Hodges 1990).

Genus *Arogalea* Wlsm.

1851 ***A. cristifasciella*** (Cham.)
PARKS AND RESERVES: BBL, BMF, HWS, OCP. **COUNTIES:** BOO, BUL, CAM, JEF, LAU, MCC, MEA, OLD. **DATES:** Mar. 27–May 3, 18, July 25, Aug. 10. **REMARKS:** Kentucky is the type locality (Chambers 1878c). Miller and Hodges (1990) report one female syntype in the MCZ. It was collected in Kentucky on May 11. Common.

Genus *Telphusa* Cham.

1855 ***T. alexandriacella*** (Cham.)
The type locality is Alexandria, Campbell Co., collected in June.

1856 ***T. fasciella*** (Cham.)
The type locality is Kentucky (Chambers 1873).

1857 ***T. latifasciella*** (Cham.)
Christian Co., PYR, June 27; Jefferson Co., May 23; McCreary Co., Yahoo Ridge, June 15; Meade Co., OCP, June 7; Rowan Co., Morehead, emerged June 1. The type locality may be Kentucky, though the original description indicates it to be Missouri (Chambers 1875b).

1858 ***T. longifasciella*** (Clem.)
Kentucky is the type locality for the synonym *curvistrigella* (Chambers 1872). Menifee Co., RRG, Indian East Fork, Kelley Branch, Apr. 13, 14; McCreary Co., Great Meadows Recreation Area, Apr. 15; Owsley Co., 4.8 km (3 mi) NE of Booneville, Apr. 14.

Genus ***Pseudochelaria*** Dietz

1862 ***P. pennsylvanica*** Dietz
Bell Co., PNM, July 10; Campbell Co., AJJ, July 15, Harlan Co., BBM, July 12; Meade Co., OCP, Aug. 10; Menifee Co., RRG, FS 9B, Indian East Fork, Kelley Branch, in black-light trap, July 19, 20; Owsley Co., near Booneville, May 22.

1864 ***P. walsinghami*** Dietz
Edmonson Co., MCV, emerged June 3–13; Jefferson Co., Anchorage, May 26; Jefferson Co., Valley Station, May 18, Aug. 14.

Genus ***Neotelphusa*** Janse

1866 ***N. querciella*** (Cham.)
Kentucky is the type locality (Chambers 1872). Miller and Hodges (1990) report 3 syntypes from Kentucky in the MCZ. Rowan Co., Morehead, reared, adults emerged June 6, 10, A. F. Braun.

Genus ***Pseudotelphusa*** Janse

1868 ***P. basifasciella*** (Zell.)
Fayette Co., Lexington cemetery, reared from green larvae which "sewed oak leaves together." Adult emerged in lab, Sept. 28, 1928, H. Garman.

1871 ***P. fuscopunctella*** (Clem.)
Powell Co., emerged May 4, A. F. Braun.

1873 ***P. pallliderosacella*** (Cham.)
Bell Co., PNM, on oak, Mar. 31, Apr. 15; Powell Co., NBP, May 16 and 20.

1874 ***P. quercinigracella*** (Cham.)
Kentucky is the type locality (Chambers 1872). The holotype male is in the MCZ (Miller and Hodges 1990). Harlan Co., BBM, Apr.; Letcher Co., Pine Mountain, Apr. 11, and Pound Gap, emerged from red oak, May 6, 12, 20; all A. F. Braun records.

Genus ***Xenolechia*** Meyr.

1876 ***X. quinquecristatella*** (Cham.)
Kentucky is possibly the type locality (Chambers 1878b). Letcher Co., Big Black Mountain, on *Vaccinium*, emerged May 29 and June 1, A. F. Braun.

Genus ***Deltophora*** Janse

1928 ***D. sella*** (Cham.)
Bullitt Co., BMF, in Malaise trap, May 26–31, June 1–6, 1976, A. J. Brownell.

Genus ***Gelechia*** Hbn.

1929 ***G. albisparsella*** (Cham.)
Kentucky is the type locality for this name and synonyms *fuscoluteella* and *platanella* (Chambers 1872). Menifee Co., RRG, Indian East Fork, Kelley Branch, July 19, 1985, J. S. Nordin.

1931 ***G. badiomaculella*** Cham.
Kentucky is the type locality. Chambers (1872) wrote "Kentucky – at light."

1936 ***G. capiteochrella*** Cham.
Kentucky is probably the type locality (A.F. Braun, *pers. comm.*), but there is no indication in the original description (Chambers 1875b).

1952 ***G. obscurella*** Cham.
Kentucky is the type locality (Chambers 1872).

1957 ***G. palpialbella*** Cham.
Kentucky is possibly the type locality (Chambers 1875b).

1961 ***G. rileyella*** (Cham.)
Kentucky is the type locality (Chambers 1872).

1965 ***G. unistrigella*** Cham.
Kentucky is the type locality (Chambers 1873).

Genus ***Gnorimoschema*** Bsk.

1968 ***G. alaricella*** Bsk.
Jefferson Co., Valley Station, Oct. 13, 1974, A. J. Brownell.

1971 ***G. ambrosiella*** (Cham.)
Kentucky is the type locality (Chambers 1875b).

1974 ***G. batanella*** Bsk.
Bell Co., Clear Creek, emerged Aug. 11, A. F. Braun.

1981 ***G. marmorella*** (Cham.)
Kentucky is the type locality (Chambers 1875b).

1985 ***G. gallaeasterella*** (Kell.)
Menifee Co., RRG, in black-light trap, Sept. 7, 8, 21, 22, 1984, J. S. Nordin.

1986 ***G. gallaesolidaginis*** (Riley)
Goldenrod Gall Moth
Menifee Co., RRG, in black-light trap, Sept. 21–30, Oct. 12, 13; Rowan Co., Clack Mtn., reared by A. F. Braun on *Solidago*, emerged Oct. 15. Research on free amino acids was conducted on larvae of this species taken between May and Oct. 1980 in Jefferson, Nelson and Oldham counties (Heady et al. 1982).

1988 ***G. grisella*** (Cham.)
The type locality is Kentucky "May" (Chambers 1872).

1989 ***G. inexpertum*** (Meyr.)
Kentucky is the type locality for homonym *simpliciella* (Chambers 1875b).

1998 ***G. saphirinella*** (Cham.)
Kentucky is the type locality (Chambers 1875b). The holotype is in the MCZ (Miller and Hodges 1990).

Genus *Phthorimaea* Meyr.

2011 ***P. operculella*** (Zell.)
Potato Tuberworm Moth
Jefferson Co., Louisville, in seed potatoes, Mar. 10; Muhlenberg Co., Bremen, June 23 and 25. A pest of potatoes in the field and in storage. Also destroys tobacco leaves and other Solonaceae.

Genus *Scrobipalpula* Povolny´

2019 ***S. polemoniella*** (Braun)
Campbell Co., mining in *Polemonium reptans*, Nov. 18, 1934, A. F. Braun.

Genus *Scrobipalpa* Janse

2030 ***S. scutellariaeela*** (Cham.)
The type locality is Boone Co., near Verona, where it was "very abundant in September and October" (Chambers 1873). Campbell Co., mining in *Scutellaria canescens*, Nov. 18, 1934, A.F. Braun.

Genus *Symmetrischema* Povolny´

2037 ***S. pallidochrella*** (Cham.)
Kentucky is the type locality. Chambers (1873) collected it in June.

Genus *Caryocolum* Gregor and Povolny´

2041 ***C. protecta*** (Braun)
Holotype male, Carter Co., CAC, adult emerged June 11. Paratypes: Carter Co., Cascades Caves, adults June 15–18; McCreary Co., Yahoo Valley, emerged May 20–31, June 1–5; Powell Co., NBP, adult June 18–30, July 1–4; Rowan Co., Morehead, emerged July 12, and Clack Mtn., emerged June 6; Whitley Co., CBF, emerged June 18, 19. Foodplant: *Silene rotundifolia* Nuttall; all A. F. Braun.

Genus *Tildenia* Povolny´

2049 ***T. glochinella*** (Zell.)
Campbell Co., Alexandria, mining in *Physalis* sp., Oct. 11, 1942, A. F. Braun.

Genus *Frumenta* Bsk.

2052 ***F. nundinella*** (Zell.)
Bullitt Co., BMF, in Malaise trap, Mar. 27–31, Apr. 3–20, May 22–30; Hardin Co., Glendale, in Malaise trap, Aug. 14–22; Jefferson Co., Apr., 1986.

Genus *Chionodes* Hbn.

2064 ***C. bicostomaculella*** (Cham.)
Rowan Co., Morehead, June 22, A. F. Braun. Referred to in Dyar ("1903" [1904]) as from Kentucky, which may be the type locality for this name and also synonym *quercifoliella* (Chambers 1872). Meade Co., OCP, June 30; Menifee Co., RRG, East Fork Gladie Cr., June 2 and Indian East Fork, Kelley Branch, July 19; Rowan Co., Morehead, June 22.

2072 ***C. discoocellella*** (Cham.)
The type locality is "Kentucky, Sept." (Chambers 1872). Fayette Co., Lexington, larvae fed on *Rumex crispus*, June 7, 1917; Hardin Co, Glendale, in Malaise trap, July 8–15, P. H. Freytag.

2076 ***C. fondella*** (Bsk.)
Menifee Co., RRG, Apr. 26, 1981, L. D. Gibson.

2077 ***C. formosella*** (Murt.)
Boone Co., BBL, May 31; Owsley Co., near Booneville, May 22; Rowan Co., Clack Mtn., emerged May 13, and Morehead, emerged June 22.

2079 ***C. fuscomaculella*** (Cham.)
Kentucky is the type locality (Chambers 1872).

2093 ***C. mediofuscella*** (Clem.)
PARKS AND RESERVES: HWS, RRG. **COUNTIES:** BOO, CLK, FAY, HRD, JEF, MEN, OLD, OWS. **DATES:** Apr. 12, 14, 28, May 10–19, July 1–8, 19–20, Sept. 6. **REMARKS:** Kentucky is the type locality for synonym *fuscoochrella* (Chambers 1872).

2099 ***C. obscurusella*** (Cham.)
Kentucky is the type locality for this name and synonym *fuscopulvella* (Chambers 1872). The holotypes labelled "Kentucky" are in the MCZ (Miller and Hodges 1990). Meade Co., OCP at bait, Oct. 10; Menifee Co., RRG near Slade, 12 Oct. and East Fork Gladie Cr., June 2, and Indian East Fork, Kelley Branch, Oct. 12; all J.S. Nordin.

2104 ***C. pereyra*** Clarke
Jefferson Co., Valley Station, July 14, 1975, A. J. Brownell; Meade Co., OCP, June 30.

2110 ***C. pseudofondella*** (Bsk.)
Meade Co., OCP, June 7; Owsley Co., near Booneville, July 5; Whitley Co., Goldbug, June 5; L. D. Gibson.

2119 ***C. thoraceochrella*** (Cham.)
Harlan Co., BBM, July 27, A. F. Braun; Meade Co., Doe Run, July 21, C.V. Covell, Jr.

2119.1 ***C.*** undescribed species
Harlan Co., PMS, July 18, 1975, A. J. Brownell.

2119.2 ***C.*** undescribed species
Fulton Co., Sept. 9, 1975, C. Cornett.

2119.3 ***C.*** undescribed species
Meade Co., OCP, Aug. 19, 1979, C.V. Covell, Jr., and Mar. 17, 1987, B.S. Nichols; Menifee Co., RRG, FS 9B, Indian East Fork, Kelley Branch, Aug. 9–10 and 17–18, 1985, J.S. Nordin; Powell Co., RRG, Tunnel Ridge, Mar. 4 and 11, 1989, D.J. Wright.

Genus *Filatima* Bsk.

2169 ***F. pseudacaciella*** (Cham.)
Kentucky is the type locality (Chambers 1872). Bell Co., PNM, July 18; Boone Co., BBL, Apr. 30, May 18; Fayette Co., Lexington, in Malaise trap, Aug. 8–11, 19–22.

2172 ***F. serotinella*** (Bsk.)
Boone Co., BBL, May 4, 18; Bullitt Co., BMF, in Malaise trap, Apr. 17–20; Meade Co., OCP, Apr. 11.

Genus *Aroga* Bsk.

2184 ***A. argutiola*** Hodges
Two female paratypes labelled "10 miles E of London, Laurel Co., Ky.; iss. July 1936; A. F. Braun." are cited in the original description (Hodges 1974).

2198 ***A. trialbimaculella*** (Cham.)
Powell Co., RRG, Tunnel Ridge, July 11, L. D. Gibson.

Genus *Fascista* Bsk.

2203 ***F. bimaculella*** (Cham.)
The type designation of Chambers (1872) is "Kentucky, Common." Kentucky is also the type locality for synonym *sylvaecolella* (Chambers 1878b).

2204 ***F. cercerisella*** (Cham.)
Redbud Leaffolder Moth
PARKS AND RESERVES: BMF, HWS, NBP, OCP, RRG. **COUNTIES:** BAR, BUL, FAY, JEF, MEA, MEN, OLD, OWS, PER, POW. **DATES:** Apr. 17–20, May 1–10, 27, June 6–14, July 11, 25, Aug. 9–19, Sept. 21. **REMARKS:** Miller and Hodges (1990) note 2 syntypes from Kentucky in the MCZ; however, Chambers (1872) did not publish a type locality.

Genus *Stegasta* Meyr.

2209 ***S. bosqueella*** (Cham.)
Fayette Co., Lexington, in Malaise trap, Aug. 11–15; Fulton Co., 3.2 km (2 mi) E of Cayce, Sept. 7 and 11; Hardin Co., Glendale, in Malaise trap, July 8–15; Henderson Co., SWR, Aug. 22.

Genus *Syncopacma* Meyr.

2218 ***S. adversa*** (Braun)
Fleming Co., Sand Lick, emerged July 11; Powell Co., emerged Dec. 31; Rowan Co., Clack Mtn., emerged July 25; all A. F. Braun.

2222 ***S. palpilineella*** (Cham.)
Kentucky is the type locality (Chambers 1875b). Miller and Hodges (1990) report 3 syntypes from Kentucky in the MCZ. Bullitt Co., BMF, June 19; Fayette Co., Lexington (Malaise trap), June 12–20, July 11–14; Hardin Co., Glendale (Malaise trap), July 29–Aug. 6; Jefferson Co., Louisville, reared from clover, July 26.

Genus *Untomia* Bsk.

2223 ***U. albistrigella*** (Cham.)
The type citation (Chambers 1872) is "Kentucky, June." Bell Co., Fern Lake, July 10; Fayette Co., Lexington, in Malaise trap, June 16–27, Aug. 15–25; Jefferson Co., WVP, July 11, 14; Letcher Co., Big Black Mountain, July 5, and Pine Mountain, July 1–14; Whitley Co., CBF, June 25.

Genus *Battaristis* Meyr.

2227 ***B. nigratomella*** (Clem.)
Kentucky is the type locality for synonym *apicistrigella* (Chambers 1872). The holotype female is in the MCZ (Miller and Hodges 1990). Bell Co., PNM, June 22; Bullitt Co., BMF, in Malaise trap, Apr. 17–20; Elliott Co., Little Sandy River, May 9.

2229 ***B. vittella*** (Bsk.)
Bullitt Co., BMF, June 10; Christian Co., PYR, June 27.

Genus *Anacampsis* Curt.

2230 ***A. agrimoniella*** (Clem.)
Bullitt Co., BMF, in Malaise trap, June 1–15; Laurel Co., London, emerged May 29, 31; Powell Co., RRG, Tunnel Ridge near Nada Tunnel, July 11.

2233 **A. conclusella** (Wlk.)
Kentucky is probably the type locality of synonym *grissefasciella* (Chambers 1875b; A.F. Braun, *pers. comm.*). Bell Co., PNM, July 13; Bullitt Co., BMF, in Malaise trap, Apr. 17–20; Fayette Co., Lexington, Aug. 19–22; Hardin Co., Glendale, in Malaise trap, July 29–31, Aug. 1–14.

2236 **A. fullonella** (Zell.)
Fulton Co., Willingham Bottoms, 4.0 km (2.5 mi) E of Cayce, Sept. 6, and Williams Rd., 3.2 km (2 mi) E of Cayce, Sept. 7; both L. D. Gibson.

2238 **A. kearfottella** (Bsk.)
McCreary Co., South Fork of Cumberland River, June 18, A. F. Braun.

2241 **A. levipedella** (Clem.)
Fayette Co., Lexington, in Malaise trap, Sept. 15–30, Oct. 1–6; Floyd Co., Prestonsburg, May 22; Jefferson Co., Louisville, May 26; McCreary Co., Yahoo Ridge, June 14.

2244 **A. nonstrigella** Bsk.
Edmonson Co., MCV, July 1; Owen Co., June 23; both A. F. Braun.

2248 **A. rhoifructella** (Clem.)
Boone Co., BBL, larvae collected on *Viburnum* on May 3 and adults emerged May 15, 1993, L. D. Gibson.

2250 **A. tephriasella** (Cham.)
Letcher Co., Big Black Mountain, July 7, A. F. Braun.

Genus *Strobisia* Clem.

2253 **S. iridipennella** Clem.
Bullitt Co., July 8; Christian Co., PYR, June 27; Letcher Co., Cumberland Valley, July 1; Owsley Co., 4.8 km (3 mi) NE of Booneville, June 30; Todd Co., Clifty Creek, June 15.

Genus *Holophysis* Wlsm.

2255 **H. emblemella** (Clem.)
Letcher Co., Colliers Creek, June 8, July 7, 8, Cumberland Valley, June 26, and Pine Mountain, June 16, 21, July 1; all A. F. Braun.

Genus *Anarsia* Zell.

2257 **A. lineatella** Zell.
Peach Twig Borer Moth
Fayette Co., Lexington, mining in tips of peach twig, emerged Apr. 28; Henderson Co., Henderson, reared but host not stated, emerged May 29.

Genus *Sitotroga* Heinemann

2260 **S. cerealella** (Olivier)
Angoumois Grain Moth
Bell Co., Beverly, July 4; Daviess Co., Stanley, from rye head, June 18; Fayette Co., Lexington, Jan. 28, Feb. 2–5, May 27, Oct. 24, Nov. 15, Dec. 24, and Muir, Sept.; Jefferson Co., Louisville, Apr. 19, May 29; Meade Co., Midway, in ear of corn, Oct. 10, 18, and Payneville, from popcorn, Nov. 7. Commonly found infesting dried ear corn – including ornamental Indian corn. Larvae eat out individual kernals, leaving a small hole in each.

Genus *Brachmia* Hbn.

2265 **B. chambersella** (Murt.)
One old specimen in the AMNH is labeled "Ky." and "Aug. Busck Collector." It was identified as *chambersella* by R. W. Hodges.

2266 **B. discoannulella** (Cham.)
Kentucky is the type locality (Chambers 1875b). Hodges (1986) returned this to *Gelechia* (the genus in which it was originally described) as an unknown species that may later be recognizable. It is retained here in its Hodges et al. (1983) combination and position for easy reference.

Genus *Helcystogramma* Zell.

2268 **H. hystricella** (Braun)
Hardin Co., Glendale, in Malaise trap, June 8–15, July 29, Aug. 5; Jefferson Co., Valley Station, May 23, 27; Woodford Co., 1.6 km (1 mi) W of SR 68 over the Kentucky River, in black-light trap, Aug. 18. Common.

Genus *Dichomeris* Hbn.

NOTE: This and the preceding genus have recently been revised by Hodges (1986). The order for *Dichomeris* species has been kept close to that of Hodges et al. (1983), and new species have been numbered to reflect as closely as possible their relative positions in the revised classification.

2277 **D. georgiella** (Wlk.)
Bullitt Co., BMF, in Malaise trap, Apr. 13–17; Christian Co., PYR, June 20; Harlan Co., BBM, Apr. 20; Marshall Co., KLK, Apr. 1.

2281 **D. ligulella** Hbn.
Palmerworm Moth
PARKS AND RESERVES: BBM, BMF, CBB, HWS, KGC, KRC, LMP, OCP, PNM, RRG. **COUNTIES:** BEL, BUL, FAY, HLN, LET, MCC, MEA, MEN, MER, MUH, NEL, OLD, OWS, PER, POW, WOL. **DATES:** Mar. 4, 17, 18, 28, Apr. 13–20, May 2, 3, 11, 14, June 6, 14, 24–28, July 1–15, Aug. 25, Sept. 6. **REMARKS:** Common. The holotype female of synonym *quercipominella* (Chambers 1872) is in the MCZ (Miller and Hodges 1990).

2282 ***D. marginella*** (F.)
Juniper Webworm Moth
Fayette Co., Lexington, May 14, 22, ("reared"); another reared probably on cedar, "adult June 5, 1935"; both in UK Collection.

2283 ***D. punctidiscella*** (Clem.)
Kentucky is the type locality of synonym *stramin-iella* (Chambers 1872). The lectotype female was selected by Hodges (1986). Chambers (1872) noted it was "captured June 16." Carter Co., CAC, Oct. 15; Jefferson Co., Valley Station, June 6; Letcher Co., Pine Mountain, June 21; Meade Co., OCP, June 7; Menifee Co., RRG, June 21–22; Powell Co., larva found on birch, emerged June 17, 18, A. F. Braun.

2286 ***D. vacciniella*** Bsk.
COUNTIES: ELL, LAU, LET, LEW, MAG, MCY, OWS, ROW. **DATES:** Apr. 6, 30, June 14–26, July 1, 5, 15. Common.

2287 ***D. ventrella*** (Fitch)
PARKS AND RESERVES: BBM, BMF, HWS, OCP, RRG. **COUNTIES:** BUL, HLN, JEF, LAU, LET, MEA, MEN, OLD. **DATES:** Mar. 27–31, Apr. 1–9, 17–20, May 4, 22–31, June 20, 28, July 21, 22, Sept. 5. **REMARKS:** Kentucky is the type locality for the synonym *querciella* (Chambers 1872). Common.

2288 ***D. punctipennella*** (Clem.)
Kentucky is the type locality of synonym *gracilella* (Chambers 1872) who wrote "a single specimen was taken May 7th, resting upon the trunk of a tree. When disturbed it fluttered around for a moment, re-alighting always on the same tree." Bell Co., PNM, June 24; Bullitt Co., BMF, Apr. 17; Jefferson Co., Valley Station, June 20; Letcher Co., head of Line Fork, June 18.

2289 ***D. ochripalpella*** (Zell.)
The name *alacella* Clemens, used in Hodges et al. (1983), was replaced as a junior homonym by Hodges (1986). Fayette Co., Lexington, July 10; Harlan Co., BBM, July 12; Owsley Co., 4.8 km (3 mi) NE of Booneville, June 30, July 1. Uncommon.

2291 ***D. bilobella*** (Zell.)
Bullitt Co., BMF, May 11–20; Menifee Co., RRG, East Fork of Gladie Creek, in Malaise trap, June 15, 16, July 3. Hodges (1986) notes that A. F. Braun reared this species on *Solidago flexicaulis* in Kentucky.

2291.1 ***D. aleatrix*** Hodges
One paratype female was collected as a larva in Rowan Co, on Clack Mtn., by A. F. Braun on *Helianthus*; it emerged June 6, 1942 (Hodges 1986).

2291.2 ***D. furia*** Hodges
One paratype male was collected in Owsley Co., 4.8 km (3 mi) NE of Booneville, July 5, 1980, L. D. Gibson.

2295 ***D. flavocostella*** (Clem.)
PARKS AND RESERVES: BBL, BMF, HWS. **COUNTIES:** BOO, BUL, FAY, HRD, JEF, OLD, OWS, ROW, WOO. **DATES:** May 19, June, July 1–8. Common.

2297 ***D. inserrata*** (Wlsm.)
Fulton Co., SR 94, Willingham Bottoms, 4.0 km (2.5 mi) E of Cayce, Aug. 3; Owsley Co., 4.8 km (3 mi) NE of Booneville, July 14.

2297.1 ***D. euprepes*** Hodges
The holotype female was reared by A. F. Braun on *Solidago flexicaulis* and collected in Letcher Co., Big Black Mountain. The adult emerged June 16, 1937 (Hodges 1986).

2298 ***D. juncidella*** Clem.
COUNTIES: FAY, JEF, LET, MCY, MER, OLD. **DATES:** May 19, June, July, Aug. 1–8, 20–27. **REMARKS:** Locally common. Kentucky is the type locality of the synonym *dubitella* (Chambers 1872). The lectotype was selected by Hodges (1986).

2300 ***D. levisella*** (Fyles)
Hodges (1986) lists this species from Kentucky, saying adults "have been collected on 12 June." More precisely, the locality is Rowan Co., Perry Branch, 1 female reared by A. F. Braun; emerged as an adult June 12, 1939.

2301 ***D. serrativittella*** (Zell.)
There are two specimens of synonym *plutella* (Chambers 1874b) in the MCZ from Kentucky (Miller and Hodges 1990). Boone Co., BBL, June 18; Bracken Co., May 8; Bullitt Co., BMF, in Malaise trap, June 1–6; Jefferson Co., Valley Station, June 4; Marshall Co., Gilbertsville, taken at lights on Sept. 10; Owsley Co., 4.8 km (3 mi) NE of Booneville, May 22, 23.

2301.1 ***D. isa*** Hodges
Male paratypes from Kentucky were listed by Hodges (1986) as follows: Bullitt Co., BMF, June 1–6; Jefferson Co., WVP, Valley Station, May 17; Owsley Co., 4.8 km (3 mi) NE of Booneville, May 22, 23.

2302 ***D. setosella*** (Clem.)
PARKS AND RESERVES: BBL, BMF, HWS, NBP, OCP. **COUNTIES:** BOO, BRA, BUL, HRD, JEF, LAU, LET, MEA, OLD, POW, ROW, WOO. **DATES:** May 4–25, June 7–16, July 8–31, Aug. 1–6, Sept. 19. **REMARKS:** Common. Kentucky is the type locality for synonyms *costolutella*

and *eupatoriella* (Chambers 1872). The lectotype male of the latter was selected by Hodges (1986).

2302.1 ***D. vindex*** Hodges
One paratype male of this species was reared from *Helianthus hirsutus* collected at Rowan Co., Clack Mtn., on Oct. 3, 1945, by A. F. Braun; adult emerged Apr. 22, 1946 (Hodges 1986).

2305 ***D. agonia*** Hodges
Formerly known as *Trichotaphe trinotella* Busck, a homonym of *Dichomeris trinotella* Coquillet. Bullitt Co., BMF, in Malaise trap, May 4–8; Hardin Co., Glendale, in Malaise trap, July 29–31, Aug. 1–6.

2305.1 ***D. offula*** Hodges
Hodges (1986) designated as paratypes two males from Menifee Co., RRG, May 13, 1981, L. D. Gibson. Menifee Co., RRG, FS 9B, Indian East Fork, Kelley Branch, in black-light trap, May 26, 27, Sept. 6, 7. Locally common.

2306 ***D. washingtoniella*** (Bsk.)
Bullitt Co., Mt. Washington, July 28; Hardin Co., Glendale, in Malaise trap, July 29–31, Aug.1–6; Owsley Co., 4.8 km (3 mi) NE of Booneville, June 14; Woodford Co., 1.6 km (1 mi) W of SR 68, along Kentucky River, June 8.

2307 ***D. nonstrigella*** (Cham.)
Kentucky is the type locality (Chambers 1878b). Bullitt Co., BMF, in Malaise trap, May 26–31, June 1–10; Owsley Co., 4.8 km (3 mi) NE of Booneville, June 10, 14.

2310 ***D. inversella*** (Zell.)
Jefferson Co., Valley Station, July 11, 14, 1975, A. J. Brownell.

Superfamily **COPROMORPHOIDEA**

Family **CARPOSINIDAE**

Carposinid Moths

Genus ***Bondia*** Newman

2319 ***B. crescentella*** (Wlsm.)
Boone Co., BBL, 2 specimens, Apr. 25, 30; Owsley Co, near Booneville, Apr 18; both L. D. Gibson.

Family **EPERMENIIDAE**

Epermeniid Moths

Genus ***Epermenia*** Hbn.

2329 ***E. cicutaella*** Kft.
Jefferson Co., Valley Station, July 18, 1975, A. J. Brownell.

2330 ***E. pimpinella*** Murt.
Letcher Co., Cumberland Valley; Perry Co., Lynn Fork, emerged July 25; both A. F. Braun.

Genus ***Ochromolopis*** Hbn.

2334 ***O. ramapoella*** (Kft.)
Edmonson Co., Mammoth Cave, on *Commandra*, May 25, A. F. Braun; Menifee Co., RRG, May 11–12.

Family **GLYPHIPTERIGIDAE**

Glyphipterigid Moths

Genus ***Glyphipterix*** Hbn.

2337 ***G. circumscriptella*** Cham.
Meade Co., OCP, at light, May 3, 1987, B. S. Nichols.

2337.1 ***G. nordini*** Heppner
Heppner (1997) selected as holotype a male from Menifee Co., RRG, Indian East Fork, Kelley Br., FS 9B, 220 m (720 ft) elev., collected on the night of May 11–12, 1985. The allotype female was collected from Menifee Co., RRG, along FS 23 near SR 77, May 21, 1988 by L.D. Gibson. A paratype male was named with the same data as the holotype.

2338 ***G. quadragintapunctata*** Dyar
Menifee Co., RRG, FS 9b, Indian East Fork, Kelley Branch, 220 m (720 ft) elevation, June 15 and 16, 1985, J. S. Nordin.

2343.1 ***G. chambersi*** Heppner
Heppner (1985) in the original description cited 9 paratypes and 4 other specimens from Kenton Co. (probably Covington). All were collected by V. T. Chambers.

Genus ***Diploschizia*** Heppner

2346 ***D. impigritella*** (Clem.)
The holotype of synonym *exoptatella* (Chambers 1875b) is from Kenton Co., Linden Grove Cemetery, Covington, taken in June (Miller and Hodges 1990). Boone Co., BBL, May 31; Hardin Co., Glendale, in Malaise trap, June 19–30; Gallatin Co, Markland Dam, May 29; Jefferson Co., Valley Station, June 4.

Superfamily **YPONOMEUTOIDEA**

Family **PLUTELLIDAE**

Diamondback Moths

Genus ***Homadaula*** Lower

2353 ***H. anisocentra*** Meyr.
Mimosa Webworm Moth
Boone Co., CER, Aug. 17; Gallatin Co., Markland

Dam, July 17; Meade Co., OCP, May 15, Aug. 10; Rowan Co., Morehead, June 28, Sept. 1; Woodford Co., along Kentucky River, 1.6 km (1 mi) W of SR 68, Aug. 28. Reported as a pest of mimosa trees in Boone and Kenton counties, and Kentucky in general (*Kentucky Cooperative Economic Insect Report* 1975).

Genus *Plutella* Schr.

2366 ***P. xylostella*** (L.)
Diamondback Moth
PARKS AND RESERVES: BBL, BMF, MCV, PYR, RNR, RRG. **COUNTIES:** BOO, BUL, CHR, EDM, FAY, FUL, JEF, MEN. **DATES:** Apr. 4, 14–17, May 12–30, June 10–13, 27, July 27, Aug. 16. **REMARKS:** A pest of cabbage and other members of the mustard family. Common.

Family YPONOMEUTIDAE

Ermine Moths

Genus *Atteva* Wlk.

2401 ***A. punctella*** (Cham.)
Ailanthus Webworm Moth
PARKS AND RESERVES: BEC, BLA, BMF, CBF, GBT, HNK, HWS, LBL, LCP, NBP, OCP, PNM, RBF. **COUNTIES:** BEL, BRE, BUL, CAL, CHR, CRL, FAY, FRA, FUL, JEF, MAD, MCC, MCY, MEA, OLD, POW, PUL, ROW, RUS, SPE, TRI, WAR, WHI. **DATES:** Mar. 11, 15, 29, Apr. 5–Oct. 6, 26, Nov. 4, 7, 19. Common.

Genus *Lactura* Wlk.

2405 ***L. pupula*** (Hbn.)
Trigg Co., LBL, June 12; Warren Co., Apr. 6, June 18, July 30, Aug. 12. Uncommon.

Genus *Swammerdamia* Hbn.

2413 ***S. caesiella*** (Hbn.)
A. F. Braun (*pers. comm.*) gave the following information for synonym *castaneae* Busck.: Letcher Co., Cumberland Valley, emerged in Aug., and Pine Mountain, on *Castanea*, emerged July 8; Powell Co., RRG, Tunnel Ridge Rd., May 10, 21, emerged May 31.

Genus *Yponomeuta* Latr.

2416 ***Y. atomocella*** Dyar
Brown-bordered Ermine Moth
Woodford Co., 1.6 km (1 mi) W of SR 68 along Kentucky River, May 25, 1985, in black-light trap, J. S. Nordin.

2417 ***Y. euonymella*** Cham.
Kentucky is the type locality for *euonymella* (Chambers 1872) and its synonym, *orbimaculella* (Chambers 1873). *Y. euonymella* was described as "very common" by Chambers (1872) in his original description. Rockcastle Co., Renfro Valley, June 20, 21; Rowan Co., Morehead, June 20–30, July 3.

2420 ***Y. multipunctella*** Clem.
American Ermine Moth
PARKS AND RESERVES: BBL, BMF, HWS, JWS, LCP, NBP, RRG. **COUNTIES:** BOO, BUL, FLO, JEF, LEE, LET, MEN, NEL, OLD, OWS, POW, ROW, RUS. **DATES:** June 3–14. **REMARKS:** Locally common. Reared from *Betula nigra*.

Genus *Zelleria* Staint.

2427 ***Z. haimbachi*** Bsk.
Owsley Co., 4.8 km (3 mi) NE of Booneville, June 14; Whitley Co., Goldbug, June 5.

2431 ***Z. retiniella*** Fbs.
Bell Co, PNM, July 19; Powell Co., RRG, Tunnel Ridge, July 23. Forbes (1923) indicated the author was Kearfott, but Hodges et al. (1983) attributed authorship to Forbes.

Family ARGYRESTHIIDAE

Argyresthiid Moths

Genus *Argyresthia* Hbn.

2434 ***A. affinis*** Braun
Hart Co., Green River bluffs, May 20; Nelson Co., Balltown, May 18. Both are A. F. Braun paratypes.

2435 ***A. alternatella*** Kft.
Boone Co., BBL, May 8, 1980, L. D. Gibson.

2438 ***A. apicimaculella*** Cham.
Kentucky ("in oak woods, in June and July") is the type locality (Chambers 1874a).

2441 ***A. austerella*** Zell.
Kenton Co., Covington, Aug. 1905; Letcher Co., Pine Mountain, June 21; both A. F. Braun.

2455 ***A. freyella*** Wlsm.
Hart Co., Green River bluffs, adults emerging around evergreens, May 27, A. F. Braun.

2467 ***A. oreasella*** Clem.
Boone Co., BBL, July 10; Menifee Co., RRG, May 15, 16.

2481 ***A. thuiella*** (Pack.)
Arborvitae Leafminer Moth
The UK Collection has a series of this species reared from larvae boring in leaves of *arborvitae* near Fayette Co., Lexington. Adults emerged Apr. 29, 1925.

2484 ***A. undulatella*** Cham.
Kentucky is the type locality (Chambers 1874a).

Family **ACROLEPIIDAE**

Acrolepiid Moths

Genus ***Acrolepiopsis*** Gaedike

2490 ***A. incertella*** (Cham.)
Letcher Co., Cumberland Valley, June 26, A. F. Braun.

Family **HELIODINIDAE**

Heliodinid Moths

Genus ***Cycloplasis*** Clem.

2493 ***C. panicifoliella*** Clem.
Blue Lick (county unrecorded; either Lincoln or Nicholas); emerged in July from *Panicum*, A. F. Braun.

Genus ***Heliodines*** Staint.

2497 ***H. bella*** (Cham.)
Kentucky is the type locality (Chambers 1873), who found it nectaring on *Celastrus scandens*. Listed from Kentucky by Forbes (1923).

Genus ***Schreckensteinia*** Hbn.

2509 ***S. festaliella*** (Hbn.)
Harlan Co., Pine Mountain, July 5; Letcher Co., Colliers Creek, June 8; both A. F. Braun.

Superfamily **SESIOIDEA**

Family **SESIIDAE**

Clear-winged Moths

Genus ***Paranthrene*** Hbn.

2522 ***P. asilipennis*** (Bdv.)
Oak Stump Borer Moth
Laurel Co., Cave Creek, SR 121, May 30; McCreary Co., E of Alum Fork, May 19; Wolfe Co., SR 715, June 16; all collected by J. S. Nordin.

2523 ***P. dollii*** (Neum.)
Two specimens in the AMNH are labelled as follows: "Kentucky," "No. 15778 Hy. Edwards" and "cotype" (with the name form *castaneum* Beutenmüller); and "No. 15778 Collection Hy. Edwards" and cotype label for "form *castaneum* Beutenmüller." This name is a synonym of *dollii*.

2527 ***P. simulans*** (Grt.)
Hornet Clearwing
Reported by Hay (1974) as *P. palmii* Henry Edwards, a synonym, attacking wounds in red oaks in vicinity of London, Laurel Co. Fayette Co., Lexington, May 29, 1943; Jefferson Co., Louisville, at pheromone bait, June 6–16; McCreary Co., 1.6 km (1 mi) E of Alum, May 19.

Genus ***Vitacea*** Engelh.

2530 ***V. polistiformis*** (Harr.)
Grape Root Borer Moth
Fayette Co., Lexington, July 15, Aug. 20, 1936; Jefferson Co., Louisville, July 4; Kenton Co., Covington, July 13.

2531 ***V. scepsiformis*** (Hy. Edw.)
Carter Co., 6.4 km (4 mi) W of Gregoryville, Aug. 7, 1966.

Genus ***Euhagena*** Hy. Edw.

2534 ***E. emphytiformis*** (Wlk.)
Bullitt Co., Pine Creek Rd., 0.8 km (0.5 mi) N of SR 480, swept during day from *Gaura filipes*, Aug. 30, 1990, J. A. Bess.

Genus ***Melittia*** Hbn.

2536 ***M. cucurbitae*** (Harr.)
Squash Vine Borer Moth
COUNTIES: FAY, FLE, JEF, MCC, OLD, WAR. **DATES:** June 1, 18–29, July 1, 12–29, Aug. 10, 11, 20–25. **REMARKS:** I have collected adults from milkweed flowers and leaves of squash plants in my garden in Louisville. The larvae bore within stalks and kill squash plants.

Genus ***Synanthedon*** Hbn.

2546 ***S. acerrubri*** Engelh.
Fayette Co., Lexington, June 9–21, July 10. The label on the June 21, 1904 specimen states that it was caught "on linden."

2549 ***S. scitula*** (Harr.)
Dogwood Borer Moth
Fayette Co., Lexington, in pheromone trap, June 6; Hickman Co., Clinton, "from woody gall of red oak, taken Nov. 2, 1911, emerged Feb. 1912"; Jefferson Co., Louisville, May 25.

2550 ***S. pictipes*** (G. & R.)
Lesser Peachtree Borer Moth
Fayette Co., Lexington, May 18, 20, 30, June 19; Fulton Co., Willingham Bottoms, 4.0 km (2.5 mi) E of Cayce, Aug. 9; Jefferson Co., Louisville, July 3, 19, 21, Aug. 24; Kenton Co., Taylor Mill, June 30, July 1; Madison Co., Richmond, May 10. A pest of peach and other *Prunus* species.

2552 ***S. rileyana*** (Hy. Edw.)
Riley's Clearwing
Bullitt Co., Shepherdsville, July 25; Fayette Co., Lexington, July 24, 30, Aug. 1–24, Sept. 9; Fulton Co., Hickman, Oct. 2; Jefferson Co., Louisville, Aug. 5, 22; Kenton Co., Taylor Mill, Aug. 12. Common.

2554 ***S. acerni*** (Clem.)
Maple Callus Borer Moth
PARKS AND RESERVES: BBM, BUC, PNM, PYR, SWR. **COUNTIES:** BEL, BRE, CAD, FAY, GAL, HEN, HLN, HND, JEF, OWS, PER, ROW. **DATES:** May 12–July 8, Aug. 10–22, Sept. 10. **REMARKS:** Common; our only clearwing that comes to lights.

2562 ***S. dominicki*** Duckworth and Eichlin
Fulton Co., Willingham Bottoms, 4.0 km (2.5 mi) E of Cayce, day capture, Sept. 7, 1985, C. V. Covell, Jr. Determination verified by T. Eichlin.

2565 ***S. pyri*** (Harr.)
Fayette Co., Lexington, June 2, 1938.

2575 ***S. arkansasensis*** Duckworth and Eichlin
Boone Co., BBL, at light, July 26, 29; Russell Co., LCP, in light trap, Aug. 14.

2583 ***S. exitiosa*** (Say)
Peachtree Borer Moth
PARKS AND RESERVES: BBM, BLA, KGC, OCP. **COUNTIES:** BOO, FAY, HLN, JEF, KEN, MEA. **DATES:** June 3, 11–17, 27, July 1, 9, 16, Aug. 6, 17, 24, Sept. 4, 5, 19. **REMARKS:** A pest on peach; more common than our records indicate.

Genus *Podosesia* Mösch.

2588 ***P. aureocincta*** Purrington and Nielsen
Banded Ash Clearwing
Boone Co., Florence, Sept. 6; Fayette Co., Lexington, in pheromone traps operated by Daniel Potter, Aug. 15, 21, Sept. 5–19, 1979. This recently described species is on the wing in late summer, while the similar *P. syringae* flies in early summer.

2589 ***P. syringae*** (Harr.)
Lilac Borer Moth
PARKS AND RESERVES: OCP, RRG. **COUNTIES:** FAY, FUL, JEF, KEN, MEA, MEN, WOO. **DATES:** Apr. 28–July 15. **REMARKS:** Collected regularly in pheromone traps (Daniel Potter, *pers. comm.*).

Genus *Carmenta* Hy. Edw.

2596 ***C. bassiformis*** (Wlk.)
Eupatorium Borer Moth
COUNTIES: FAY, FUL, JEF, JES, MER, OLD. **DATES:** June 29, July 30, Aug. 1, 23, 29, Oct. 2.

2597 ***C. corni*** (Hy. Edw.)
Fayette Co., Lexington, June 7, 1920; Laurel Co., Hazel Patch, June 5; Jefferson Co., Louisville, July 6.

2600 ***C. ithacae*** (Beutenmüller)
Boone Co., BCN, at light, July 5; Bullitt Co., BMF, in Malaise trap, June 6–10, 22–30, July 8–18; Kenton Co., Taylor Mill, Sept. 4.

2608 ***C. pyralidiformis*** (Wlk.)
Fayette Co., Clay's Mill, Aug. 31, 1894, H. Garman; Jefferson Co., Louisville, Caperton Swamp, Sept. 5.

Genus *Alcathoe* Hy. Edw.

2623 ***A. caudata*** (Harr.)
Fayette Co., Lexington, South Elkhorn Creek, June 26, 1983, J. S. Nordin.

Family **CHOREUTIDAE**

Choreutid Moths

Genus *Brenthia* Clem.

2627 ***B. pavonacella*** Clem.
Laurel Co., Fariston, May; Letcher Co., Cumberland Valley, May; McCreary Co., Yahoo Creek, emerged in July 13, 1935; all A. F. Braun. Menifee Co., RRG, collected in daytime, May 21; Owsley Co., Booneville, Aug 20; Rowan Co., Morehead, emerged from *Clematis* Sept. 24. Chambers (1878a) created a *nomen nudum*, *Microaethia amphicarpeaeana*, in his discussion of *B. pavonacella*; but he had specimens of it from Kentucky, and also sent one to the NHM in 1877 under the name *amphicarpeaeana* (Miller and Hodges 1990).

Genus *Prochoreutis* Diakonoff & Heppner

2629 ***P. inflatella*** (Clem.)
Bullitt Co., BMF, in Malaise trap, July 8–12; Henderson Co., SWR, Sept. 11; Owsley Co., 4.8 km (3 mi) NE of Booneville, at lights, Sept. 8; Rowan Co., Clack Mtn., May, and Morehead, July 3.

2632 ***P. extrincicella*** (Dyar)
Woodford Co., 1.6 km (1 mi) W of SR 68, 198 m (650 ft) elevation, June 2, 1983, J. S. Nordin.

Genus *Tebenna* Billberg

2647 ***T. gnaphaliella*** (Kft.)
Powell Co., NBP, June 22; Rowan Co., Clack Mtn., mining in *Gnaphalium obtusifolium*, Aug. 20, both A. F. Braun.

Superfamily **COSSOIDEA**

Family **COSSIDAE**

Carpenterworm and Leopard Moths

Genus ***Inguromorpha*** Edw.

2659 ***I. basalis*** (Wlk.)
Black-lined Carpenterworm Moth
Bullitt Co., PCP, June 23; Jefferson Co., Valley Station, May 26, June 22, 30, July 4, Aug. 3, 15; Meade Co., OCP, June 21. Uncommon.

Genus ***Givira*** Wlk.

2668 ***G. anna*** (Dyar)
Anna Carpenterworm Moth
Boone Co., BBL, July l0; Bullitt Co., BMF, June 17, and PCP, June 23, July 22; Jefferson Co., WVP, May 26, June 25, 30, July 15, Aug. 8, 22, 27; McCracken Co., Ray Black farm, 9.7 km (6 mi) SW of Paducah, July 26.

Genus ***Prionoxystus*** Grt.

2693 ***P. robiniae*** (Peck)
Carpenterworm Moth
PARKS AND RESERVES: BMF, CBB, LBL, NBP, PNM. **COUNTIES:** BEL, BOD, BRA, BUL, CAE, FAY, JEF, LAU, MCC, OLD, OWS, POW, ROW, SCO, TRI, WHI. **DATES:** Apr. 12–27, May 5, 13, 23–29, June, July 1–25, Aug. 10. **REMARKS:** Common. Hay (1974) reported this species invading wounded red oak in Laurel Co. during experiments in 1957–1961. He found this species more common than *P. macmurtrei*.

2694 ***P. macmurtrei*** (Guér.)
Little Carpenterworm Moth
PARKS AND RESERVES: BIP, CBB, JWS, PNM. **COUNTIES:** BEL, BRA, FLO, JEF, LAU, OLD, PIK. **DATES:** Apr. 24, 30, May 4, 14, 23, June 12, July 8. **REMARKS:** Hay (1974) reported this species attacking wounds in red oak in Laurel Co. in 1957, 1959, and 1961.

Genus ***Zeuzera*** Latr.

2700 ***Z. pyrina*** (L.)
Rockcastle Co., July 15, 20, 21, 1955 (Beebe 1955).

Superfamily **TORTRICOIDEA**

Family **TORTRICIDAE**

Tortricid Moths

Subfamily **Olethreutinae**

Genus ***Episimus*** Wlsm.

2701 ***E. argutanus*** (Clem.)
PARKS AND RESERVES: BBL, OCP. **COUNTIES:** BAR, BOO, FAY, GAL, JEF, LAU, MEA, OWS. **DATES:** Apr. 18, May 4–31, June 9–12, 22, 23, Aug. 11–15. Common.

2703 ***E. tyrius*** Heinr.
Jefferson Co., WVP, June 20, 1975, A. J. Brownell.

Genus ***Bactra*** Steph.

2706 ***B. furfurana*** (Haw.)
Boone Co., BBL, June 26, Aug. 4; Gallatin Co., Markland Dam, June 21; Henderson Co., SWR, Aug. 22; Jefferson Co., Valley Station, July 19; Letcher Co., Cumberland Valley, June 30.

2707 ***B. verutana*** Zell.
PARKS AND RESERVES: PCP. **COUNTIES:** BRA, BUL, FAY, HAR, LAU, MCC. **DATES:** June 23, July 1–17, 29–Aug. 6, 14–22, Sept. 1, 19, Oct. 6. Common.

Genus ***Endopiza*** Clem.

2711 ***E. liriodendrana*** (Kft.)
PARKS AND RESERVES: BBL, BBM, KGC, OCP, PYR, RRG, WVP. **COUNTIES:** BOO, CHR, HLN, JEF, MEA, MEN, POW. **DATES:** Apr. 18, June 1–27, July 8–13, Aug. 25. Common.

2712 ***E. viteana*** Clem.
Grape Berry Moth
Boone Co., BBL, Apr. 25; Fayette Co., Lexington, Aug. 3, 11, 31, Sept. 2, 5; Harlan Co., KGC, July 13; Jefferson Co., Louisville, reared from grape, adults emerging July 4–18.

2716 ***E. rhoifructana*** (Kft.)
Powell Co., June 19, 1922, A. F. Braun.

2718 ***E. spiraeifoliana*** (Heinr.)
Boone Co., BBL, May 5; Harlan Co., KGC, July 13, Owsley Co., 4.8 km (3 mi) NE of Booneville, May 10; all L. D. Gibson.

2720 ***E. palliolana*** (McD.)
Powell Co., RRG, Tunnel Ridge, July 8, 1989, D. J. Wright.

2724 ***E. aruncana*** (Kft.)
Lewis Co., emerged in Apr., A. F. Braun.

2727 ***E. cyclopiana*** (Heinr.)
Laurel Co., ridge top SW of Ned Branch, July 28, 1992, L. D. Gibson.

Genus ***Lobesia*** Gn.

2728 ***L. carduana*** (Bsk.)
Boone Co., BBL, May 31; Fayette Co., Lexington, at black-light, June 1; Jefferson Co., Louisville, May 30.

Genus ***Endothenia*** Steph.

2732 ***E. montanana*** (Kft.)
Boone Co., BBL, Aug. 4; Gallatin Co., Markland Dam, July 17, Aug. 1, 3; Harlan Co., BBM, May 23, July 7.

2733 ***E. heinrichi*** McD.
Boone Co., BBL, July 27, Aug. 4; Gallatin Co., Markland Dam, June 7; both L. D. Gibson.

2738 ***E. hebesana*** (Wlk.)
PARKS AND RESERVES: BBL, BBM, BMF, KGC, OCP, PNM, RRG, WVP. **COUNTIES:** BEL, BOO, BUL, CAR, FAY, FUL, GAL, HLN, JEF, LET, MEA, MEN, OLD, OWS. **DATES:** Apr. 13–23, May 5–12, 23–30, June, July 1–26, Aug. 3, 20–25, Sept. 6, 19, 30. Common.

2740 ***E. conditana*** (Wlsm.)
Boone Co., BBL, Aug. 4, Gallatin Co. Markland Dam, Aug. 1; both L. D. Gibson.

2742 ***E. infuscata*** Heinr.
Gallatin Co., Markland Dam, Aug. 3, 1988, L. D. Gibson.

2743.1 ***E. nubilana*** (Clem.)
Boone Co., BBL, Sept. 30; Bracken Co., SR 1159 at Locust Creek, Sept. 25; both L. D. Gibson.

Genus ***Taniva*** Heinr.

2745 ***T. albolineana*** (Kft.)
Boone Co., BBL, May 18, 1989; Oldham Co., Goshen, May 15. Transferred from *Endothenia* by Dang (1990).

Genus ***Hulda*** Heinr.

2747 ***H. impudens*** (Wlsm.)
Menifee Co., RRG, Aug. 23, 25, Sept. 5; Owsley Co., 4.8 km (3 mi) NE of Booneville, May 10, 16, July 24; Powell Co, RRG, Sept. 12, 13; Rowan Co., Morehead, June 27. Transferred from *Endothenia* by Dang (1990).

Genus ***Aterpia*** Gn.

2748 ***A. approximana*** (Heinr.)
Gallatin Co., Markland Dam, June 9, 11, July 17; Menifee Co., RRG, May 13; both L. D. Gibson.

Genus ***Eumarozia*** Heinr.

2749 ***E. malachitana*** (Zell.)
PARKS AND RESERVES: HWS, LBL, OCP. **COUNTIES:** FAY, FUL, JEF, LAU, MCC, MEA, OLD, OWS, TRI, WHI. **DATES:** May 25, June 5, 16–30, July, Aug. 6, 10, 22–25, Sept. 6, Oct. 2, Nov. 14, 19. Common.

Genus ***Zomaria*** Heinr.

2750 ***Z. interruptolineana*** (Fern.)
Laurel Co., Holly Bay Campground, Apr. 19; Owsley Co., 4.8 km (3 mi) NE of Booneville, May 10, 28, July 20; Powell Co., RRG, Tunnel Ridge, at lights, May 3, 14, 21; all L. D. Gibson.

Genus ***Orthotaenia*** Steph.

2770 ***O. undulana*** (D. and. S.)
Menifee Co., RRG, June 21, 22, 1985, in light trap, J. S. Nordin.

Genus ***Phaecasiophora*** Grt.

2771 ***P. confixana*** (Wlk.)
PARKS AND RESERVES: BBL, BMF, HWS, LCP, NBP, OCP, RGR, RRG. **COUNTIES:** BOO, BUL, GAL, GRY, JEF, LET, MEA, MEN, MER, OLD, OWS, POW, RUS. **DATES:** Apr. 26, May 9–24, June 2–20, Aug. 9–14. Common.

2772 ***P. niveiguttana*** Grt.
PARKS AND RESERVES: BBL, OCP, PYR, RRG, WVP. **COUNTIES:** BEL, BOO, BUL, CAD, CHR, JEF, MCC, MEA, MEN, OLD, OWS, POW, ROW, WHI. **DATES:** Apr. 11–24, May 1, 10–18, 27–June 29, July 10–31, Sept. 1, 6, 27. **REMARKS:** Common. One larva collected on Sept. 7 at Valley Station, Jefferson Co. reared on *Sassafras albidum*, pupated Sept. 15, adult emerged Sept. 27.

Genus ***Olethreutes*** Hbn.

2774 ***O. monetiferana*** (Riley)
Boone Co, BCN, May 29, June 5, 8; Bullitt Co., BMF, in Malaise trap, May 11–20; Gallatin Co., Markland Dam, June 11; Mercer Co., Shaker's Landing, May 9; Woodford Co., May 25.

2775 ***O. nitidana*** (Clem.)
Laurel Co., DBNF, ridge S of Ned Branch, July 28; Meade Co., OCP, June 20, Aug. 14, 20; Owsley Co., 4.8 km (3 mi) NE of Booneville, July 20; Powell Co., NBP, July 9; Russell Co., LCP, Aug. 14.

2776 ***O. furfurana*** (McD.)
Boone Co., BBL, July 27, Aug. 4; Laurel Co., June 19; Trigg Co., LBL, July 10.

2778 ***O. olivaceana*** (Fern.)
Owsley Co., 4.8 km (3 mi) NE of Booneville, June 27; Rowan Co., Morehead, June 20, 28.

2780 ***O. subnubilus*** (Heinr.)
Letcher Co., Pine Mountain, July 1, A. F. Braun.

2784 ***O. footiana*** (Fern.)
Letcher Co., Big Black Mountain, June 13–20; Lewis Co., Big Sulfur, June 27; Owsley Co., 4.8 km (3 mi) NE of Booneville, July 20; Powell Co., NBP, June 19; Whitley Co., Goldbug, at lights, June 30. Reared on *Hamamelis virginiana*.

2785 ***O. atrodentana*** (Fern.)
Bath Co., CAV, Aug. 17; Owsley Co., 4.8 km (3 mi) NE of Booneville, Aug. 26; Laurel Co, ridge SW of Ned Branch, July 28; all L. D. Gibson.

2787 ***O. connectus*** (McD.)
Carter Co., Cascades Cave, emerged June 9; Laurel Co., Fariston, emerged June 5; Lewis Co., Kinniconick, emerged from *Cornus* May 30; Menifee Co., Leatherwood Fork, FS 9A, July 6; Rowan Co., Brushy Creek, emerged in May.

2788 ***O. inornatana*** (Clem.)
Bullitt Co., near Shepherdsville, July 26, larvae collected on *Pinus serotina* May 4; Harlan Co., BBM, July 11; Laurel Co., July 28; Meade Co., OCP, Aug. 14; Whitley Co., no date.

2789.1 ***O. mysteriana*** W.E. Miller
A paratype male of this species (Miller 1979) was reared from Bullitt Co., BMF, on *Celtis occidentalis* by A. J. Brownell, emerging May 21, 1977. Boone Co., BBL July 9; Gallatin Co., Markland Dam, June 24; Meade Co., OCP, June 20; Menifee Co., RRG, July 6 and 8; Powell Co., RRG, Tunnel Ridge, July 8.

2790 ***O. mediopartitus*** (Heinr.)
Meade Co., OCP, May 15, 1982, L. D. Gibson.

2796 ***O. sciotana*** (Heinr.)
Bracken Co., Chatham, emerged May 25; Franklin Co., emerged May 13 from *Aesculus glabra*; Jefferson Co., emerged in Louisville, May 13; all A. F. Braun; "Ky. near Cincinnati, Ohio," emerged May 23.

2797 ***O. appalachiana*** (Braun)
Carter Co., Carter City, emerged May 24–29, June 2; Johnson Co., Paintsville, emerged May 22–25; Laurel Co., Rockcastle Recreational Area, DBNF, reared from *Aesculus octandra*, emerged May 15–25; Pulaski Co., DBNF, May 24; Whitley Co., Buck Branch, emerged May 27–30.

2799 ***O. clavana*** (Wlk.)
Bath Co., Cave Run Road to Zilpo, Aug. 17; Jefferson Co., BEC, June 13, July 16; Laurel Co., DBNF, June 11; Powell Co., RRG, Tunnel Ridge, July 8.

2800 ***O. nigrana*** (Heinr.)
PARKS AND RESERVES: BEC, CAC, KGC, LCP, OCP, PMS. **COUNTIES:** CAR, HLN, JEF, LAU, LET, MEA, OLD, RUS, WHI. **DATES:** June 3–7, 24, July 8, 16, 31, Aug. 9–16. **REMARKS:** Common. Reared on *Acer saccharum*.

2801 ***O. viburnana*** (McD.)
Fayette Co., Lexington, June 1; Oldham Co., HWS, June 7; Trigg Co., LBL, July 10.

2802 ***O. hippocastana*** (Kft.)
Letcher Co., Gilley, June 13, A. F. Braun.

2803 ***O. merrickana*** (Kft.)
Bell Co., Beverly, July 4, A. F. Braun.

2804 ***O. hamameliana*** (McD.)
PARKS AND RESERVES: BBM, NBP, PNM, RRG. **COUNTIES:** BEL, CAR, HLN, LAU, LET, MEN, OWS, POW. **DATES:** May 27–June 14. Reared from *Hamamelis virginiana*.

2805 ***O. corylana*** (Fern.)
Campbell Co., Schmudde-Phillips Creek Rd., near Schmudde-Tarvin Rd., reared from *Corylus americana* (John Merkle), emerged May 21; Letcher Co., Big Black Mountain, larva on *Cornus aeternifolia*, emerged July 12; Meade Co., OCP, June 7; Owsley Co., 4.8 km (3 mi) NE of Booneville, at lights, May 29.

2806 ***O. ochrosuffusana*** (Heinr.)
PARKS AND RESERVES: BBM, BCN, HWS, RRG. **COUNTIES:** BOO, GAL, HLN, KEN, LAU, LET, MEN, OLD, PEN. **DATES:** May 27–July 1. **REMARKS:** L. D. Gibson has reared this species from larvae on *Aesculus glabra* collected in Boone, Kenton and Pendleton counties.

2807 ***O. brunneopurpuratus*** (Heinr.)
Harlan Co., Pine Mountain, July 5; Letcher Co., Colliers Creek, June 27; both A. F. Braun.

2808 ***O. ferrugineana*** (Riley)
Boone Co., BBL, May 18; Gallatin Co., Markland Dam, June 2; Rowan Co., Morehead, emerged June 26, A. F. Braun.

2809 ***O. fagigemmeana*** (Cham.)
Kentucky is the type locality (Chambers 1878b). Edmonson Co., MCV, July 1; Laurel Co., FS 9A near Bolton Branch, June 17; Meade Co., OCP, June 20; Owsley Co., 4.8 km (3 mi) NE of Booneville, June 14 and 25; Whitley Co., CBF, June 24, 25.

2810 ***O. sericorana*** (Wlsm.)
Letcher Co., Pine Mountain, June 24, A. F. Braun.

2817 ***O. permundana*** (Clem.)
Boone Co., BBL, June 7–23; Harlan Co., KGC, July 11; Jefferson Co., Valley Station, July 3; Owsley Co., 4.8 km (3 mi) NE of Booneville, June 14; Powell Co., Clay City, June 30.

2820 ***O. malana*** (Fern.)
Laurel Co., Fariston, emerged June 4, A. F. Braun.

2821 ***O. appendicea*** (Zell.)
PARKS AND RESERVES: AJJ, BAN, BBL, BBM, OCP, RRG. **COUNTIES:** BAR, BOO, CAM, HLN, JEF, LAU, LET, MCY, MEA, MEN, POW. **DATES:** May 24, 27, June 6–26, July 8. **REMARKS:** Common. L. D. Gibson reared this from larvae found on *Gaylussacia baccata*, *Fagus grandifolia*, and *Acer rubrum*.

2822 ***O. concinnana*** (Clem.)
PARKS AND RESERVES: AJJ, BBL, BCN, OCP, RRG. **COUNTIES:** BAT, BOO, CAM, GAL, JEF, KEN, LAU, LET, MEA, MEN, OWS, WHI. **DATES:** May 10, 13, 27, June 5–8, July 7, 17, Aug. 7. Common.

2823 ***O. fasciatana*** (Clem.)
PARKS AND RESERVES: BAN, BMF, NBP, PNM, WVP. **COUNTIES:** BAR, BEL, BUL, FAY, GAL, HRT, JEF, LET, MCY, MEA, OWS, POW, WHI. **DATES:** May 16–July 11. Common.

2824 ***O. troglodana*** (McD.)
Laurel Co., DBNF, May 30, June 6; Owsley Co., 4.8 km (3 mi) NE of Booneville, May 28–29; all L. D. Gibson.

2825 ***O. exaeresima*** (Heinr.)
Bell Co., Beverly, July 4, and PNM, July 14; Letcher Co., Big Black Mountain, July 7; Perry Co., Lynn Fork, July 11; all A.F. Braun.

2826 ***O. lacunana*** (Free.)
PARKS AND RESERVES: KGC, RRG. **COUNTIES:** BAT, HLN, LAU, MEN, OWS, POW. **DATES:** May 29, June 30–July 1–14, 28–30, Aug. 17. Common.

2827 ***O. ferriferana*** (Wlk.)
PARKS AND RESERVES: BBM, BMF, KGC, RRG. **COUNTIES:** BUL, HLN, JEF, LAU, MEN, OWS. **REMARKS:** Common. Reared by L.D. Gibson from *Hydrangea arborescens*.

2828 ***O. griseoalbana*** (Wlsm.)
PARKS AND RESERVES: CAV, OCP, PNM, RRG. **COUNTIES:** BAT, BEL, JEF, LAU, MEA, MEN, OWS. **DATES:** May 10, 22–31, June 6–14, 22, July 14, Aug. 27, 28. Common.

2829 ***O. osmundana*** (Fern.)
Bath Co., CAV, Aug. 26, 28; Laurel Co., DBNF, June 6; McCreary Co., near Whitley City, Aug. 30 and Pine Knot, June 14; Owsley Co., 4.8 km (3 mi) NE of Booneville, June 30, Aug. 15; Powell Co., June 3, 6, 25, 26.

2831 ***O. agilana*** (Clem.)
Bullitt Co., PCP, July 8; Jefferson Co., Valley Station, Apr. 7; Letcher Co., Colliers Creek, June 8, Pine Mountain, June 16, 21; Menifee Co., RRG, May 10; Owsley Co., 4.8 km (3 mi) NE of Booneville, May 10, 28.

2837 ***O. astrologana*** (Zell.)
PARKS AND RESERVES: BCN, BMF, BUC, HWS, PNM, RRG. **COUNTIES:** BEL, BOO, BRE, BUL, GAL, JEF, MCC, MEN, OLD, OWS, PER, POW. **DATES:** May 6, 24 –June 19. Common.

2838 ***O. coruscana*** (Clem.)
Boone Co., BBL, May 26, and BCN, May 29; Bullitt Co., SR 480 E of SR 61, May 28; Gallatin Co., Markland Dam, May 27, 29, June 2; Letcher Co., Big Black Mountain, June 8, Colliers Creek, June 27, Gilley, June 13, and Pine Mountain, June 24.

2838.1 ***O. ferrolineana*** (Wlk.)
Boone Co., BBL, May 24–29; Bullitt Co., BMF, Apr. 8–11, May 11–June 15, and PCP, in Malaise trap, May 28, June 10–15; Gallatin Co., Markland Dam, May 25–June 2; Whitley Co., Goldbug, June 5. Common.

2859 ***O. cespitana*** (Hbn.)
PARKS AND RESERVES: AJJ, BBL, CBB, OCP. **COUNTIES:** BOO, BUL, CAM, FAY, GAL, JEF, LAU, MCC, MEA, OLD, OWS. **DATES:** May 14–31, June 2–9, 26, 29, July 5–28, Aug. 1, 29, Sept 14. Common.

Genus ***Hedya*** Hbn.

2860 ***H. separatana*** (Kft.)
PARKS AND RESERVES: BBL, BCN, CAV, PYR, RRG. **COUNTIES:** BAT, BOO, CHR, FAY, GAL, HRD, JEF, LAU, OLD, OWS, POW. **DATES:** Apr. 21, May 4, 18–28, June 11, 19–27, July 10–Aug. 17. Common.

2861 ***H. ochroleucana*** (Frölich)
Cited from Kentucky by Heinrich (1926). Possibly Lexington, from an old specimen in the UK Collection labelled "336." H. Garman code book entry #336 states "Insect from rose bud emerged May 25, 1891". Boone Co., BCN, June 15, 1984, L. D. Gibson.

2864 ***H. cyanana*** (Murt.)
Boone Co., BBL, July 10, and BCN, May 29; Bracken Co., SR 1159 at Locust Creek, Aug. 23; McCracken Co., Ray Black farm, 9.7 km (6 mi) SW of Paducah, Sept. 6; Owsley Co., 4.8 km (3 mi) NE of Booneville, June 22.

Genus ***Evora*** Heinr.

2866 ***E. hemidesma*** (Zell.)
Cited from Kentucky by Heinrich (1926).

Genus ***Rhyacionia*** Hbn.

2867 ***R. buoliana*** (D. & S.)
Jefferson Co., Louisville, June 7; Fayette Co., Lexington, June 8, 9; Oldham Co., Goshen, June 14.

2868 ***R. rigidana*** (Fern.)
Pitch Pine Tip Moth
Harlan Co., KGC, July 13; Jefferson Co., Louisville, Apr. 2; Laurel Co., DBNF, Apr. 6, 11;

Owsley Co., 4.8 km (3 mi) NE of Booneville, Apr. 22, May 10, 22, June 30, July 3. Eastern Kentucky was indicated in distribution map for this species in Powell and Miller (1978), but a precise locality was not given.

2879 ***R. busckana*** Heinr.
Jefferson Co., Louisville, Apr. 4; Menifee Co., RRG, Tunnel Ridge, Mar. 11–12; Owsley Co., 4.8 km (3 mi) SE of Booneville, Mar. 15; Powell Co., RRG, Mar. 4, 11.

2882 ***R. frustrana*** (Comstock)
Fayette Co., Lexington, reared from Scotch Pine twig, emerged June 25, July 1; Jefferson Co., Louisville, Apr. 4, and Shively, reared from pine twig, emerged Feb. 28, Mar. 4; McCracken Co., Paducah, emerged from pine twig, July 24; Ohio Co., Feb. 7–14; Owsley Co., 4.8 km (3 mi) NE of Booneville, July 14.

Genus ***Retinia*** Gn.

NOTE: W.E. Miller (1987) indicated this was the correct generic name for moths placed in *Petrova* Heinr. in Hodges et al. (1983).

2889 ***R. comstockiana*** (Fern.)
Laurel Co., June 6; Menifee Co., RRG, May 26, 27; Powell Co., RRG, Tunnel Ridge, May 21; Whitley Co., Goldbug, June 5.

2891 ***R. wenzeli*** (Kft.)
Jefferson Co., Valley Station, May 5; McCreary Co., DBNF, Blue Heron Campground, Apr. 27; Menifee Co., RRG, Apr. 27, May 11; Owsley, Co., 4.8 km (3 mi) NE of Booneville, May 10, 23; Powell Co., RRG, Tunnel Ridge, May 3–21.

2898 ***R. gemistrigulana*** (Kft.)
Bullitt Co., PCP, June 23; Laurel Co., DBNF, May 31, June 6; Menifee Co., RRG, June 15, 16; Powell Co., RRG, Tunnel Ridge, May 21, June 9–19, July 8; Whitley Co., Goldbug, June 5.

2902 ***R. houseri*** W.E. Miller
Laurel Co., DBNF, FS 775, July 10, FS 4176 on ridge top SW of Ned Branch, June 6, 11, FS 121 near Bolton Branch, June 11; all L.D. Gibson.

Genus ***Phaneta*** Steph.

2910 ***P. essexana*** (Kft.)
Boone Co., BBL, May 4; Campbell Co, AJJ, May 1990.

2913 ***P. umbrastriana*** (Kft.)
PARKS AND RESERVES: BBL, BCN, MUR, RRG. **COUNTIES:** BOO, BRA, GAL, HIC, JEF, LAU, MCY, MEN. **DATES:** May 4–31, June 8, July 24. Common.

2925 ***P. autumnana*** (McD.)
Boone Co., BBL, Sept. 30, 1981, L. D. Gibson.

2926 ***P. verna*** W.E. Miller
Boone Co., BBL, Apr. 10, 26, 30, May 1; Bracken Co., Meldahl Dam, May 9; Bullitt Co., BMF, in Malaise trap, Apr. 3–7.

2927 ***P. ochrocephala*** (Wlsm.)
Boone Co., BBL, Aug. 25, East Bend, 8.0 km (5 mi) W of BBL, Aug. 21; Bracken Co., Meldahl Dam, Aug. 24, Sept. 1; all L. D. Gibson.

2928 ***P. raracana*** (Kft.)
Boone Co., East Bend, Aug. 21; Bracken Co., SR 1159 at Locust Creek, Aug. 23, and Meldahl Dam, Aug. 24, Sept. 1; Letcher Co., Pine Mountain, near Edia, Aug. 28; McCracken Co., Ray Black farm, 9.7 km (6 mi) SW of Paducah, Sept. 6; Owsley Co., Aug. 22–Sept. 1.

2929 ***P. ochroterminana*** (Kft.)
Boone Co., BBL, Aug. 27, Sept. 5; Bracken Co., Meldahl Dam, Sept. 1; Bullitt Co., BMF, in Malaise trap, Apr. 13–17, and Bearcamp Rd., May 6; Menifee Co., NE of Slade, FS 9A, Indian Creek, Sept. 7; Owsley Co., 4.8 km (3 mi) NE of Booneville, Aug. 27, Sept. 1.

2933 ***P. marmontana*** (Kft.)
Fayette Co., Lexington, in Malaise trap, Aug. 22–25; Harlan Co., BBM, 1,204 m (3,950 ft) elevation, at black-light, Aug. 14.

2936 ***P. tomonana*** (Kft.)
PARKS AND RESERVES: BBL, PCP, RRG. **COUNTIES:** BOO, BRA, BUL, LAU, MEN, WOO. **DATES:** June 2, Aug. 31–Sept. 8. Uncommon.

2937 ***P. parmatana*** (Clem.)
PARKS AND RESERVES: BBL, BCN, CER, RRG. **COUNTIES:** BOO, BRA, GAL, JEF, MCC, MEN, OWS, ROW. **DATES:** May 23–June 2, 30, Aug. 15–Sept. 8. **REMARKS:** Common. This species has until recently been known by the name of *crispana* Clemens.

2950 ***P. kokana*** (Kft.)
Jefferson Co., Eastwood, Oct. 28; Rowan Co., DBNF, FS 1274, 3.2 km (2 mi) W of SR 519, Oct 8.

2951 ***P. ornatula*** (Heinr.)
Boone Co., BBL, June 28, July 10; Bracken Co., SR 1159 at Locust Creek, Aug. 23, Meldahl Dam, Aug. 24, Sept. 1; Campbell Co., AJJ, July 15; Gallatin Co., Markland Dam, Aug. 3, 17; Owsley Co., 4.8 km (3 mi) NE of Booneville, June 30; all L. D. Gibson.

2973 ***P. striatana*** (Clem.)
Edmonson Co., Park Mammoth Resort, Park City, May 16; Hart Co., May 20; Menifee Co., RRG, FS 9B, Indian East Fork, Kelley Branch, 220 m (720 ft) elevation, May 10, 11; Owsley Co., 4.8 km (3 mi) NE of Booneville, June 14.

2982 ***P. kiscana*** (Kft.)
Boone Co., BBL, May 22, and BCN, June 15, 19; Owsley Co., 4.8 km (3 mi) NE of Booneville, May 10–16; all L. D. Gibson.

3005.1 ***P. ambodaidaleia*** W.E. Miller
Paratype male, Oldham Co., HWS, Apr. 6, 1972, C. V. Covell, Jr. (Miller 1983a). Bullitt Co., N side of SR 480, 11.1 km (6.9 mi) E of SR 61, Mar. 30.

3005.2 ***P. canusana*** Wright
Paratypes are cited as follows in Wright et al. (1997): Bullitt Co., SR 480, 11.3 km (7 mi) E of Shepherdsville, Mar. 30, and N side of SR 580, 11.1 km (6.9 mi) E of SR 61, Mar. 30; Rowan Co., E side of SR 1274, 3.2 km (2 mi) W of SR 519, Mar. 13.

Genus ***Eucosma*** Hbn.

3008 ***E. quinquemaculana*** (Rob.)
Laurel Co., DBNF, Craig Creek Rd., Sept. 19; Powell Co., Sept. 17; Whitley Co., SR 8, NE of CBF, Sept. 26. Uncommon.

3009 ***E. robinsonana*** (Grt.)
PARKS AND RESERVES: HWS. **COUNTIES:** LAU, MCY, OLD, OWS, ROC, ROW. **DATES:** May 8, 10, June 10–22, Sept. 1, 8. Common.

3014 ***E. ridingsana*** (Rob.)
Barren Co., Diamond Caverns, Oct. 11; Bullitt Co., PCP, Sept. 6, 8; Hart Co., Pine Ridge, Aug. 20.

3037 ***E. agricolana*** (Wlsm.)
Bullitt Co., PCP, May 28, July 8 and BMF, July 19; Fleming Co., June 4; Jefferson Co., Valley Station, July 14; Lewis Co., Aug. 5; Taylor Co., July 5. Our subspecies is *argentialbana* (Wlsm.), which includes *pergandeana* (Fern.), a synonym.

3042 ***E. vagana*** McD.
Boone Co., BBL, July 26, 27; Bullitt Co., PCP, July 22, S side of SR 480, 11.1 km (6.9 mi) E of SR 61, July 8; Henderson Co., SWR, Aug. 22; all L. D. Gibson.

3043 ***E. albiguttana*** (Zell.)
Boone Co., BBL, Aug. 25; Bullitt Co., PCP, July 8, 22; Christian Co., PYR, Aug. 14, 15; McCreary Co., Greenwood, South Fork, and Stearns, June 18, 21, Aug. 28; Powell Co., NBP, Aug. 21.

3044 ***E. graciliana*** Kft.
Bullitt Co., S side of SR 480, 11.1 km (6.9 mi) E of SR 61; Metcalf Co., Knob Creek, Aug. 12; McCreary Co., Pine Knott, June 14, and "South Fork," June 18.

3045.1 ***E.*** species near ***galenapunctana*** Kft.
Bullitt Co., S side of SR 480, 11.1 km (6.9 mi) E of SR 61, June 23, July 8, and PCP, July 22; all L. D. Gibson.

3051 ***E. glomerana*** (Wlsm.)
Boone Co., BBL, Aug. 25; Bracken Co., SR 1159 at Locust Creek, Sept. 12; Lewis Co., Aug. 6; Metcalf Co., Oct. 12; Oldham Co., HWS, Sept. 6, Goshen, Aug. 27, Sept. 1–5.

3052 ***E. circulana*** Hbn.
Christian Co., PYR, Aug. 22, 26; Henderson Co., SWR, Aug. 12; both L. D. Gibson. Our subspecies is *gemellana* Heinr.

3072 ***E. cocana*** Kft.
Laurel Co., Holly Bay camping area, Apr. 19; McCreary Co., Blue Heron Campground, Apr. 27; Owsley Co., 4.8 km (3 mi) NE of Booneville, May 10; Powell Co., RRG, Tunnel Ridge, at light, Apr. 28, May 3–14; Whitley Co., Goldbug, June 5.

3073 ***E. monitorana*** Heinr.
Bullitt Co., PCP, May 28; McCreary Co., Blue Heron Campground, Apr. 27; Owsley Co., 4.8 km (3 mi) NE of Booneville, May 10, 22; Powell Co., RRG, Tunnel Ridge, Apr. 28, May 3, 14–21.

3074 ***E. tocullionana*** Heinr.
McCreary Co., Blue Heron Campground, Apr. 27; Menifee Co., RRG, FS 9B Indian East Fork, Kelley Branch, May 16–17, RRG, Salt Fork of Gladie Creek, 366 m (1,200 ft) elevation, June 25; Owsley Co., 4.8 km (3 mi) NE of Booneville, May 16; Powell Co., RRG, Tunnel Ridge, May 3, 14–21, June 1, 9.

3091 ***E. matutina*** (Grt.)
Bullitt Co., SR 480, 8.8 km (5.5 mi) E of SR 61, July 22, 31, Sept. 6; Christian Co., PYR, Aug. 14, 15, 26; Rowan Co., DBNF, E side of FS 1274, Aug. 26. Locally common.

3098 ***E. giganteana*** (Riley)
Bracken Co., Meldahl Dam, July 30, Aug. 9; Bullitt Co., 10.0 km (6.2 mi) E of Shepherdsville, July 8, 22, Aug. 18, and PCP, July 22; Gallatin Co., Markland Dam, July 17; Jefferson Co., Valley Station, Aug. 9; Owsley Co., 4.8 km (3 mi) NE of Booneville, July 1.

3110 ***E. gomonana*** Kft.
Boone Co., BBL, Apr. 26; Laurel Co., FS 615A/B, Apr. 18; McCreary Co., Blue Heron Campground, Apr. 27; Owsley Co., 4.8 km (3 mi) NE of Booneville, Apr. 20; Powell Co., RRG, Tunnel Ridge, May 10; all L. D. Gibson.

3112 ***E. nandana*** Kft.
Bath Co., CAV, Sept. 5, 1987, L. D. Gibson.

3116 ***E. dorsisignatana*** (Clem.)
COUNTIES: BRA, JEF, LAU, MCC, OWS, ROC, ROW. **DATES:** Sept. 5, 9, 19–Oct. 6, 15. Locally common.

3116.1 ***E. similiana*** (Clem.)
PARKS AND RESERVES: BAN, BBM, CAV, RRG. **COUNTIES:** BAR, BAT, BOO, FUL, HLN, JEF, MCC, MEN, OWS. **DATES:** July 27, Aug. 8, 27, Sept. 5–Oct. 10. **REMARKS:** Recently elevated to species status by Miller (1985); spelling incorrect ("*similana*") in Hodges et al. (1983).

3120 ***E. derelicta*** Heinr.
PARKS AND RESERVES: AJJ, BBL, SWR. **COUNTIES:** BOO, CAM, CHR, FAY, GAL, HEN, HLN, HND, MEN, OWS, ROW. **DATES:** July 23–26, Aug. 8–31. Common.

3122 ***E. wandana*** Kft.
PARKS AND RESERVES: AJJ, BBL, OCP, WVP. **COUNTIES:** BOO, BUL, CAM, GAL, JEF, LAU, MEA, OLD, OWS, ROW. **DATES:** May 15, 20, June 6, 20–Aug. 4. **REMARKS:** Common. See Gibson and Miller (1994) for illustrations, synonomies and records.

3124 ***E. fulminana*** (Wlsm.)
Bullitt Co., 11.3 km (7 mi) E of I-65, July 8, 22–31, and PCP, July 22, 31; Christian Co., PYR, Aug. 26; Rowan Co., SR 1274, 2 miles W of SR 519, July 1, Aug. 26.

3125 ***E. rusticana*** Kft.
Bullitt Co., PCP, at lights, July 8, 21; Fulton Co., SR 94, Willingham Bottoms, 4.0 km (2.5 mi) E of Cayce, Sept. 7; Rowan Co., DBNF, E side of SR 1274, July 1.

3127 ***E. sombreana*** Kft.
PARKS AND RESERVES: BBL, PCP, PYR, RRG, SWR. **COUNTIES:** BOO, BUL, CHR, GAL, HND, MEN, OLD, ROW. **DATES:** June 15, July 22, Aug. 3, 4, 12–15, 26, Sept. 2. Common.

3128 ***E. pandana*** Kft.
Bullitt Co., N side of SR 480, 11.1 km (6.9 mi) E of SR 61, July 22, 1989, L.D. Gibson.

3142 ***E. cataclystiana*** (Wlk.)
"Kentucky, Aug." (specimens in the NMNH); Bullitt Co., S side of SR 480, 11.1 km (6.9 mi) E of SR 61, June 23, July 8; Fleming Co., Aug. 21; Gallatin Co., Markland Dam, Aug. 3; Meade Co., OCP, Aug. 16; Menifee Co., RRG, June 29.

Genus *Pelochrista* Led.

3151 ***P. scintillana*** (Clem.)
Boone Co., BBL, June 7, 24–28, July 7, and BCN, June 19; Bracken Co., Meldahl Dam, Aug. 24; Fayette Co., S. Elkhorn Creek S of Higbee Mill Rd., July 20; Gallatin Co., Markland Dam, July 15; Rowan Co., Clack Mtn., June 20.

3153 ***P. pallidipalpana*** (Kft.)
McCreary Co., June 14, A. F. Braun.

3168 ***P. zomonana*** (Kft.)
PARKS AND RESERVES: BBL, BCN, RRG. **COUNTIES:** BOO, BUL, FAY, FUL, GAL, MEN, OWS. **DATES:** May 29, June 15–25, July 22, Aug. 25, 26, Sept. 17. Common.

3169 ***P. womonana*** (Kft.)
Bullitt Co., PCP, July 22, S side of SR 480, 11.1 km (6.9 mi) E of SR 61, July 8; Campbell Co., AJJ, July 23; Henderson Co., SWR, Aug. 22; Laurel Co., DBNF, S side of SR 3497 near Ned Branch Trail, Aug. 8; all L. D. Gibson.

Genus *Epiblema* Hbn.

3171 ***E. boxcana*** (Kft.)
PARKS AND RESERVES: BBL, BMF, HWS. **COUNTIES:** BOO, BUL, BRA, GAL, LAU, OLD. **DATES:** May 4–9, 17–25, June 2–6.

3172 ***E. strenuana*** (Wlk.)
PARKS AND RESERVES: BCN, CER, HWS, PNM, RNR, SWR, WVP. **COUNTIES:** BEL, BOO, FAY, FUL, GAL, HEN, HND, JEF, LET, OLD, ROW. **DATES:** May 12, 27, 29, June 15–26, July 20, Aug. 13–29, Sept. 8, 14. Common.

3173 ***E. abruptana*** (Wlsm.)
Boone Co., BBL, June 24; Gallatin Co., Markland Dam, June 15–24, July 15; Meade Co., OCP, June 20; all L. D. Gibson.

3174 ***E. numerosana*** (Zell.)
Jefferson Co., WVP, June 6, July 3; Oldham Co., Goshen, June 14, 24, July 3–12, 23; Pulaski Co., Somerset, July 9.

3174.1 ***E. luctuosana*** Blanchard
Boone Co., BBL, at lights, July 26, 1989, L. D. Gibson.

3184 ***E. tripartitana*** (Zell.)
Boone Co., BBL, July 27; Rowan Co., DBNF, E side of SR 1274, 3.2 km (2 mi) W of SR 519, Aug. 18, 28; all L.D. Gibson.

3186 ***E. scudderiana*** (Clem.)
PARKS AND RESERVES: BBL, LMP. **COUNTIES:** BAR, BOO, FUL, GAL, MCC, MUH, OLD.

DATES: Apr. 30, May 12–25, June 2, July 4. Common.

3189 ***E. obfuscana*** (Dyar)
Boone Co., BBL, May 4, 18, June 18; Fayette Co., Lexington, June 11; Gallatin Co., Markland Dam, May 25, June 2–11; Jefferson Co., Louisville, July 11; Owsley Co., 4.8 km (3 mi) NE of Booneville, May 10.

3190 ***E. desertana*** (Zell.)
Boone Co., BBL, May 4, 24, June 24; Bullitt Co., PCP, May 28; Oldham Co., Goshen, May 3.

3192 ***E. carolinana*** (Wlsm.)
Boone Co., BBL, July 31, 8.0 km (5 mi) W of BBL, East Bend area, Aug. 21; Bullitt Co., S side of SR 480, June 23, July 8; Gallatin Co., Markland Dam, June 11; Menifee Co., NE of Slade, FS 9A, Indian Creek, June 6, 15, Aug. 9, 10, RRG, June 29, Aug. 31.

3198 ***E. iowana*** McD.
Boone Co., BBL, May 1; Bracken Co., May 8; Bullitt Co., BMF, in Malaise trap, Apr. 6–9.

3201 ***E. infelix*** Heinr.
PARKS AND RESERVES: BBL, BMF, CAC, RRG. **COUNTIES:** BOO, BUL, CAR, LAU, OWS, POW, ROW. **DATES:** Apr. 13–18, 28, May 2, 10, June 10, 14, July 17. Common.

3201.1 ***E.*** undescribed species near ***infelix*** Heinr.
Bullitt Co., PCP, July 22; Menifee Co., RRG, Leatherwood Fork of Indian Creek, FS 9A, July 6; Owsley Co., 4.8 km (3 mi) NE Booneville, July 20, 24, Aug. 15; Rowan Co., DBNF, E side of SR 1274, 3.2 km (2 mi) W of SR 19, July 1, 16. Locally common.

3202 ***E. otiosana*** (Clem.)
Bidens Borer Moth
PARKS AND RESERVES: BBL, CBF, HWS, SWR. **COUNTIES:** BOO, GAL, HND, HRD, JEF, OLD, OWS, WHI. **DATES:** May 5, 29, June, July 1–17, 31, Aug., Sept. 10, 12. Common.

3203 ***E. brightonana*** (Kft.)
Bullitt Co., PCP, 0.8 km (0.5 mi) N of SR 480, 7.2 km (4.5 mi) E of I-65, at light, July 22 and S side of SR 480, July 8; Gallatin Co., Markland Dam, at light, June 22, Aug. 3, 4, 17; Rowan Co., DBNF, E side of SR 1274, 3.2 km (2 mi) W of SR 19, Aug. 26; all L. D. Gibson.

3204 ***E. tandana*** (Kft.)
Gallatin Co., Markland Dam, June 22; Menifee Co., RRG, June 27, 28; Owsley Co., 4.8 km (3 mi) NE Booneville, June 25; L. D. Gibson.

3205 ***E. resumptana*** (Wlk.)
PARKS AND RESERVES: BBL, CBF. **COUNTIES:** BOO, LAU, LET, NEL, OWS, WHI. **DATES:** Apr. 18, 26, 29, May 4, 6, 16.

3206 ***E. dorsisuffusana*** (Kft.)
Campbell Co., AJJ, July 15; Boone Co., June 30; Gallatin Co., Markland Dam, June 9.

Genus ***Notocelia*** Hbn.

3208 ***N. trimaculana*** (Haw.)
Boone Co., BCN, June 8; Gallatin Co., Markland Dam, June 2, 11, 15, 27; Oldham Co., Goshen, May 26–28, June 3, HWS, June 7; Owsley Co., 4.8 km (3 mi) NE of Booneville, May 10, 11, 29.

3210 ***N. illotana*** (Wlsm.)
Rowan Co., July l, A. F. Braun.

3211 ***N. culminana*** (Wlsm.)
Boone Co., BBL, Aug. 25; Bullitt Co., N side of SR 480, 11.1 km (6.9 mi) E of SR 61, Sept. 6; Oldham Co., Goshen, Oct 1.

Genus ***Suleima*** Heinr.

3212 ***S. helianthana*** (Riley)
Campbell Co., AJJ, July 15; Owsley Co., 4.8 km (3 mi) NE of Booneville, July 15, 20; Rowan Co., Morehead, May 22, DBNF, E side of SR 1274, July 16.

3217 ***S. cinerodorsana*** Heinr.
Boone Co., BCN, Aug. 15; Bracken Co., Meldahl Dam, Sept. 1; Owsley Co., 4.8 km (3 mi) NE Booneville, July 20, Aug. 22; all L. D. Gibson.

Genus ***Sonia*** Heinr.

3218 ***S. constrictana*** (Zell.)
Menifee Co., RRG, East Fork Indian Creek, Kelley Branch, July 8; Gallatin Co., Markland Dam, Aug. 3.

3218.1 ***S. paraplesiana*** Blanchard
Bullitt Co., PCP, July 22; Menifee Co., RRG, July 6; Owsley Co., 4.8 km (3 mi) NE of Booneville, June 22; all L. D. Gibson.

3219 ***S. canadana*** McD.
Bullitt Co., PCP, July 8, 1988, D. J. Wright and L. D. Gibson.

3221.1 ***S. divaricata*** W. E. Miller
Laurel Co., DBNF, FS 121 near Bolton Branch, May 18, June 21; Powell Co., RRG, Tunnel Ridge, one paratype female, May 14, 1988; all L. D. Gibson.

Genus ***Gypsonoma*** Meyr.

3226 ***G. haimbachiana*** (Kft.)
Gallatin Co., Markland Dam, at light, May 27, June 7–29, July 15, Aug. 17, 1989, L. D. Gibson and D. J. Wright.

3228 ***G. salicicolana*** (Clem.)
Boone Co., BBL, June 18, 26; Gallatin Co., Markland Dam, June 9; Menifee Co., RRG, Sept. 5; Owsley Co., 4.8 km (3 mi) NE of Booneville, May 10; all L. D. Gibson.

Genus ***Proteoteras*** Riley

3230 ***P. aesculana*** Riley
PARKS AND RESERVES: BBL, BBM, BCN, OCP, RRG. **COUNTIES:** BOO, BRA, BUL, CLI, FAY, GAL, HLN, HRD, HRT, MCY, MEA, POW, PUL. **DATES:** Mar. 25, Apr. 11–26, May 1, 18–29, June 9, 15, 24, July 5, 8. **REMARKS:** Common. "Bores into maple twigs..." (*Kentucky Cooperative Plant Pest Report* 1977).

3232 ***P. willingana*** (Kft.)
Boone Co., BCN, June 15, 19; Bullitt Co., SR 480, 11.3 km (7 mi) E of Shepherdsville, July 8; Gallatin Co., Markland Dam, May 29, June 2, 11; Jefferson Co., Caperton Swamp, Apr. 6; Owsley Co., 4.8 km (3 mi) NE of Booneville, June 22.

3233 ***P. crescentana*** Kft.
Boone Co., BBL, June 25, and BCN, June 15, July 5; Campbell Co., AJJ, July 15; Jefferson Co., Valley Station, July 25; Meade Co., OCP, Apr. 11; Menifee Co., RRG, June 8.

3234 ***P. naracana*** Kft.
Bullitt Co., S side of SR 480, 11.1 km (6.9 mi) E of SR 61, May 28; Gallatin Co., Markland Dam, May 25, 29, June 2; Meade Co., OCP, June 7, 11; Oldham Co., HWS, June 7; Whitley Co., Goldbug, June 5.

3235 ***P. moffatiana*** (Fern.)
PARKS AND RESERVES: BBL, BBM, BMF, CAH, LCP, OCP, PNM, PYR, RRG. **COUNTIES:** BEL, BOO, BUL, CAD, HLN, JEF, LAU, MEA, MEN, OLD, OWS, ROW, RUS. **DATES:** Apr. 17, May 12, June 3–July 28, Aug. 2–16, 29. Common.

3237 ***P. obnigrana*** Heinr.
Gallatin Co., Markland Dam, May 29; Laurel Co., DBNF, FS 121, May 18; Powell Co., near Nada Tunnel, June 9; all L.D. Gibson.

Genus ***Zeiraphera*** Tr.

3238 ***Z. claypoleana*** (Riley)
Boone Co., BCN, reared from *Aesculus glabra*, emerged May 18; Laurel Co., DBNF, Rockcastle Recreation Area, reared from *Aesculus octandra*, emerged May 17–21, and 31; Letcher Co., head of Line Fork, June 18; Meade Co., OCP, June 11.

3241 ***Z. improbana*** (Wlk.)
Carter Co., CAC, Apr. 29; Floyd Co., JWS, Apr. 22; Jefferson Co., Louisville, Mar. 13; Rowan Co., Bluestone, Apr. 16, 19.

Genus ***Pseudexentera*** Heinr.

3246 ***P. cressoniana*** (Clem.)
PARKS AND RESERVES: BBL, BMF, CAC, GBT, HWS, LMP, NBP, OCP, RRG, WVP. **COUNTIES:** BOO, BUL, CAR, CRL, JEF, MEA, MUH, OLD, OWS, POW. **DATES:** Mar. 2–4, 13–28, Apr. 1–17, May 12. Common.

3247 ***P. mali*** Free.
Boone Co., BBL, Mar. 3; Meade Co., OCP, Mar. 16.

3251 ***P. spoliana*** (Clem.)
PARKS AND RESERVES: AJJ, BBL, CAC, CBF, GBT, LMP, RRG. **COUNTIES:** BOO, CAM, CAR, CRL, JEF, LAU, MCY, MEN, MUH, OWS, POW, PUL, WHI. **DATES:** Mar. 4, 17, 25–28, Apr. 2–17, 25. Common.

3252 ***P. haracana*** (Kft.)
Laurel Co., DBNF, FS 615A/B, Apr. 6; Menifee Co., RRG, Mar. 12; Powell Co., RRG, Tunnel Ridge, Apr. 17, 28, 1989, L. D. Gibson and D. J. Wright.

3253 ***P. faracana*** (Kft.)
Carter Co., CAC, Apr. 12; Menifee Co., RRG, Mar. 23.

3253.1 ***P. sepia*** W. E. Miller
Campbell Co., AJJ, Apr. 16; Carter Co., CAC, Apr. 12; Jefferson Co., Louisville, Mar. 27; Menifee Co., RRG, Mar. 12, 23; Owsley Co., 4.8 km (3 mi) NE of Booneville, Apr. 3.

3253.2 ***P. hodsoni*** W.E. Miller
PARKS AND RESERVES: AJJ, CAC, LMP, OCP, RRG. **COUNTIES:** CAM, CAR, LAU, MEA, MUH, POW. **DATES:** Mar. 2–Apr. 17, 28.

3254 ***P. maracana*** (Kft.)
Boone Co., BBL, Mar. 25, Apr. 3; Rowan Co., Morehead, emerged Apr. 1.

3254.1 ***P. vacinii*** W. E. Miller
Laurel Co., DBNF, FS 615 A/B, Apr. 6–18; Powell Co., RRG, Tunnel Ridge, Apr. 28, May 3.

3257 ***P. costomaculana*** (Clem.)
PARKS AND RESERVES: BBM, RRG. **COUNTIES:** ELL, FRA, HLN, LAR, LAU, LEW, MCY, MEN, POW. **DATES:** Mar. 30, Apr. 6–30, May 3, 4, 19, July 14. Common.

3258 ***P. virginiana*** (Clem.)
Carter Co., CAC, Apr. 12; Laurel Co., on *Hamamelis*, Apr. 6; Menifee Co., RRG, Mar. 12, 23, Apr. 26; Owsley Co., 4.8 km (3 mi) NE of Booneville, Apr. 3; Powell Co., RRG, Tunnel Ridge, Apr. 17.

Genus *Gretchena* Heinr.

3259 ***G. deludana*** (Clem.)
PARKS AND RESERVES: BBL, BMF, CAC, HWS, OCP, RRG. **COUNTIES:** BOO, BUL, CAR, GAL, JEF, MCY, MEA, MEN, OLD, OWS, POW. **DATES:** Apr. 11–May 29. Common.

3260 ***G. concubitana*** Heinr.
Laurel Co., FS 615A/B, Apr. 6; Powell Co., RRG, Tunnel Ridge, May 10; both L. D. Gibson.

3261 ***G. watchungana*** (Kft.)
Campbell Co., AJJ, Apr. 16; Laurel Co., DBNF, FS 615A/B, Apr. 6, 11; Meade Co., OCP, May 10; Powell Co., NBP, Apr. 25, and RRG, Tunnel Ridge, Apr. 17, 28, May 10.

3263 ***G. bolliana*** (Slingerland)
Boone Co., BBL, Apr. 3, 30, May 10, 18; Laurel Co., DBNF, FS 615A/B Apr. 18; Owsley Co., 4.8 km (3 mi) NE of Booneville, May 10; Rowan Co., 3.2 km (2 mi) W of SR 519 on SR 1274, Oct. 15.

3264 ***G. amatana*** Heinr.
PARKS AND RESERVES: BBL, BMF, CAV, HWS, RRG. **COUNTIES:** BAT, BOO, BRA, BUL, JEF, MEN, OLD, ROW. **DATES:** Apr. 2–7, 26–June 8, Aug. 17. Common.

3265 ***G. delicatana*** Heinr.
Edmonson Co., MCV, May 3; Fleming Co., May 7; Harlan Co., BBM, May 23; McCreary Co., May 10; all A. F. Braun.

3266.1 ***G. nymphana*** Blanchard & Knudson
One paratype female, Laurel Co., Fariston, May 2, 1935, A. F. Braun. See original description by Blanchard and Knudson (1983); specimen in the ANSP.

3268 ***G. concitatricana*** (Heinr.)
Boone Co., BBL, Apr. 3, 10, 26, May 4, June 18; Campbell Co., AJJ, Apr. 16; Meade Co., OCP, Apr. 11; Menifee Co., RRG, Apr. 26; all L. D. Gibson.

Genus *Chimoptesis* Powell

3272 ***C. gerulae*** (Heinr.)
Laurel Co., DBNF, FS 615A/B, Apr. 6, 11; Powell Co., RRG, Tunnel Ridge, Mar. 4, 11; L. D. Gibson and D. J. Wright.

3273 ***C. pennsylvaniana*** (Kft.)
PARKS AND RESERVES: BBL, BMF, CAC, GBT, HWS, KGC, KLK, LMP, NBP, OCP, RRG. **COUNTIES:** BOO, BUL, CAR, CRL, HLN, JEF, LAU, MEA, MEN, MRL, MUH, OLD, OWS, POW. **DATES:** Mar. 4, 16–Apr. 10, 20–29. Common.

Genus *Rhopobota* Led.

3276 ***R. naevana*** (Hbn.)
Harlan Co., BBM, emerged, June 12–16; Laurel Co., FS 4176, ridge top SW of Ned Branch, June 11, 19.

3277 ***R. dietziana*** (Kft.)
Harlan Co., BBM, June 3; Laurel Co., DBNF, FS 4176, Apr. 6–19, 27, July 28; McCreary Co., Blue Heron Campground, Apr. 27; all collected by L. D. Gibson.

3278 ***R. finitimana*** (Heinr.)
Elliott Co., Little Sandy River, June 8–19, A. F. Braun.

Genus *Catastega* Clem.

NOTE: Numbers for the two following species are out of normal sequence because of reinstatement of *Catastega* by Brown ("1986" [1987]).

(3333) ***C. timidella*** Clem.
Boone Co., BBL, May 24; McCreary Co., Blue Heron Campground, Apr. 27; Powell Co., RRG, Tunnel Ridge, Apr. 28.

(3334) ***C. aceriella*** Clem.
Maple Trumpet Skeletonizer
This species was moved to *Catastega* from *Epinotia* by Brown ("1986" [1987]). Logan Co., pest of yard trees in Sept., 1976 (*Annual Summary of Insect Conditions in Kentucky* 1976); Oldham Co., Goshen, May 6.

Genus *Epinotia* Hbn.

3286 ***E. medioviridana*** (Kft.)
Bath Co., CAV, FS 918, Aug. 28; Menifee Co., RRG, Sept. 1, 2; Owsley Co., 4.8 km (3 mi) NE of Booneville, Sept. 8; Rowan Co., DBNF, E side of SR 1274, Sept. 15.

3291 ***E. celtisana*** (Riley)
Boone Co., BBL, Mar. 25, Apr. 10, L. D. Gibson and D. J. Wright. This species was formerly known as *E. laracana* Kft.

3294 ***E. zandana*** (Kft.)
Boone Co., BBL, Mar. 21–25; Jefferson Co., Louisville, Mar. 14.

3295 ***E. xandana*** (Kft.)
Boone Co., BBL, 1 female, at light, Mar. 25, 1989, L. D. Gibson.

3310 ***E. transmissana*** (Wlk.)
Harlan Co., BBM, July 14, and KGC, July 11, 13; Menifee Co., RRG, July 6; Powell Co., RRG, Tunnel Ridge, July 8.

3330 ***E. unica*** Heinr.
Edmonson Co., MCV, reared from *Clitoria*; Harlan Co., Pine Mountain, July 22, 23; Laurel Co., S side of SR 3497 at Ned Branch Trail, Aug. 8; Letcher Co., Pine Mountain, Mar. 17, 23, July 11, 15.

3335 ***E. nonana*** (Kft.)
Gallatin Co., Markland Dam, July 15, Aug. 1, 3, 17, Sept. 20; L. D. Gibson.

3351 ***E. lindana*** (Fern.)
Menifee Co., RRG, in black-light trap, Oct. 12–14, in a bait trap, Nov. 14, L.D. Gibson and J. S. Nordin.

3351.1 ***E. sotipena*** R. L. Brown
Boone Co., BBL, Mar. 21, 25; Fayette Co., Lexington, Apr. 3; Jefferson Co., Louisville, Apr. 2; Menifee Co., RRG, Mar. 12.

Genus ***Ancylis*** Hbn.

3354 ***A. nubeculana*** (Clem.)
PARKS AND RESERVES: RRG. **COUNTIES:** LAU, LEW, MCC, MEN, OWS, POW. **DATES:** Apr. 27–30, May 10–June 5. Common.

3359 ***A. metamelana*** (Wlk.)
PARKS AND RESERVES: AJJ, BBL, BMF, RRG. **COUNTIES:** BOO, BUL, CAM, FAY, GAL, HRD, MEA, MEN, OWS, ROW. **DATES:** Mar. 27–May 22, June 9–12, 22, July 8–15, Oct. 15. **REMARKS:** Common. Some captures were in Malaise traps.

3361 ***A. semiovana*** (Zell.)
PARKS AND RESERVES: PCP, PNM. **COUNTIES:** BEL, BUL, LAU, LET, OWS, ROW. **DATES:** May 21–28, June 3, 11, 13, July 1, 22. Common.

3365 ***A. spiraeifoliana*** (Clem.)
Bullitt Co., BMF, in Malaise trap, Apr. 4–8; Carter Co., Olive Hill, June 10; Owsley Co., 4.8 km (3 mi) NE of Booneville, June 14.

3366 ***A. laciniana*** (Zell.)
PARKS AND RESERVES: AJJ, BCN, RRG. **COUNTIES:** BOO, CAM, MCC, MEN, OWS, POW. **DATES:** May 4–June 6, 14, 18, July 5, 11. Common.

3367 ***A. burgessiana*** (Zell.)
PARKS AND RESERVES: BBL, BMF, HWS, KGC, OCP, PNM, RRG. **COUNTIES:** BEL, BOO, BUL, HLN, LAU, MEA, MEN, OLD, OWS. **DATES:** May 4–June 6, 14, 18, July 5, 11. Common.

3369 ***A. fuscociliana*** (Clem.)
Menifee Co., RRG, FS 9B, Indian East Fork, Kelley Branch, 220 m (720 ft) elevation, May 17–18, 1986, J. S. Nordin.

3370 ***A. platanana*** (Clem.)
PARKS AND RESERVES: BBL, OCP, RRG. **COUNTIES:** BOO, CAR, GAL, JEF, MEA, MEN, OWS. **DATES:** Apr. 30, May 2, 4, 13–31, June 7, July 5, 6, 15, 19, 20, Aug. 9, 10. Common; found in daytime as well as at light.

3374 ***A. comptana*** (Frölich)
Bullitt Co., BMF, in Malaise trap, June 22–30; Fayette Co., Lexington, in Malaise trap, July 7–11, Aug. 15–19; Harlan Co., BBM, May 3, July 21; Laurel Co., DBNF, FS 121, June 17; Owsley Co., 4.8 km (3 mi) NE of Booneville, June 30, Aug. 20.

3375 ***A. divisana*** (Wlk.)
PARKS AND RESERVES: AJJ, BCN, BMF, OCP, RRG. **COUNTIES:** BOO, BUL, CAM, MEA, OWS, POW, ROW. **DATES:** Apr. 4–8 (Malaise trap), May 8–June 7, 15–22. Common.

3376 ***A. apicana*** (Wlk.)
Jefferson Co., Valley Station, June 20, 1975, A. J. Brownell.

3377 ***A. muricana*** (Wlsm.)
Boone Co., BBL, Apr. 30; Bullitt Co., BMF, in Malaise trap, Apr. 4–8, May 11–20; Gallatin Co., Markland Dam, July 15.

3379 ***A. diminutana*** (Haw.)
Menifee Co., RRG, Apr. 26; Owsley Co., June 22; both L. D. Gibson.

Genus ***Hystrichophora*** Wlsm.

3396 ***H. ochreicostana*** (Wlsm.)
Boone Co., BBL, July 10, 1980, L. D. Gibson.

Genus ***Dichrorampha*** Gn.

3404 ***D. simulana*** (Clem.)
Boone Co., BBL, May 31; Owsley Co., 4.8 km (3 mi) NE of Booneville, May 16–28, Sept. 8; Whitley Co., Goldbug, June 5; all L. D. Gibson.

3406 ***D. bittana*** (Bsk.)
Boone Co., BBL, May 18, 31; Bullitt Co., BMF, in Malaise trap, June 1–6; Gallatin Co., Markland Dam, May 27; Menifee Co., RRG, May 26–31, June 1.

3414 ***D. leopardana*** (Bsk.)
Bell Co., PNM, June 24; Boone Co., BBL, Aug. 4, and BCN, June 5, and CER, Aug. 17.

Genus *Talponia* Heinr.

3417 ***T. plummeriana*** (Bsk.)
Henderson Co., Henderson, reared on pawpaw blooms, May 22, 1950, T.D. Hodge; Laurel Co., DBNF, FS 615A/B, Apr. 6, 11; Menifee Co., RRG, at light, Apr. 4.

Genus *Pammene* Hbn.

3419 ***P. felicitana*** Heinr.
Gallatin Co., Markland Dam, May 25–June 11; Jefferson Co., Louisville, June 7.

Genus *Eucosmomorpha* Obr.

3422.1 ***E. albersana*** (Hbn.)
Boone Co., BBL, July 9, 1991, Aug. 4, 1989, L. D. Gibson. These records extend the known North American range of this recently discovered European moth southward from the state of Michigan (Miller 1983b).

Genus *Larisa* W. E. Miller

3423 ***L. subsolana*** W. E. Miller
Boone Co, BBL, May 24, 31; Christian Co., PYR, June 27; Gallatin Co., Markland Dam, June 7; Laurel Co., DBNF, FS 4176, June 6; Powell Co., RRG, 1.5 km (0.9 mi) E of Nada Tunnel, June 9, July 8, Tunnel Ridge, June 1.

Genus *Sereda* Heinr.

3425 ***S. tautana*** (Clem.)
Boone Co., BBL, Mar. 25; Campbell Co., AJJ, Apr. 16; Laurel Co., DBNF, FS 615A/B, Apr. 6; Meade Co., OCP, Mar. 28; Menifee Co., RRG, Apr. 4; Powell Co., RRG, Tunnel Ridge, Mar. 4, 11, Apr. 5, 27.

Genus *Grapholita* Tr.

3426 ***G. molesta*** (Bsk.)
Oriental Fruit Moth
Fayette Co., Lexington, June 7, 1938, L. H. Townsend; Fayette Co., Lexington, reared from peach twigs, adult July 20, 1927; Monroe Co., pest on nectarine, May 1976, (*Kentucky Cooperative Plant Pest Report* 1977).

3428 ***G. packardi*** Zell.
Boone Co., BBL, May 18, 24, and BCN, Aug. 15; Gallatin Co., Markland Dam, May 27, 29; all L. D. Gibson.

3434 ***G. fana*** (Kft.)
Boone Co., BBL, May 10; Bullitt Co., BMF, in Malaise trap, June 1–6; Menifee Co., Leatherwood Fork, July 6; Owsley Co., 4.8 km (3 mi) NE of Booneville, May 10, June 16; Powell Co., May 23; Pulaski Co., Cumberland, North Fork, Apr. 25.

3438 ***G. eclipsana*** Zell.
PARKS AND RESERVES: BBL, BBM, KGC, OCP, RRG. **COUNTIES:** BOO, BUL, CAR, GAL, HLN, JEF, MEA, OWS, POW, ROW. **DATES:** Mar. 2, Apr. 6–11, 20, 25, May 2–10, 18, 29. Common.

3439 ***G. interstinctana*** (Clem.)
Bullitt Co., BMF, May 11–20 in Malaise trap, S side of SR 480, July 8; Fayette Co., Lexington, reared from clover head, Aug. 29; Menifee Co., RRG, Sept. 5; Oldham Co., Goshen, July 12; Owsley Co., 4.8 km (3 mi) NE of Booneville, Aug. 10.

3440 ***G. edwardsiana*** (Kft.)
Rowan Co., Morehead, June 26, 1962, T. N. Freeman.

3443.1 ***G. delineana*** (Walker)
Miller (1982) distinguished the introduced *G. delineana*, which feeds on hemp (*Cannabis sativa*) from the native *G. tristrigana*, which does not. Records of *tristrigana* reared from hemp at the University of Kentucky, Lexington, are *delineana*. Earliest records seem to be specimens labelled "borer in hemp, Lex., Ky., Sept. 2, 1915, H.H. J[ewett]." Jewett apparently took this adult in nature: "Collected on hemp grown for seed, Lex., Ky., Sept. 16, 1916." Reared specimens emerged in June, July, and Aug. *Cannabis* was grown legally in Kentucky in the early part of this century for its fiber, which was used in making rope. A series in the NMNH was collected by M.L. Didlake "on hemp in field," Lexington, Ky., June 25–30, 1943. Recent records: Menifee Co., Leatherwood Fork, May 31; Owsley Co., 4.8 km (3 mi) NE Booneville, May 14, June 25, July 14.

Genus *Corticivora* Clarke

3446 ***C. clarki*** Clarke
Whitley Co., Goldbug, June 5, 1981, L. D. Gibson.

Genus *Cydia* Hbn.

3457 ***C. garacana*** (Kft.)
Gallatin Co., Markland Dam, at light, May 29, June 2, 7, 22, Aug. 3, L. D. Gibson.

3461 ***C. albimaculana*** (Fern.)
Laurel Co., DBNF, FS 615A/B, May 4, 1996, L. D. Gibson.

3464 ***C. lacustrina*** (W.E. Miller)
Gallatin Co., Markland Dam, at light, May 25, 1989, L. D. Gibson.

3469 ***C. candana*** (Fbs.)
Jefferson Co., Louisville, Apr. 26; Laurel Co., DBNF, FS 615A/B, Apr. 18; Oldham Co., Goshen, Aug. 11; Powell Co., RRG, Tunnel Ridge, Apr. 17.

3471 ***C. caryana*** (Fitch)
PARKS AND RESERVES: AJJ, BBL, KGC, RRG. **COUNTIES:** BOO, CAM, EDM, GAL, HLN, JEF, LAU, LET, MCY, OWS, POW, ROW. **DATES:** Apr. 19, 27, May 16–27, June 15–28, July 13–26, Aug. 9. Common.

3479 ***C. gallaesaliciana*** (Riley)
Boone Co., BBL, June 18; Owsley Co., 4.8 km (3 mi) NE of Booneville, May 28; both L. D. Gibson.

3486 ***C. toreuta*** (Grt.)
Gallatin Co., Markland Dam, June 12–22; Laurel Co., DBNF, FS 4176, June 6, 19; Muhlenberg Co., LMP, July 15; Owsley Co., 4.8 km (3 mi) NE of Booneville, June 14; Whitley Co., Goldbug, June 5.

3492 ***C. pomonella*** (L.)
The Codling Moth
Boone Co., BCN, June 19, and Florence, June 20; Campbell Co., AJJ, July 20; Fayette Co., Lexington, University of Kentucky, Apr. 23, May 6–9, June 6, 7, series labelled by H. Garman, emerged May 26, 1930 and May 6, 1931; Jefferson Co., Louisville, June 1–11, Valley Station, July 11; Owsley Co., 4.8 km (3 mi) NE of Booneville, July 24. A pest of apple; much more widespread than these records indicate.

3494 ***C. latiferreanus*** (Wlsm.)
Filbertworm Moth
PARKS AND RESERVES: BBL, BUC, HWS, LCP, OCP, PCP, PYR, RRG. **COUNTIES:** BOO, BRE, BUL, CAD, GRY, JEF, MCC, MEA, MEN, MRL, OLD, OWS, PER, POW, RUS. **DATES:** May 3, 21, 28, June, July 6, 10, 22, 31, Aug. 1–10, 19, 21, Sept. 1–15, Oct. 7. Very common.

Genus *Ecdytolopha* Zell.

3495 ***E. punctidiscana*** (Dyar)
PARKS AND RESERVES: BBL, BMF, HWS, LBL, OCP, PNM, PYR, RRG, WVP. **COUNTIES:** BEL, BOO, BUL, CHR, GAL, JEF, MCC, MEA, OLD, OWS, POW, TRI. **DATES:** Apr. 3, May 7, 23, 26, June 1–11, 24–30, July 15, 16, Aug. 14, 26, Sept. 19, 23. Common.

3497 ***E. insiticiana*** Zell.
Locust Twig Borer Moth
PARKS AND RESERVES: BBL, BMF, HWS, OCP. **COUNTIES:** BOO, BUL, FAY, GAL, JEF, MEA, OLD, OWS. **DATES:** Mar. 25, May 15–24, June 11–15, July 11, 27, Aug. 6, 18, Sept. 8, 23. Common.

Genus *Pseudogalleria* Rag.

3500 ***P. inimicella*** (Zell.)
Jefferson Co., Valley Station, May 29; Owsley Co., 4.8 km (3 mi) NE of Booneville, May 10; Powell Co., RRG, Tunnel Ridge, May 14; Russell Co., LCP, Aug. 14.

Subfamily **Tortricinae**

Genus *Croesia* Hbn.

3502 ***C. albicomana*** (Clem.)
Bullitt Co., BMF, June 19; Jefferson Co., May 22; Meade Co., OCP, June 11. Also reported damaging oak in Fleming, Nicholas, Montgomery, and Boyd counties for the week ending July 27, 1975 (*Kentucky Cooperative Economic Insect Report* 1975).

3503 ***C. semipurpurana*** (Kft.)
Oak Leaftier Moth
Jefferson Co., Louisville, June 4; Meade Co., OCP, June 7; Owsley Co., 4.8 km (3 mi) NE of Booneville, May 22. Larvae of this species were reported defoliating hardwoods in Lyon and Trigg counties, LBL, May 7, 1981 (Carothers 1981a) and in parts of Breathitt, Estill, Lee, Menifee, Powell, and Wolfe counties on June 2–5, 1981 (Carothers 1981b). Locally common.

3504 ***C. curvalana*** (Kft.)
PARKS AND RESERVES: BMF, OCP, PNM, RRG. **COUNTIES:** BEL, BUL, LET, MEA, OWS, POW, WHI. **DATES:** May 21–June 24. **REMARKS:** Common. L. D. Gibson reared it from *Gaylussacia baccata*, with emergence May 10–11.

Genus *Acleris* Hbn.

3506 ***A. macdunnoughi*** Obr.
Menifee Co., RRG, Oct. 3; Owsley Co., 4.8 km (3 mi) NE of Booneville, Oct. 10, 1980; both L. D. Gibson.

3509 ***A. ptychogrammos*** (Zell.)
Powell Co., 1.3 km (0.8 mi) E of Nada Tunnel, along Martin Fork, at bait, Nov. 15, 1991, L. D. Gibson.

3510 ***A. nivisellana*** (Wlsm.)
Gallatin Co., Markland Dam, July 17; Harlan Co., PMS, July 8.

3514 ***A. cervinana*** (Fern.)
Boone Co., BBL, Nov. 14; Letcher Co., Colliers Creek, June 27, 28; Meade Co., OCP, June 20; Menifee Co., RRG, FS 9B, Indian East Fork, Kelley Branch, 220 m (720 ft) elevation, June 27;

Owsley Co., 4.8 km (3 mi) NE of Booneville, Mar. 15, June 14, 22, Oct. 8.

3517 ***A. subnivana*** (Wlk.)
PARKS AND RESERVES: AJJ, BBL, BCN, BMF, CAC, NBP, OCP. **COUNTIES:** BOO, BUL, CAM, CAR, MEA, OWS, POW. **DATES:** Jan. 26, Mar. 17, 21, Apr. 11, May 18, June 10–15, July 15, Aug. 17, Nov. 14, 23. Common.

3521 ***A. semiannula*** (Rob.)
Boone Co., BBL, Nov. 14, and BCN, Mar. 17, and Florence, Jan. 26; Henry Co., E of Defoe, Nov. 18.

3523 ***A. cornana*** (McD.)
Boone Co., BBL, May 18; Carroll Co., GBT, reared from *Cornus drummondii*, emerged May 10, 11; Jefferson Co., Louisville, May 30.

3526 ***A. negundana*** (Bsk.)
Boone Co., BCN, Nov. 18, and Florence, May; Campbell Co., AJJ, reared from *Acer negundo*, emerged May 22–24; Jefferson Co., CCR, Feb. 15; Meade Co., OCP, Sept. 25; Powell Co., NBP, Nov. 22. Common.

3527 ***A. schalleriana*** (L.)
Boone Co., BBL, May 24, 25, reared from *Viburnum prunifolium*, emerged May 12–21; Carroll Co., GBT, reared from *V. prunifolium*, emerged May 12–15; Jefferson Co., Louisville, emerged from *Viburnum*, July 18, 27; Owsley Co., 4.8 km (3 mi) NE of Booneville, Sept. 26, Oct. 14; Rowan Co., emerged from *Viburnum*, May 5.

3531 ***A. hastiana*** (L.)
Boone Co., BBL, Mar. 23, Apr. 3, May 26, Oct. 26, Nov. 14; Harlan Co., BBM, July 12; Meade Co., OCP, Apr. 11; Owsley Co., 4.8 km (3 mi) NE of Booneville, July 14.

3536 ***A. robinsoniana*** (Fbs.)
Boone Co., BBL, Mar. 23, Florence, Nov. 15; Jefferson Co., Louisville, Apr. 2. A highly variable species.

3539 ***A. chalybeana*** (Fern.)
Boone Co., BBL, Nov. 17, Florence, Mar. 18; Bullitt Co., BMF, in Malaise trap, Apr. 6–9; Jefferson Co., Valley Station, Apr. 1; Powell Co., NBP, Apr. 20. Uncommon.

3540 ***A. logiana*** (Cl.)
PARKS AND RESERVES: KGC, NBP, RRG. **COUNTIES:** HLN, LET, MCC, MEN, POW, ROW, TOD. **DATES:** May 3, 10, June 11, July 2–11, 21, 22, Aug. 28, Oct. 15, Nov. 15, 23, 24. **REMARKS:** The North American subspecies is *placidana* (Rob.). Common.

3542 ***A. flavivittana*** (Clem.)
Boone Co., BBL, Apr. 3, 10, 22, May 26, Oct. 26, and BCN, Mar. 24, and Florence, Mar. 19, July 20, Nov. 15; Bullitt Co., BMF, in Malaise trap, June 1–6; Jefferson Co., Jeffersontown, Oct. 28; Meade Co., OCP, at bait, Sept. 24; Owsley Co., 4.8 km (3 mi) NE of Booneville, Apr. 22.

3543 ***A. maculidorsana*** (Clem.)
Boone Co., BBL, Mar. 23; Bullitt Co., BMF, in Malaise trap, June 1–6, N side of SR 480, Mar. 30; Laurel Co., DBNF, FS 615A/B, Apr. 11; McCracken Co., Ray Black farm, 9.7 km (6 mi) SW of Paducah, Feb. 5.

3548 ***A. variana*** (Fern.)
Eastern Blackheaded Budworm Moth
Menifee Co., RRG, FS 6B, Indian East Fork, Kelley Branch, Aug. 12, Sept. 6, 7, and RRG, FS 9A, Leatherwood Fork, Aug. 12.

3549 ***A. maccana*** (Tr.)
Harlan Co., BBM, May 17, 1993, L. D. Gibson.

3551 ***A. inana*** (Rob.)
Jefferson Co., CCR, Feb. 15; McCracken Co., Ray Black farm, 9.7 km (6 mi) SW of Paducah, Feb. 24.

Genus *Decodes* Obr.

3573 ***D. basiplaganus*** (Wlsm.)
Rowan Co., 3.2 km (2 mi) W of SR 519 on SR 1274, Oct. 15, 1993, D. J. Wright.

Genus *Pandemis* Hbn.

3593 ***P. lamprosana*** (Rob.)
Boone Co., BCN, June 5; Bullitt Co., BMF, reared from *Celtis*, emerged May 5; Gallatin Co., Markland Dam, May 27; Grayson Co., ASC, Sept. 11; Harlan Co., BBM, at light, July 12, 14, another emerged as adult on June 7; Henderson Co., SWR, Sept. 10; Oldham Co., HWS, June 14.

3594 ***P. limitata*** (Rob.)
Three-lined Leafroller Moth
PARKS AND RESERVES: BBL, BMF, HWS, KLK, LCP, OCP, PNM. **COUNTIES:** BEL, BOO, BUL, FUL, GAL, JEF, LEW, MAR, MEA, MRL, OLD, RUS. **DATES:** May 18, 27–June 22, Aug. 3, Sept. 5, 6, 17. Common.

Genus *Argyrotaenia* Steph.

3597 ***A. velutinana*** (Wlk.)
Red-banded Leafroller Moth
PARKS AND RESERVES: BBL, BMF, CER, HWS, OCP, RBF. **COUNTIES:** BOO, BRE, BUL, CAD, FAY, JEF, MCC, MEA, OLD, OWS.

DATES: Mar. 20–27, Apr. 6–20, 30–July 11, 23–Aug. 25, Sept. 5–9. **REMARKS:** Specimens reared from *Celtis laevigata* and *C. occidentalis* by A. J. Brownell emerged May 15–June 6 and Oct. 1. A locally common pest of fruit and other trees.

3599 ***A. floridana*** Obr.
Christian Co., PYR, June 27; Owsley Co., 4.8 km (3 mi) NE of Booneville, June 22; Powell Co., RRG, Tunnel Ridge, May 21, July 8, 11; all L. D. Gibson.

3602 ***A. pinatubana*** (Kft.)
Pine Tube Moth
Jefferson Co., Louisville, June 24, July 3; Laurel Co., DBNF, Holly Bay Campground, Apr. 19; Powell Co., RRG, Tunnel Ridge, Apr. 28.

3603 ***A. tabulana*** Free.
Bullitt Co., PCP, Sept. 6; Laurel Co., DBNF, FS 615A/B, Apr. 6, and Holly Bay Campground, Apr. 19; McCracken Co., Ray Black farm, 9.7 km (6 mi) SW of Paducah, July 4; Nelson Co., Leslie farm near Boston, July 21; Owsley Co., 4.8 km (3 mi) NE of Booneville, May 10.

3607 ***A. occultana*** Free.
Laurel Co., DBNF, Holly Bay Campground, Apr. 19, 1991, L. D. Gibson.

3621 ***A. quadrifasciana*** (Fern.)
Jefferson Co., Louisville, May 30; Kenton Co., Taylor Mill, June 9.

3622 ***A. juglandana*** (Fern.)
Hickory Leafroller Moth
PARKS AND RESERVES: BMF, KGC, OCP. **COUNTIES:** BUL, CAM, GAL, HLN, LAU, MEA, NEL, ROW, WOL. **DATES:** May 15–21, June 6–16, July 11, 15. **REMARKS:** Reared from a pupa found on *Celtis* at BMF, emerged May 21, A. J. Brownell.

3623 ***A. quercifoliana*** (Fitch)
PARKS AND RESERVES: BBL, BBM, BMF, HWS, KRC, OCP, PYR, WVP. **COUNTIES:** BOO, BUL, CHR, HLN, JEF, LEE, LET, MCC, MEA, OLD, OWS, ROW, WOL. **DATES:** May 21–31, June, July 10–15, Sept. 6. Locally common.

3624 ***A. alisellana*** (Rob.)
PARKS AND RESERVES: BMF, CBF, OCP, PNM, PYR, RBF. **COUNTIES:** BEL, BRE, BUL, CAD, JEF, MCC, MEA, OWS, WHI. **DATES:** May 11, June, July 3, Aug. 8. Locally common.

3625 ***A. mariana*** (Fern.)
Graybanded Leafroller Moth
PARKS AND RESERVES: BBL, BIP, HWS, JWS, KGC, NBP, OCP, RBF, RRG. **COUNTIES:** BOO, BRE, FLO, HLN, JEF, MEA, OLD, OWS, PIK, POW. **DATES:** Apr. 10–26, May 3, 5, 17, 21, June 18. Common.

Genus ***Choristoneura*** Led.

3632 ***C. fractivittana*** (Clem.)
PARKS AND RESERVES: BBL, BMF, CBB, LCP, OCP, RRG. **COUNTIES:** BOO, BUL, GAL, JEF, MEA, OLD, POW, RUS. **DATES:** May 14–31, June 2, 11. Common.

3633 ***C. parallela*** (Rob.)
Spotted Fireworm Moth
PARKS AND RESERVES: AJJ, BBL, BBM, BMF, BUC, HWS, LBL. **COUNTIES:** BOO, BRE, BUL, CAD, CAM, FAY, GAL, HIC, HLN, HRD, JEF, MCC, OLD, PER, ROW, TRI, WHI. **DATES:** June 2–12, 20–30, July 5, 13, 23–31, Aug., Sept. 4, 11. Common.

3635 ***C. rosaceana*** (Harr.)
Obliquebanded Leafroller Moth
PARKS AND RESERVES: BBL, BCN, BMF, GBT, HWS, KRC, LBK, LBL, LCP, NBP, OCP, PYR, RGR, RRG, SWR, WVP. **COUNTIES:** BOO, BRE, BRK, BUL, CAD, CRL, FUL, GAL, GRY, HND, HOP, JEF, LET, MCC, MEA, MEN, OLD, OWS, POW, PUL, ROW, RUS, SPE, TRI, WHI, WOL. **DATES:** May 6, 18–31, June 2–26, July 13–27, Aug. 3–19, 29–30, Sept. 1–10, 18, 29, Oct. 6. **REMARKS:** Specimens reared by A. J. Brownell from *Celtis laevigata* and *C. occidentalis* emerged May 6–31, June 1–3, July 21. Common; a pest of many plants.

3638 ***C. fumiferana*** (Clem.)
Spruce Budworm Moth
Bell Co., PNM, June 8, 9; Bullitt Co., BMF, May 22, June 10–22; Caldwell Co., PYR, June 11, 12; Laurel Co., DBNF, FS 121, June 11, 17, and FS 4176, June 6; Nelson Co., June 23; Russell Co., LCP, June 10. This species is one of America's most destructive defoliating pests of conifers.

3643 ***C. pinus*** Free.
Jack Pine Budworm Moth
PARKS AND RESERVES: BMF, PNM, RRG. **COUNTIES:** BEL, BUL, LAU, OWS, POW, ROW. **DATES:** June 9–22, July 1. **REMARKS:** Our subspecies is *maritima* Free.

Genus ***Archips*** Hbn.

3648 ***A. argyrospila*** (Wlk.)
Fruit-tree Leafroller Moth
PARKS AND RESERVES: AJJ, BCN, BMF, CAC, HWS, LCP, MCV, OCP, PNM, PYR. **COUNTIES:** BEL, BOO, BUL, CAD, CAM, CAR, EDM, FAY, GAL, JEF, MCC, MCY, MEA, OLD, OWS, ROW, RUS. **DATES:** May 14, 29,

June, July 25. **REMARKS:** Specimens reared from *Celtis laevigata* emerged May 14, 17, June 4. Common.

3648.1 ***A. nigriplagana*** Franc.
Two paratypes are from Kentucky (Franclemont 1986): Bell Co., PNM, 1 paratype female, June 9, 1981, C. Cornett; and Owsley Co., 4.8 km (3 mi) NE of Booneville, 1 paratype male, June 14, 1980, L. D. Gibson. Additional material includes: Bell Co., PNM, 10 specimens, June 9, 10; Bullitt Co., BMF, May 18; Jefferson Co., Valley Station, June 14, July 7; Laurel Co., DBNF, FS 121, June 11; Meade Co., OCP, June 7; Russell Co., LCP, June 9, 10.

3653 ***A. semiferana*** (Wlk.)
Oak Leafroller Moth
PARKS AND RESERVES: BBM, BMF, HWS, LBL, OCP, PNM, PYR. **COUNTIES:** BEL, BRE, BUL, CAD, EST, HLN, JEF, LAU, LEE, LYO, MCC, MEA, MEN, OLD, POW, TRI, WHI, WOL. **DATES:** May 6–9, 29, June 7–22, July 14 (larvae collected May 7, June 2–5). **REMARKS:** Larvae of this species were reported (Carothers 1981a) defoliating hardwoods in Lyon and Trigg counties, LBL, May 7, 1981, and in parts of Breathitt, Estill, Lee, Menifee, Powell, and Wolfe counties, June 2–5, 1981.

3655 ***A. fervidana*** (Clem.)
Oak Webworm Moth
Harlan Co., KGC, July 13, 1979, C. V. Covell, Jr.

3656 ***A. georgiana*** (Wlk.)
McCreary Co., Pine Knot and Yahoo Ridge, emerged June 14, 15, 26; Rowan Co., Clack Mtn., on *Vaccinium stamineum*, emerged May 1; all A. F. Braun.

3657 ***A. magnoliana*** (Fern.)
PARKS AND RESERVES: BBM, NBP, PNM, RRG. **COUNTIES:** BEL, HLN, LAU, LET, MEN, POW. **DATES:** June 7–20, July 6–14. **REMARKS:** Common. L. D. Gibson has reared this species from *Magnolia macrophylla* and *M. acuminata*; adults emerged May 24–June 6.

3658 ***A. purpurana*** (Clem.)
Boone Co., Florence, July 3, 20; Harlan Co., BBM, July 9, 12.

3659 ***A. infumatana*** (Zell.)
Bullitt Co., S side of SR 480, June 23; Harlan Co., BBM, July 14, and KGC, July 13; Meade Co., OCP, June 20; Owsley Co., 4.8 km (3 mi) NE of Booneville, June 25, July 3; Rowan Co., Morehead, June 27.

3660 ***A. grisea*** (Rob.)
Bullitt Co., BMF, reared on *Celtis*, emerged May 14, June 4; Campbell Co., AJJ, June 12; Laurel Co., DBNF, FS 4176, June 6, 11; Meade Co., OCP, June 7; Oldham Co., HWS, June 14.

3661 ***A. cerasivorana*** (Fitch)
Laurel Co., DBNF, FS 121 at Bolton Branch, June 17, 1994, L. D. Gibson and D. J. Wright.

3662 ***A. rileyana*** (Grt.)
Laurel Co., DBNF, FS 121 near Bolton Branch, June 17, 1994, L. D. Gibson.

Genus *Syndemis* Hbn.

3672 ***S. afflictana*** (Wlk.)
Campbell Co., AJJ, Apr. 16; Harlan Co., BBM, Apr. 21, KGC, Apr. 20; Menifee Co., RRG, Apr. 26; Pike Co., BIP, Apr. 22. Locally common in eastern Kentucky.

Genus *Clepsis* Gn.

3679 ***C. listerana*** (Kft.)
Letcher Co., Pine Mountain, July 1, 4; McCreary Co., Natural Bridge, June 22; Powell Co., June 28; Todd Co., Clifty Creek, June 15; all A. F. Braun.

3682 ***C. persicana*** (Fitch)
White Triangle Tortrix
Harlan Co., BBM, several specimens taken at light, July 12, 14, C. V. Covell, Jr., and L. D. Gibson.

3684 ***C. clemensiana*** (Fern.)
Freeman (1958) includes "Ky." in the range of this species. Boone Co., BBL, May 29, June 25, and BCN, May 29; all L. D. Gibson.

3686 ***C. melaleucana*** (Wlk.)
PARKS AND RESERVES: BBL, BBM, CBB, HWS, NBP, OCP, PNM, RRG. **COUNTIES:** BEL, BOO, HLN, JEF, MEA, MER, OLD, POW. **DATES:** May 9–June 11. Common.

Genus *Ptycholoma* Steph.

3688 ***P. peritana*** (Clem.)
Garden Tortrix
PARKS AND RESERVES: BMF, CER, OCP, PNM, RRG. **COUNTIES:** BEL, BOO, BUL, FAY, GAL, HOP, HRT, JEF, MCC, MEA, MEN, OLD, OWS, WOO. **DATES:** Apr. 26, May 10, 18–30, June 2, 10–26, July 5–14, Aug. 1–24, Sept. 1–13. Common.

3689 ***P. virescana*** (Clem.)
Harlan Co., BBM, Aug. 27; Jefferson Co., Louisville, Apr. 30; Oldham Co., Goshen, May 16–30, June 1–13, Aug. 16–23, Sept. 5, 12. Locally common.

Genus *Adoxophyes* Meyr.

3690 ***A. furcatana*** (Wlk.)
Boone Co., BBL, June 24, and BCN, June 15; Fulton Co., SR 94, Willingham Bottoms, 4.0 km (2.5 mi) E of Cayce, Sept. 11; Jefferson Co., Valley Station, May 29; Menifee Co., RRG, Aug. 12, 25. Uncommon.

Genus *Xenotemna* Powell

3693 ***X. pallorana*** (Rob.)
PARKS AND RESERVES: BBL, BMF, FAO, GBT, LCP, PYR. **COUNTIES:** BOO, BUL, CHR, CRL, GAL, HND, HRD, JEF, MEA, RUS. **DATES:** May 18–26, June 2, 11–13, July 8, 24, Aug. 10–21, Sept. 16. Common.

Genus *Sparganothis* Hbn.

3695 ***S. sulfureana*** (Clem.)
Sparganothis Fruitworm Moth
PARKS AND RESERVES: RRG. **COUNTIES:** BOO, BRA, FAY, GAL, HAR, JEF, MCY, OWS. **DATES:** May 28–June 2, 11, 21, July 1–14, Aug. 19, Sept. 10, Oct. 27. Common.

3696 ***S. belfrageana*** (Zell.)
Carroll Co., GBT, Sept. 18; Fayette Co., Lexington, Sept. 6–12; "Fort Garrett, Kentucky," Sept. 7; Jefferson Co., Surry Hills Farm, May 19, 24, Sept. 12, Oct. 1; Oldham Co., HWS, May 26.

3699 ***S. tristriata*** Kft.
Jefferson Co., Valley Station, Aug. 30; Laurel Co., DBNF, Craig Creek Rd., Sept. 19; Menifee Co., RRG, N end of FS 9B, Sept. 5; Owsley Co., 4.8 km (3 mi) NE of Booneville, Sept. 8; Rowan Co., E side of SR 1274, 3.2 km (2 mi) W of SR 519, Aug. 26. Uncommon.

3700 ***S. caryae*** (Rob.)
McCracken Co., 14.5 km (9 mi) SW of Paducah, June 29, 1981, L. D. Gibson.

3701 ***S. pulcherrimana*** (Wlsm.)
Fayette Co., Lexington, June 3, 11; Jefferson Co., May 16.

3703 ***S. demissana*** (Wlsm.)
Bell Co., PNM, June 22, 24; McCreary Co., Yahoo Ridge, June 14; both A. F. Braun.

3704 ***S. distincta*** (Wlsm.)
PARKS AND RESERVES: KDV, SWR. **COUNTIES:** HEN, HIC, HND, JEF, MAR, MRL, ROW, WHI. **DATES:** June 5, 20–29, Aug. 12, Sept. 11. **REMARKS:** A.F. Braun reared this species from *Solidago rigida*; adults emerged June 27.

3706 ***S. xanthoides*** (Wlk.)
Boone Co., BBL, July 18, Aug. 3, 21; Bullitt Co., Mt. Washington, July 28; Campbell Co., AJJ, July 20; Harlan Co., BBM, July 11; Owsley Co., 4.8 km (3 mi) NE of Booneville, July 3.

3709 ***S. striata*** (Wlsm.)
Whitley Co., CBF, Sept. 6, A. F. Braun.

3711 ***S. unifasciana*** (Clem.)
Laurel Co., DBNF, FS 121, June 17, 21; Powell Co., reared from *Kalmia latifolia*, emerged June 22, A. F. Braun.

3716 ***S. diluticostana*** (Wlsm.)
PARKS AND RESERVES: BMF, MCV, OCP, PYR, RRG. **COUNTIES:** BUL, CHR, EDM, LAU, MEA, POW. **DATES:** May 16, 19, June 6–20, July 8. **REMARKS:** A .J. Brownell reared this on *Celtis laevigata* and *C. occidentalis* from BMF; emerged June 6 and 13. L. D. Gibson reared it on *Fagus grandifolia* from Laurel Co.; adults emerged June 16–19.

3720 ***S. reticulatana*** (Clem.)
PARKS AND RESERVES: BBL, BBM, CAV, LBL, RRG. **COUNTIES:** BAT, BOO, BRA, GAL, HLN, JEF, MEN, TRI. **DATES:** June 15, 27, July 3–19, Aug. 17, 23. **REMARKS:** Common. A. F. Braun reared it from "*Aster* or *Solidago*," Harlan Co., emerged June 15.

3720.1 **S.** undescribed species
Barren Co., Park City, July 15, 1976, A.J. Brownell; Christian Co., PYR, June 20, 1980, L. D. Gibson; Marshall Co., KLK, June 16, 1973, C. Cornett. These seem identical to moths labelled "*Sparganothis* B" from Arkansas in the NMNH.

3721 ***S. albicaudana*** Bsk.
Laurel Co., DBNF, FS 121, June 11, FS 4176, June 19, and Lakeside South Trail, larvae collected by L. D. Gibson on *Acer saccharum* Apr. 23, adults emerged May 16–20; Meade Co., OCP, 3 pattern variants represented, June 11–20.

3722 ***S. directana*** (Wlk.)
PARKS AND RESERVES: AJJ, BEC, BMF, COB, OCP, RRG. **COUNTIES:** BAR, BUL, CAM, HIC, JEF, LAU, MCC, MEA, POW. **DATES:** June 11, 20, July 8–20. **REMARKS:** Locally common. A. J. Brownell reared it from *Celtis* collected at BMF, and L. D. Gibson reared it on *Hamamelis virginiana* from RRG, Powell Co.; adults in both cases emerged June 20.

3723 ***S. chambersana*** (Kft.)
Bullitt Co., BMF, reared on *Celtis laevigata*, emerged July 23; McCracken Co., 14.5 km (9 mi)

SW of Paducah, June 29; Meade Co., OCP, June 20; Todd Co., Clifty Creek, June 16; Wolfe Co., KRC, July 15.

3725 ***S. pettitana*** (Rob.)
PARKS AND RESERVES: BMF, KLK, LMP, OCP, PYR. **COUNTIES:** BUL, CAD, JEF, LAU, MEA, MRL, MUH, OWS. **DATES:** May 14, June 1–20, July 12. **REMARKS:** Common. Wings vary from yellow to pure white, with or without a darker pattern.

3727 ***S. niveana*** (Wlsm.)
Meade Co., OCP, June 11; Russell Co., LCP, June 9. Uncommon.

Genus *Platynota* Clem.

3732 ***P. flavedana*** Clem.
PARKS AND RESERVES: BBL, BMF, BUC, HWS, LBL, OCP. **COUNTIES:** BOO, BRE, BUL, FAY, JEF, LAU, MCC, MEA, OLD, OWS, PER, TRI, WHI. **DATES:** May 20–31, June 1–22, July 10–24, Aug. 15–21. Common.

3740 ***P. idaeusalis*** (Wlk.)
Tufted Apple Bud Moth
PARKS AND RESERVES: HWS, KGC, LCP, OCP, PNM, RRG. **COUNTIES:** BEL, BOO, BRA, FAY, FUL, GAL, HLN, JEF, MCC, MEA, MEN, OLD, OWS, RUS. **DATES:** May 7–30, June 4–17, July 6–30, Aug., Sept. 1–7, 20, 28, Oct. 8. Common.

3742 ***P. scotiana*** (McD.)
Meade Co., OCP, June 7; Owsley Co., 4.8 km (3 mi) NE of Booneville, June 14; both L. D. Gibson.

3743 ***P. exasperatana*** (Zell.)
PARKS AND RESERVES: BBM, KGC, OCP, PYR, RRG. **COUNTIES:** CHR, HLN, JEF, MEA, MEN, ROW. **DATES:** May 31, June 7, July 11–17, Aug. 26–28. Locally common.

Genus *Coelostathma* Clem.

3747 ***C. discopunctana*** Clem.
PARKS AND RESERVES: NBP. **COUNTIES:** FAY, JEF, MCC, OLD, OWS, POW. **DATES:** May 22, June 4, 9, 14, July 7–11, 22, 24, Aug. 1, 2, 17, Sept. 29–Oct. 6 (Malaise trap).

Genus *Amorbia* Clem.

3748 ***A. humerosana*** Clem.
PARKS AND RESERVES: BBL, BCN, BIP, BMF, CBF, LBL, NBP, OCP, PNM, RBF, RRG. **COUNTIES:** BEL, BOD, BOO, BRE, BUL, JEF, LET, MCC, MEA, MEN, OWS, PIK, POW, TRI, WHI. **DATES:** Apr. 15, 29, 30, May 11–June 11, July 15. Common.

NOTE: Species from here to the end of the Tortricidae were included as Family Cochylidae in Hodges et al. (1983). Horak and Brown (1991) have placed the group as a tribe, Cochylini, in Subfamily Tortricinae. Many species are still to be named in this group.

Genus *Aethes* Billberg

3757 ***A. deutschiana*** (Zett.)
Letcher Co., Big Black Mountain, May 4, A. F. Braun.

Genus *Cochylis* Tr.

3768.1 ***C. caulocatax*** Razowski
Hardin Co., Glendale, in Malaise trap, Aug. 22–31, Sept. 1–3, 1975, P. H. Freytag (Pogue and Friedlander 1987).

Genus *Thyraylia* Wlsm.

3769 ***T. bunteana*** (Rob.)
Boone Co., CER, Aug. 17; Bracken Co., SR 1159 at Locust Creek, Aug. 23; Clinton Co., June 20; Menifee Co., RRG, Aug. 25; Owsley Co., 4.8 km (3 mi) NE of Booneville, Sept. 8.

Genus *Henricus* Bsk.

3774 ***H. contrastanus*** (Kft.)
Jefferson Co., Valley Station, May 27, 1975, A. J. Brownell.

Genus *Carolella* Bsk.

3782 ***C. sartana*** (Hbn.)
Bullitt Co., BMF, in Malaise trap, June 6–10, July 8–12; Woodford Co., 1.6 km (1 mi) W of SR 68 bridge, Kentucky River, Sept. 2, 3.

3783 ***C. bimaculana*** (Rob.)
Bullitt Co., BMF, in Malaise trap, July 8–12; Jefferson Co., Louisville, June 29; Owsley Co., 4.8 km (3 mi) NE of Booneville, June 30.

3784 ***C. erigeronana*** (Riley)
McCracken Co., Ray Black farm, 9.7 km (6 mi) SW of Paducah, May 20, July 4, W.R. Black, Jr.

Genus *Hysterosia* Steph.

3801 ***H. birdana*** Bsk.
Grayson Co., ASC near Leitchfield, Sept. 11, 1976, C. V. Covell, Jr. (questionably this species; M.G. Pogue, *pers. comm.*).

3802 ***H. riscana*** Kft.
Boone Co., East Bend area, 8.0 km (5 mi) W of BBL, Aug. 21, 1981; Fayette Co., S of Higbee Mill near Hopemont Dr., Aug. 10.

3804 ***H. baracana*** Bsk.
Boone Co., BBL, July 10; Gallatin Co., Markland Dam, July 15; Menifee Co., RRG, Aug. 25, 1984;

all L. D. Gibson, who labelled the first and third record as "probably this species."

Cochylini species of uncertain generic placement:

NOTE: The species below are listed in the genera to which they were originally assigned, in keeping with their treatment in Hodges et al. (1983). This was done by J.A. Powell, the author of the treatment of what was family Cochylidae in Hodges et al. (1983), who was unable to place them in genera. Since they are not arranged in generic groups, the whole genus name is spelled out for each. Numbers correspond to those in Hodges et al. (1983).

3807 ***Aethes angulatana*** Rob.
Menifee Co., RRG, Aug. 25, 1984, L. D. Gibson.

3808 ***Conchylis augustana*** Clem.
Menifee Co., RRG, Aug. 25; Owsley Co., 4.8 km (3 mi) NE of Booneville, July l, 24, Aug. 22; all L. D. Gibson.

3809 ***Conchylis argentilimitana*** Rob.
PARKS AND RESERVES: BBL, OCP, PNM, WVP. **COUNTIES:** BAR, BEL, BOO, BUL, JEF, MEA, OWS, ROC. **DATES:** May 2, 3, 17, June 20, July 8, 10, Aug. 22, Sept. 11. Common.

3811 ***Phalonia aureana*** Bsk.
Harlan Co., BBM, July 22, 23; Letcher Co., Pine Mountain, June 11; Rowan Co., May 21; all A. F. Braun.

3812 ***Phalonia aurorana*** Kft.
Henderson Co., SWR, Aug. 12; Menifee Co., RRG, Aug. 25; both L. D. Gibson.

3815 ***Phalonia biscana*** Kft.
Fulton Co., SR 94, Willingham Bottoms, 4.0 km (2.5 mi) E of Cayce, Sept. 8; Harlan Co., BBM, June 30; Letcher Co, head of Line Fork, June 18.

3822 ***Conchylis floccosana*** Wlk.
Boone Co., BCN, June 8; Jefferson Co., Valley Station, May 27, June 4, 1975; Owsley Co., 4.8 km (3 mi) NE of Booneville, May 23, 28. This species name was misspelled "*flaccosana*" in Hodges et al. (1983).

3827 ***Phalonia gunniana*** Bsk.
Owsley Co., 4.8 km (3 mi) NE of Booneville, July 24, 1982, L. D. Gibson.

3828 ***Phalonia hoffmanana*** Kft.
Boone Co., BBL, May 4, 5, 22, 31, Aug. 25, L. D. Gibson.

3834 ***Phalonia lavana*** Bsk.
Boone Co., BBL, May 4–31; Owsley Co., 4.8 km (3 mi) NE of Booneville, May 28; Woodford Co., 1.6 km (1 mi) W of SR 68 along Kentucky River, May 25.

3835 ***Phalonia leguminana*** Bsk.
Boone Co., BBL, Aug. 25, and CER, Aug. 17; both L. D. Gibson.

3841 ***Phalonia obliquana*** Kft.
Barren Co., Diamond Caverns, Sept. 11; Letcher Co., Pound Gap, Pine Mountain, June 24, Aug. 26; McCreary Co., Natural Bridge, Sept. 7, Whitley City, Aug. 29, and Yahoo Ridge, June 14, Sept. 16; McCreary Co., June 15, 18; Metcalfe Co., Knob Lick, July 20, Sept. 12; Powell Co., Aug. 9; all A. F. Braun.

3842 ***Conchylis oenotherana*** Riley
PARKS AND RESERVES: OCP. **COUNTIES:** FAY, GAL, HAR, MCC, MEA, OWS. **DATES:** Apr. 27, June 19–July 22, Aug. 6–Sept. 6. **REMARKS:** Common. One specimen from Fayette Co., Lexington, was labelled "from museum cases containing wheat, June 27, 1910."

3843 ***Phalonia plummeriana*** Bsk.
Gallatin, Markland Dam, May 27; Harlan Co., BBM, July 12, and KGC, July 11; Jefferson Co., Valley Station, May 27, 29.

3846 ***Phalonia rana*** Bsk.
Boone Co., BBL, Aug. 26, East Bend area, 8.0 km (5 mi) W of BBL, Aug. 21; Bracken Co., SR 1159 at Locust Creek, Aug. 23; Hardin Co., Glendale, in Malaise trap, July 29–31, Aug. 1–6.

3854 ***Phalonia temerana*** Bsk.
Boone Co., BBL, May 4, 18; Franklin Co., Apr. 20.

3858 ***Phalonia voxcana*** Kft.
Boone Co., CER, Aug. 17, L. D. Gibson.

3859 ***Phalonia wiscana*** Kft.
Bullitt Co., BMF, in Malaise trap, Apr. 3–20; Menifee Co., RRG, Apr. 26; Powell Co., NBP, Apr. 21.

Superfamily **HESPERIOIDEA**

Family **HESPERIIDAE**

Skippers

Subfamily **Pyrginae**

Spread-wing Skippers

Genus ***Epargyreus*** Hbn.

3870 ***E. clarus*** (Cram.)
Silver-spotted Skipper
PARKS AND RESERVES. BBM, BEC, BLD, BMF, CBB, COB, HHW, HWS, IRO, JCF, KGC, LBL, LWO, NBP, OCP, PNM, PYR, RRG, WVP. **COUNTIES:** ADA, ALL, BAL, BAR, BEL, BOL, BOO, BRE, BUL, BUT, CAD, CAE, CAL, CAR,

CHR, CLA, CLI, CRI, CRL, CUM, DAV, EDM, ELL, FAY, FLE, FLO, FUL, GRA, GRE, GRV, GUP, HIC, HLN, HND, HOP, HRD, HRT, HSN, JAC, JEF, JES, JOH, KNO, LAR, LAU, LAW, LEE, LES, LET, LEW, LIN, LIV, LOG, LYO, MAG, MAR, MAS, MCC, MEA, MEN, MON, MOR, MRL, MUH, NEL, OHI, OLD, OWE, OWS, PER, PIK, POW, PUL, ROW, RUS, SCO, SHE, SIM, SPE, TOD, TRI, TRM, UNI, WAR, WAY, WEB, WHI, WOO, WOL. **DATES:** Apr. 2, 7, 15–Sept. 11, 25–Oct. 8, 20, Nov. 6. **REMARKS:** Common to very common throughout the state.

Genus ***Urbanus*** Hbn.

3886 ***U. proteus*** (L.)
Long-tailed Skipper
Fulton Co., Hickman, Sept. 4, 1997, W. R. Black, Jr.; Jefferson Co., Anchorage, Sept. 16, 1946, B. L. Monroe, Jr., and Louisville Oct. 11, 1998, nectaring on *Buddleia*, C.V. Covell, Jr.; McCracken Co., Paducah, nectaring on frost asters at Tennessee St., Oct. 18, 1995, W. R. Black, Jr. A rare migrant into our area from much farther south.

Genus ***Autochton*** Hbn.

3902 ***A. cellus*** (Bdv. & LeConte)
Golden-banded Skipper
PARKS AND RESERVES: MCV, RRG, WVP. **COUNTIES:** EDM, HRT, JAC, JEF, LAU, MCY, MEN, PIK, POW, PUL, WAY, WOL. **DATES:** Apr. 24, May 20, June 3, 5, 15, July 2–16, Aug. 6. **REMARKS:** Sometimes locally common in eastern mountains.

Genus ***Achalarus*** Scudder

3904 ***A. lyciades*** (Gey.)
Hoary Edge
PARKS AND RESERVES: BMF, CAC, DHW, KGC, LBL, LCP, MCV, PYR. **COUNTIES:** ALL, BAR, BUL, CAD, CAR, CHR, CLA, CLI, CRI, CUM, EDM, GRE, GRV, GRY, HLN, HRD, HRT, JEF, KNO, LAR, LAU, LEE, LES, MAR, MCY, MEA, MEN, MUH, NEL, OHI, OWE, OWS, POW, PUL, RUS, SIM, SPE, TOD, TRI, WAR, WAY, WHI. **DATES:** Apr. 26, May 2–June 12, 23–July 3, 16–Aug. 21, 30, Sept. 10, Nov. 3. **REMARKS:** Widespread and frequent, but rarely common.

Genus ***Thorybes*** Scudder

3909 ***T. bathyllus*** (J. E. Smith)
Southern Cloudywing
PARKS AND RESERVES: BBL, BMF, IRO, LCP, OCP, RRG, WKM. **COUNTIES:** BEL, BOO, BUL, CHR, CLI, CRI, CUM, DAV, EDM, FUL, GRA, GRE, GRY, HLN, HOP, HRD, HRT, JEF, JES, LAR, LET, LEW, LOG, MAR, MCC, MCY, MEA, MEN, MUH, NEL, OHI, OLD, PEN, ROW, RUS, SPE, TOD, TRI, WAR, WAY, WHI, WOL. **DATES:** Apr. 26, May 8, 16–29, June 11–26, July 3, 11–Aug. 23, Sept. 5. **REMARKS:** Locally common.

3910 ***T. pylades*** (Scudder)
Northern Cloudywing
PARKS AND RESERVES: BMF, HWS, MCV, RRG, WKM. **COUNTIES:** BUL, BUT, CHR, EDM, GRY, HLN, HRD, JEF, JES, LAR, LAU, MAG, MAR, MCC, MCY, MEA, MEN, MRL, NEL, OHI, OLD, OWS, PER, PUL, ROW, SPE, TRI, TRM, WAY, WHI. **DATES:** Apr. 29, May 7–31, June 3–27, July 9, 13, 24, 30, Aug. 12. **REMARKS:** Locally common.

3913 ***T. confusis*** Bell
Confused Cloudywing
PARKS AND RESERVES: LCP. **COUNTIES:** FLE, HRT, JEF, LOG, MUH, PUL, ROW, RUS, TOD, WOL. **DATES:** May 4–9, June 13, 29, July 2, Aug. 6, 22, Sept. 11. **REMARKS:** Uncommon to rare.

Genus ***Staphylus*** Godm. & Salvin

3932 ***S. hayhurstii*** (Edw.)
Hayhurst's Scallopwing
PARKS AND RESERVES: BBL, FAO, HWS. **COUNTIES:** BOO, BUL, CLI, FAY, FUL, HRD, JEF, KEN, LAR, MEA, NEL, OLD, POW, TRM. **DATES:** Apr. 16, May 16–29, June 16, July 22–Aug. 14, 27, 29, Sept. 9, 20, 24. **REMARKS:** Local and uncommon.

Genus ***Erynnis*** Schrank

3945 ***E. icelus*** (Scudder & Burgess)
Dreamy Duskywing
PARKS AND RESERVES: BBL, BEC, BMF, CAC, CBB, HWS, IRO. **COUNTIES:** BOO, BRE, BUL, CAR, CLI, FAY, FLO, GAL, HLN, JEF, JOH, KEN, KNO, LAR, LES, MAG, MCY, MEA, MEN, MOR, NEL, OLD, POW, ROC, ROW, SCO, SPE, TRM, WAY, WHI. **DATES:** Apr. 3–30, May 1–16, 28–June 8, 19–30, July 3–12, 21–29, Aug. 31. **REMARKS:** Common to abundant.

3946 ***E. brizo*** (Bdv. & Leconte)
Sleepy Duskywing
PARKS AND RESERVES: BMF, CAC, CBB, HWS, IRO, NBP, OCP, RRG. **COUNTIES:** ALL, BRE, BUL, CAR, CRI, FLO, HAN, HLN, JAC, JEF, JOH, KNO, LAU, LAW, LEE, MAG, MAR, MCY, MEA, MEN, MOR, NEL, OLD, OWS, POW, PUL, ROC, ROW, WAR, WAY, WHI.

DATES: Mar. 3, 21–Apr. 29, May 7, 8, 16. **REMARKS:** Very common to abundant.

3947 ***E. juvenalis*** (F.)
Juvenal's Duskywing
PARKS AND RESERVES: BBM, BMF, CAC, CBB, IRO, MCV, NBP, OCP, RRG. **COUNTIES:** BRE, BUL, CAR, CRI, CUM, EDM, EST, FLO, GRY, HLN, JAC, JEF, JES, JOH, KNO, LAR, LAU, LAW, MAG, MCY, MEA, MEN, MOR, NEL, OLD, PER, POW, PUL, ROC, ROW, SCO, TRM, WAR, WHI. **DATES:** Mar. 19–31, Apr., May 1–16, June 6, July 6, 7, 23, Sept. 9. **REMARKS:** Common to abundant statewide.

3952 ***E. horatius*** (Scudder & Burgess)
Horace's Duskywing
PARKS AND RESERVES: BMF, CBB, HHW, HWS, IRO, OCP, PNM, SWR, WVP. **COUNTIES:** BEL, BUL, CAD, CHR, CRL, GRV, HAN, HIC, HLN, HND, HRT, JAC, JEF, LAU, LEE, LIV, MCC, MCY, MEA, NEL, OHI, OLD, OWE, PUL, SPE, TRM, UNI, WAR, WAY. **DATES:** Apr. 1–8, 16–28, May 11, June 9–30, July 2–23, Aug. 1–18, Sept. 5–11. **REMARKS:** Sometimes locally common.

3954 ***E. martialis*** (Scudder)
Mottled Duskywing
PARKS AND RESERVES: FKN, MCV, PYR. **COUNTIES:** CHR, CLI, EDM, LAR, LIV, MAG, MEA, PUL, WAR. **DATES:** Apr. 23, 29, May 7, 8 20, 30, July 14–22, Aug. 24. **REMARKS:** Uncommon.

3956 ***E. zarucco*** (Luc.)
Zarucco Duskywing
Marshall Co., KDV, Sept. l, 1951, B. L. Monroe, Jr. Rare.

3957 ***E. funeralis*** (Scudder & Burgess)
Funereal Duskywing
Henderson Co., SWR, Sept. 10; Muhlenberg Co., LMP, Apr.; Trigg Co., LBL, Oct. 4. Rare.

3958 ***E. lucilius*** (Scudder & Burgess)
Columbine Duskywing
Boone Co., Aug. 13; Estill Co., 11.3 km (7 mi) W of Irvine on SR 52, Apr. 20; Floyd Co., Apr. 25; Lawrence Co., Apr. 25; Powell Co., 1.9 km (1.2 mi) S of Stanton on SR 213, Apr. 21 (first state records, G. B. Straley and W. H. Wagner); Trimble Co., 3.2 km (2 mi) NW of Providence, along SR 316, 4 males and 1 female on columbine, June 20. Rare.

3959 ***E. baptisiae*** (Fbs.)
Wild Indigo Duskywing
PARKS AND RESERVES: BEC, BMF, HWS, OCP. **COUNTIES:** BOO, BUL, CAD, HLN, HND, JEF, KEN, LAU, MAD, MCC, MCY, MEA, OLD, POW, PUL, SPE, TOD, WAR, WHI. **DATES:** Apr. 6, 22–28, May 14, June 24, July 4–10, 19–23, Aug. 2, 18, 30, Sept. 2–30, Oct. 2, 29, Nov. 1. **REMARKS:** Increasingly common and widespread in recent years. It seems to be using crown vetch planted along highways as a foodplant, but may be found in gardens far from vetch.

Genus *Pyrgus* Hbn.

3962 ***P. centaureae*** (Rambur)
Grizzled Skipper
Harlan Co., BBM, several specimens collected on Apr. 25, 1980, L. C. Koehn. It has not been found on subsequent visits there. Listed as a threatened species in Kentucky (KSNPC 1996). Our subspecies is *wyandot* (Edw.).

3966 ***P. communis*** (Grt.)
Common Checkered-Skipper
PARKS AND RESERVES: BEC, CBB, FAO, HHW, HWS, SWR. **COUNTIES:** AND, BAL, BEL, BOL, BOO, BUL, CHR, CLK, CRL, CUM, FAY, FUL, HND, HRD, JEF, KEN, LAU, MAD, MCC, MCY, MEA, MER, NEL, OLD, OWE, SHE, TRI, TRM, UNI, WAR, WHI. **DATES:** Apr. 18, 26, May 17, June 16, July 7, 15–25, Aug. 11–27, Sept., Oct. 1, 2, 10–31, Nov. 1–18. **REMARKS:** Widespread; sometimes locally common.

Genus *Pholisora* Scudder

3977 ***P. catullus*** (F.)
Common Sootywing
PARKS AND RESERVES: BEC, BLD, CBB, FAO, HWS, PYR, SWR, WVP. **COUNTIES:** AND, BAL, BOL, BOO, BRA, BUL, BUT, CAR, CHR, CLI, CLK, CRI, CRL, EDM, FAY, FUL, GRA, GRE, HLN, HND, HRD, HRT, JEF, JES, KNO, LAR, LIV, LOG, MAD, MAR, MAS, MCC, MEA, MRL, NEL, OHI, OLD, OWE, SCO, SHE, SIM, SPE, TOD, TRI, TRM, WAY, WOL. **DATES:** Apr. 29–Sept. 11, 20. **REMARKS:** Common.

Subfamily **Hesperiinae**

Grass Skippers

Genus *Nastra* Evans

3993 ***N. lherminier*** (Latr.)
Swarthy Skipper
PARKS AND RESERVES: BBL, HHW, HWS, IRO, OCP. **COUNTIES:** BOO, BUL, FUL, GAL, GRA, GRE, HRD, JEF, LAU, MAR, MCC, MCY, MEA, OLD, OWE, RUS, UNI, WAR, WHI. **DATES:** May 22–31, June 3, 19–30, July 16–29,

Aug. 6–27, Sept. 8–13, 27. **REMARKS:** Common, but easily overlooked, often staying low among tall grasses and other plants.

Genus ***Lerema*** Scudder

3998 ***L. accius*** (J. E. Smith)
Clouded Skipper
PARKS AND RESERVES: LBL, PNM, PYR. **COUNTIES:** BEL, CHR, FUL, JEF, MAD, MCC, SPE, TRI. **DATES:** July 13, 20, Aug. 22–Sept. 15, Oct. 4, 7, 15–26, Nov. 6, 23. **REMARKS:** Sometimes locally common in the western counties.

Genus ***Ancyloxypha*** C. Felder

4004 ***A. numitor*** (F.)
Least Skipper
PARKS AND RESERVES: BBL, BLA, BMF, CAC, CBB, FAO, HHW, JCF, LBL, RGR, SWR, WKM. **COUNTIES:** BAL, BAT, BOO, BRK, BUL, CAD, CAL, CAR, CHR, CLA, CLI, CLK, CRI, EDM, EST, FAY, FLO, FUL, GAL, GRE, GRV, GRY, GUP, HIC, HLN, HND, HOP, HRD, HSN, JEF, JES, LAR, LET, LIV, MAD, MAR, MCC, MEA, MEN, MER, MUH, NEL, OHI, OLD, OWE, PUL, ROW, SHE, SPE, TOD, TRI, TRM, UNI, WAS, WAY, WHI, WOO. **DATES:** May 6–29, June 8–Oct. 2. **REMARKS:** Common in wet, grassy habitats. It was particularly abundant in Fulton Co., Aug., 1989, 1995, and 1997.

Genus ***Thymelicus*** Hbn.

4012 ***T. lineola*** (Ochs.)
European Skipper
COUNTIES: BOO, BOU, BUL, CLK, FAY, FRA, LET, MER, MON, MTG, NEL, PER, ROC, SCO, WOO. **DATES:** May 27, June, July 4, 5, 20, 21, Aug. 21, 27, Sept. 16. **REMARKS:** An introduced species, the first Kentucky capture was by Ralph Beebe in Rockcastle Co., Aug. 21, 1955 (Beebe 1955). Locally common in unmowed grass habitats, such as along interstate highways. The late captures seem to be unusual; most records are from the first half of June.

Genus ***Hylephila*** Billberg

4013 ***H. phyleus*** (Drury)
Fiery Skipper
PARKS AND RESERVES: BBL, FAO, HHW, KDV, LBL, SWR, WVP. **COUNTIES:** BAL, BAR, BOO, CLI, FAY, FUL, GRV, HND, HRD, JEF, KNX, LAU, MAD, MAS, MEA, MRL, OLD, OWE, OWS, PUL, TRI, UNI, WAY. **DATES:** July 13–26, Aug. 3, 13, 22–Oct. 12, 22, 28, Nov. 16. **REMARKS:** Locally common, especially late in the season.

Genus ***Hesperia*** F.

4023 ***H. leonardus*** Harr.
Leonard's Skipper
PARKS AND RESERVES: CAV, KGC, OCP, PCP, PYR, WVP. **COUNTIES:** BAT, BUL, CHR, CUM, HLN, HRD, JEF, LAR, LAU, LIV, MEA, PUL, WHI. **DATES:** Aug. 20, 30–Sept. 27, Oct. 13. **REMARKS:** Adults nectar on tall fall flowers such as ironweed, thistles, and sunflower group. May be locally common.

4027 ***H. metea*** Scudder
Cobweb Skipper
PARKS AND RESERVES: KGC, OCP, PCP. **COUNTIES:** BUL, HLN, HRD, JEF, LES, MCC, MCY, MEA, PUL, WHI. **DATES:** Apr. 6, 14–30. **REMARKS:** Sometimes locally common. Prefers open grassy patches in cedar glades and power transmission corridors where its foodplant, *Andropogon*, is abundant.

4033 ***H. sassacus*** Harr.
Indian Skipper
Clay Co., July, 1938. One male from F. W. Case Collection in the CMNHS. Rare, if still present in Kentucky.

Genus ***Polites*** Scudder

4036 ***P. peckius*** (Kby.)
Peck's Skipper
PARKS AND RESERVES: BEC, BMF, CBB, FKN, HHW, HWS, OCP, PYR, SWR, WKM, WVP. **COUNTIES:** AND, BAL, BAR, BEL, BOO, BRK, BUL, CAE, CAM, CHR, CRI, DAV, EDM, FAY, FRA, GRA, GRE, GRV, HAN, HIC, HLN, HND, HOP, HRD, HRT, HSN, JEF, JES, KNX, LAR, LET, LIV, LOG, MAD, MAR, MCC, MCL, MCY, MEA, MEN, MTG, MUH, NEL, OHI, OLD, OWE, POW, ROW, SHE, SPE, TAY, TOD, TRM, UNI, WEB, WHI, WOL, WOO. **DATES:** May 2–Oct. 2, 10, 17, Nov. 2. **REMARKS:** Common to abundant throughout the state.

4041 ***P. themistocles*** (Latr.)
Tawny-edged Skipper
PARKS AND RESERVES: BEC, BLA, BMF, CBB, HHW, HWS, OCP, PYR, SWR. **COUNTIES:** AND, BAL, BAR, BEL, BOD, BOO, BRA, BRK, BUL, BUT, CAD, CHR, CLI, CRI, FAY, FRA, FUL, GRA, GRE, GRY, GUP, HLN, HND, HOP, HRD, HRT, JEF, JES, KNX, LAR, LAU, LET, LEW, LIV, LOG, MAD, MAR, MAS, MCC, MCY, MEA, MUH, NEL, OHI, OLD, OWE, PEN, POW, PUL, ROW, SCO, SHE, SIM, SPE, TOD, UNI, WHI, WOL, WOO. **DATES:** May 9–Oct. 3.

REMARKS: Abundant statewide, especially late in the season.

4042 ***P. origenes*** (F.)
Crossline Skipper
PARKS AND RESERVES: BBL, BMF, CBB, HWS, MCV, OCP, SWR. **COUNTIES:** ALL, BOO, BUL, CAL, CHR, CUM, EDM, FAY, FUL, GRV, HND, HRD, HRT, JEF, LAU, LOG, LYO, MAD, MCC, MCY, MEA, MRL, NEL, OLD, OWE, POW, PUL, ROW, TOD, TRI, WAS, WHI. **DATES:** May 9, 22, 29, June 1, 15–24, July 2–9, 19, 26, Aug. 4–30, Sept. 3–28. **REMARKS:** Moderately common; resembles tawny-edged skipper, but a bit larger, and stigma of male forewing angled less steeply.

Genus *Wallengrenia* Berg

4046 ***W. otho*** (J. E. Smith)
Southern Broken-Dash
One old specimen labelled "Louisville" in the NMNH has no date or collector's name. Burns (1985) reported it as a valid Kentucky record.

4047 ***W. egeremet*** (Scudder)
Northern Broken-Dash
PARKS AND RESERVES: BBL, BMF, CBB, HWS, OCP, SWR, WVP. **COUNTIES:** ALL, AND, BAL, BAR, BOO, BRA, BUL, BUT, CAD, CAL, CHR, CLI, CRI, FAY, FRA, FUL, GRA, GRE, GRV, HLN, HND, HOP, HRD, JEF, JES, KEN, LAR, LAU, LET, LEW, MCC, MCY, MEA, MEN, MER, MUH, NEL, OHI, OLD, OWE, OWS, POW, PUL, ROW, SCO, SPE, TOD, TRI, TRM, WAR, WHI, WOO. **DATES:** May 20–Sept. 28. **REMARKS:** Very common. Burns (1985) lists data on 16 specimens, mostly from the ULK Collection.

Genus *Pompeius* Evans

4048 ***P. verna*** (Edw.)
Little Glassywing
PARKS AND RESERVES: BMF, HWS, OCP, PYR, WVP. **COUNTIES:** BOO, BRA, BUL, BUT, CAM, CHR, CLI, FLE, FUL, HLN, HOP, JEF, LAR, LOG, MAD, MEA, MUH, NEL, OHI, OLD, PUL, ROW, SCO, SHE, SPE, TOD, TRM, WHI. **DATES:** May 18, 20, 27, June 4–26, July 2–23, Aug. 2–23, Sept. 3, 15, 28. **REMARKS:** Widespread, but uncommon.

Genus *Atalopedes* Scudder

4049 ***A. campestris*** (Bdv.)
Sachem
PARKS AND RESERVES: BEC, BMF, CBB, FAO, HHW, HWS, LBL, OCP, PYR, SWR, WVP. **COUNTIES:** ALL, BAL, BAR, BEL, BOL, BOO, BOU, BRA, BUL, BUT, CAD, CAE, CAL, CHR, CLA, CLI, CLK, CRI, CUM, DAV, ELL, FAY, FLO, FRA, FUL, GAR, GRE, GRV, GRY, GUP, HIC, HLN, HND, HOP, HRD, HRT, JAC, JEF, KEN, LAR, LAU, LET, LIV, LOG, LYO, MAD, MAR, MCC, MCL, MCY, MEA, MEN, MER, MET, MRL, MUH, NEL, OHI, OLD, OWS, POW, PUL, ROC, ROW, SCO, SIM, SPE, TOD, TRI, TRM, UNI, WAR, WAY, WHI, WOL, WOO. **DATES:** May 4, 7, 16–30, June 3–11, 21–23, July 5–Nov. 3. **REMARKS:** Common to abundant throughout the state; the dominant skipper in autumn.

Genus *Anatrytone* Dyar

4051 ***A. logan*** (Edw.)
Delaware Skipper
PARKS AND RESERVES: BEC, BMF, HWS, SWR. **COUNTIES:** BAL, BRA, BUL, CHR, FAY, FUL, GRV, HND, HRD, JEF, LAR, LOG, MAD, MCC, MEA, NEL, OLD, OWE, PUL, ROC, SPE, TOD, TRM, WHI, WOL. **DATES:** May 28, June 1–24, July 2–16, Aug. 11–23, 31–Sept. 21, Oct. 2. **REMARKS:** Sometimes locally common; thistles are a favored nectar source.

Genus *Poanes* Scudder

4059 ***P. hobomok*** (Harr.)
Hobomok Skipper
PARKS AND RESERVES: BBM, BMF, CBB, OCP, PYR, RRG, WVP. **COUNTIES:** BRK, BUL, CHR, EDM, EST, GRY, HLN, JEF, JES, MCY, MEA, MEN, NEL, OLD, OWS, POW, ROW, WHI. **DATES:** Apr. 10, 18, May 6–July 4, Aug. 21, 23. **REMARKS:** Locally common in eastern Kentucky.

4060 ***P. zabulon*** (Bdv. & Leconte)
Zabulon Skipper
PARKS AND RESERVES: BBL, BEC, BMF, CBB, HHW, HWS, KGC, OCP, PYR, RRG, SWR, WVP. **COUNTIES:** ADA, AND, BAL, BEL, BOO, BRA, BUL, BUT, CAM, CAR, CHR, CRI, EST, FAY, FLO, FUL, GAL, GRA, GRE, GRY, HIC, HLN, HND, HOP, HRD, HRT, JAC, JEF, JES, LAR, LAU, LOG, MAR, MCC, MCY, MEA, MEN, MER, MOR, MUH, NEL, OHI, OLD, OWS, POW, PUL, RUS, SPE, TOD, TRI, TRM, UNI, WAR, WEB, WHI, WOL. **DATES:** Apr. 20–Sept. 16. **REMARKS:** Locally common. Males habitually perch on tree leaves at forest edges and in clearings.

4063 ***P. yehl*** (Skin.)
Yehl Skipper
PARKS AND RESERVES: HHW, RNR, SWR. **COUNTIES:** FUL, GRV, HND, LIV, MCC, UNI.

DATES: June 18–21, Aug. 23–Sept. 21, Oct. 7. **REMARKS:** Two broods. A southern wetlands species; may be locally common.

4064 ***P. viator*** (Edw.)
Broad-winged Skipper
Fulton Co., 4.8 km (3 mi) SE of Bondurant, May 17, 1972 (first state record, B.L. Monroe, Jr.), May 28, 29, June 5, 21. Rare and very local in stands of *Phragmites*, its foodplant.

Genus *Euphyes* Scudder

4072 ***E. dion*** (Edw.)
Dion Skipper
Fulton Co., RNR, May 28, Sept. 16, SR 94, Willingham Bottoms, 4.0 km (2.5 mi) E of Cayce, Aug. 8, 22, 27, Sept. 4–12, and SR 94, 3.4 km (2.1 mi) W of Cayce, June 16, 21, Aug. 23, Sept. 5–15; Graves Co., Thomas Bell farm, Sept. 10–15; Henderson Co., SWR, Sept. 11; McCracken Co., Paducah, Sept. 6, 14. Two broods. Locally common in western Kentucky wetlands.

4074 ***E. dukesi*** (Linds.)
Duke's Skipper
Fulton Co., 4.0 km (2.5 mi) E of Cayce in Willingham Bottoms, along Little Bayou de Chien, Sept. 5, 6; Marshall Co., Ramada Inn near KDV, June 2, 1995, M. S. Monroe; McCracken Co., Paducah, two localities, Aug. 29–Sept. 13, 1998, at least 12 individuals recorded, W.R. Black, Jr. One male was found in the student collections at Western Kentucky University labelled "Webster, 7–6–72, M. Johnson." I feel certain that the specimen was taken in Webster County. The collector, according to former entomology professor Dr. Herb Shadowen at Western Kentucky University, was Melinda Johnson, whom I have not been able to contact for further details. Webster County has extensive areas of grassy, sedgy, wet meadows suitable for *dukesi*. Other colonies will certainly be discovered in appropriate habitats in western counties.

4078 ***E. vestris*** (Bdv.)
Dun Skipper
PARKS AND RESERVES: BBL, BEC, CBB, HHW, OCP, PNM, SWR. **COUNTIES:** BEL, BOO, BUL, CLA, FAY, FLE, FUL, GRA, GRV, HIC, HLN, HND, HOP, HRD, JEF, LES, LOG, MEA, MUH, NEL, OLD, OWE, POW, PUL, ROC, SCO, SHE, SPE, TRM, UNI, WAY, WEB, WOL. **DATES:** May 20, June 12–30, July 2–23, Aug. 12–30, Sept. 3–18. **REMARKS:** Our subspecies is *metacomet* (Harr.). Common.

Genus *Atrytonopsis* Godm.

4080 ***A. hianna*** (Scudder)
Dusted Skipper
PARKS AND RESERVES: CBF, OCP, PCP, PYR. **COUNTIES:** BUL, CHR, GRY, HRD, MEA, OWS, WAR, WHI. **DATES:** Apr., May 6–30, June 10. **REMARKS:** Widespread but rarely seen.

Genus *Amblyscirtes* Scudder

4096 ***A. hegon*** (Scudder)
Pepper and Salt Skipper
PARKS AND RESERVES: BMF, JWS, MCV, OCP, PYR, RRG. **COUNTIES:** BRA, BRE, BUL, CHR, EDM, FLO, JEF, KNO, LAR, LAU, LAW, LES, LEW, MAR, MCC, MCY, MEA, MEN, NEL, OWS, POW, PUL, TRM, WAY, WHI. **DATES:** Apr. 15–30, May 2–8, 16–26, June 1–23, July 9. **REMARKS:** Locally common to uncommon.

4099 ***A. aesculapius*** (F.)
Lace-winged Roadside-Skipper
PARKS AND RESERVES: BBL, PYR, RRG. **COUNTIES:** BEL, BOO, BUL, CAM, CHR, FLO, FUL, GRV, JEF, MAG, MCC, MEN, MOR, OHI, OWS, POW. **DATES:** Apr. 23, 24, May 7, 20–31, June 6, 15–Aug. 25, Sept. 5, 8. **REMARKS:** Local; uncommon to rare. Individual specimens were nectaring on ironweed near cane thickets at Willingham Bottoms, Fulton Co., in September of several years.

4105 ***A. vialis*** (Edw.)
Common Roadside-Skipper
PARKS AND RESERVES: HWS, NBP, OCP, PYR, RRG. **COUNTIES:** BUL, CAR, CHR, FLO, FUL, JEF, LAR, LAU, LEE, MAR, MCC, MCY, MEA, MEN, MET, OLD, OWS, POW, PUL, TOD, TRI, TRM, WAY, WHI, WOL. **DATES:** Apr. 15–29, May 2–16, June 14, 18, July 21, 22, Aug. 21, Sept. 8, 18. **REMARKS:** Locally uncommon to common.

4107 ***A. belli*** H. A. Freeman
Bell's Roadside-Skipper
PARKS AND RESERVES: OCP, PYR. **COUNTIES:** BRA, CAL, CHR, CLA, FUL, GRE, HRT, JEF, MCC, MEA, OHI. **DATES:** May 4, 20, 24, June 10, July 26, 28, Aug. 6–24, Sept. 8. **REMARKS:** Uncommon; captures have usually been single, but PYR record was multiple.

Genus ***Lerodea*** Scudder

4111 ***L. eufala*** (Edw.)
Eufala Skipper
Fulton Co., Willingham Bottoms, SR 94, 4.0 km (2.5 mi) E of Cayce, 1 fresh male, Sept. 5; Union Co., HHW, SR 56 near Morganfield, Sept. 9. Rare.

Genus ***Panoquina*** Hemming

4119 ***P. ocola*** (Edw.)
Ocola Skipper
PARKS AND RESERVES: CBB. **COUNTIES:** FUL, HRD, JEF, MCC, OLD, TRI. **DATES:** Sept. 2–13, 26, Oct. 10–20. **REMARKS:** Uncommon, but regular in Fulton Co. A late colonizer most years from farther south.

Superfamily **PAPILIONOIDEA**

Family **PAPILIONIDAE**

Swallowtails

Genus ***Battus*** Scop.

4157 ***B. philenor*** (L.)
Pipevine Swallowtail
PARKS AND RESERVES: BBL, BBM, BLA, BMF, CBB, CBF, HHW, HLN, HWS, KGC, LBL, MUR, NBP, PNM, RRG, SWR. **COUNTIES:** ALL, BAL, BAT, BEL, BOO, BRE, BUL, BUT, CAD, CAR, CAS, CHR, CLA, CLI, CRI, CUM, EDM, ELL, EST, FAY, FLE, FLO, FUL, GRA, GRE, GRV, GRY, GUP, HAN, HIC, HLN, HND, HOP, HRD, HRT, JAC, JEF, JES, JOH, KNO, LAR, LAU, LAW, LEE, LES, LET, LEW, LIV, LOG, LYO, MAD, MAG, MAR, MCY, MEA, MEN, MET, MOR, MRL, MUH, MUN, NEL, OHI, OLD, OWE, OWS, POW, PUL, ROW, RUS, SHE, SPE, TOD, TRI, TRM, UNI, WAR, WAY, WHI, WOL. **DATES:** Apr. 6–Oct. 14. **REMARKS:** Common, and may be abundant, especially in the mountains during late summer. Presumably the model for a complex of Batesian mimicry among several large blackish-blue butterfly species.

4158 ***B. polydamas*** (L.)
Polydamas Swallowtail
Carter Co., (Beebe 1960); Rockcastle Co., Mt. Vernon, July 25, 1959, sight record by Ralph Beebe after Hurricane Debra, on flowers of bouncing bet, *Saponaria officianalis*. Another was reported as sighted by John D. Greenhill near Grayson, in Sept., 1985, after Hurricane Elena. A rare stray from the southernmost United States, or the tropics.

Genus ***Papilio*** L.

4159 ***P. polyxenes*** F.
Black Swallowtail
PARKS AND RESERVES: BAN, BBL, BMF, CBB, FAO, HHW, HWS, KGC, LBL, PYR, RRG, SGS, SWR. **COUNTIES:** ALL, AND, BAL, BAR, BOL, BOO, BRA, BUL, CAD, CAL, CAR, CHR, CLI, CLK, CRL, CUM, EDM, EST, FAY, FUL, GRA, GRE, GRV, HIC, HLN, HND, HOP, HRD, HRT, HSN, JAC, JEF, KNX, LAR, LAU, LOG, LYO, MAD, MCC, MCY, MEA, MEN, MER, MET, MUH, NEL, OLD, OWE, OWS, POW, PUL, ROW, RUS, SCO, SHE, SIM, SPE, TOD, TRI, TRM, UNI, WAR, WAS, WAY, WHI, WOL. **DATES:** Mar. 1, Apr. 6–Oct. 10, 26, Dec. 7. **REMARKS:** Our subspecies is *asterius* Stoll. Common, especially in open field habitats.

4160 ***P. joanae*** J. R. Heitzman
Ozark Swallowtail
J. R. Heitzman (*pers. comm.*) reported a male of this species collected in Oldham Co. by Judy Maxwell on Aug. 5, 1979. Rare, but easily overlooked because of close resemblance to black swallowtail. See Heitzman (1974) for features.

Genus ***Heraclides*** Hbn.

4170 ***H. cresphontes*** (Cram.)
Giant Swallowtail
PARKS AND RESERVES: BBL, BLA, BMF, GBT, HWS, SGS. **COUNTIES:** BOO, BRA, BUL, CLI, CRL, CUM, FAY, FRA, GRA, GRE, GRV, GUP, HLN, HRT, JEF, JES, LAR, MEA, MER, MRL, NEL, OLD, OWE, POW, SCO, SPE, TOD, TRI, TRM, WAR, WAY, WHI. **DATES:** Mar. 17, Apr. 22, May 5, 13–31, June 18–26, July 11–30, Aug., Sept. 9–16. **REMARKS:** Widespread, but generally uncommon. Our largest butterfly.

Genus ***Pterourus*** Scop.

4176 ***P. glaucus*** (L.)
Eastern Tiger Swallowtail
PARKS AND RESERVES: BEC, BLA, BMF, CBB, CBF, HHW, HWS, IRO, JCF, KGC, LBL, LCP, NBP, OCP, PNM, PYR, RRG, SGS, SWR, WKM, YEL, YHF. **COUNTIES:** ALL, BAL, BAR, BAT, BEL, BOO, BRE, BRK, BUL, CAD, CAE, CAL, CAR, CAS, CHR, CLA, CLI, CRI, CRL, CUM, EDM, EST, FAY, FLE, FLO, FUL, GAR, GRE, GRV, GRY, HAN, HIC, HLN, HND, HOP, HRD, HRT, HSN, JAC, JEF, JOH, KNO, LAR, LAU, LAW, LES, LET, LIV, LOG, MAD, MAG, MAR, MCC, MCL, MCY, MEA, MEN, MET, MOR, MRL, MUH, NEL, OHI, OLD, OWE, OWS, POW, PUL, ROC, ROW, RUS, SIM,

SPE, TAY, TOD, TRI, TRM, UNI, WAR, WAS, WAY, WEB, WHI, WOL. **DATES:** Mar. 21, 31–Sept. 30. **REMARKS:** Common throughout the state. Yellow and dark females occur in about equal numbers. Females with mixed yellow and dark scales prevalent after drought in July 1988, and in April 1989. A strikingly aberrant male was collected in the DBNF (county unknown) April 18, 1971 by the late Siegfried Scholz.

4181 ***P. troilus*** (L.)
Spicebush Swallowtail
PARKS AND RESERVES: BBL, BEC, BLA, BLD, BMF, CAC, CBB, COB, FKN, GBT, HHW, HWS, IRO, JCF, KGC, LBL, NBP, OCP, PNM, PYR, RRG, SGS, SWR. **COUNTIES:** ALL, BAL, BAR, BAT, BEL, BOL, BOO, BRA, BRE, BUL, BUT, CAD, CAE, CAL, CAR, CHR, CLA, CLI, CRI, CRL, DAV, EDM, EST, FLO, FRA, FUL, GAL, GRA, GRE, GRV, GRY, GUP, HAN, HIC, HLN, HND, HOP, HRD, HRT, JAC, JEF, KNO, LAR, LAU, LAW, LES, LET, LEW, LIV, LOG, LYO, MAD, MAR, MAS, MCY, MEA, MEN, MON, MOR, MUH, OHI, OLD, OWE, OWS, PEN, POW, PUL, ROC, ROW, RUS, SHE, SIM, SPE, TAY, TOD, TRI, TRM, UNI, WAR, WAY, WEB, WHI, WOL. **DATES:** Apr. 1–Oct. 15. **REMARKS:** Probably our most common of broadly distributed swallowtails, although *B. philenor* is more abundant in some eastern areas in July through September.

4182 ***P. palamedes*** (Drury)
Palamedes Swallowtail
Fulton Co., Willingham Bottoms 4.0 km (2.5 mi) E of Cayce, Sept. 10, 1993; Hickman Co., COB, July 30, 1962; Meade Co., OCP and again near Brandenburg, 1970; Nelson Co., between New Haven and Boston, Aug. 21; Whitley Co., Woodbine, Aug. 7. Rare; a stray or temporary summer resident in western Kentucky.

Genus ***Eurytides*** Hbn.

4184 ***E. marcellus*** (Cram.)
Zebra Swallowtail
PARKS AND RESERVES: BBM, BEC, BMF, CAC, CBB, CBF, HWS, IRO, LBL, LCP, PNM, RRG, YHF. **COUNTIES:** ADA, BEL, BRA, BRE, BUL, BUT, CAD, CAR, CHR, CLI, CUM, EDM, EST, FAY, FLE, FLO, FUL, GRE, GRV, GRY, GUP, HLN, HRD, HRT, HSN, JEF, JES, JOH, KNO, KNX, LAR, LAU, LAW, LES, LET, LEW, LYO, MAD, MAG, MAR, MAS, MCC, MCY, MEA, MEN, MET, MOR, NEL, OHI, OLD, OWE, OWS, PEN, PIK, POW, PUL, ROB, ROC, ROW, RUS, SCO, SPE, TAY, TRI, TRM, UNI, WAR, WAS, WEB, WHI. **DATES:** Mar. 26–Sept. 24. **REMARKS:** May be locally common, as it was unusually so in the springs of 1988 and 1989. Two broods; individuals of the second brood notably larger than those of the first.

Family **PIERIDAE**

Sulphurs and Whites

Subfamily **Pierinae**

Whites

Genus ***Pontia*** F.

4193 ***P. protodice*** (Bdv. & Leconte)
Checkered White
PARKS AND RESERVES: BBL, CBB, DHW, FKN, HHW, HWS, SWR. **COUNTIES:** ALL, AND, BAL, BAR, BOD, BOL, BOO, BRA, BUL, CAD, CHR, CLI, CLK, CUM, EDM, FAY, FUL, GRE, HIC, HND, HOP, HRD, JEF, KEN, LAR, LES, LIN, LIV, LOG, MAD, MAR, MEA, MER, NEL, OHI, OLD, PER, SHE, TOD, TRI, TRM, UNI, WAR, WAS, WAY, WOO. **DATES:** Mar. 2–Nov. 7. **REMARKS:** May be locally common, especially in open field and roadside habitats.

Genus ***Pieris*** Schr.

4196 ***P. virginiensis*** (Edw.)
West Virginia White
PARKS AND RESERVES: BBM, BMF, CBF, KGC, OCP, PNM, RBF, RRG, YHF. **COUNTIES:** AND, BAL, BAR, BEL, BRE, BUL, CAR, CLK, EST, FAY, FLO, FUL, GUP, HLN, JEF, KNO, LAU, LAW, LEE, MAD, MCC, MCY, MEA, MEN, NEL, OWS, PER, POW, PUL, ROC, SPE, WHI, WOL, WOO. **DATES:** Mar. 17, 20, Apr. 1–May 9, 21, June 22, Aug. 3. **REMARKS:** Probably common throughout the eastern mountains, and in isolated populations westward. Our first records were from J. K. Ettman at PNM, 1973. Westernmost occurrence was discovered in April 1981. First Jefferson Co. records were at one site on March 17 and three different sites in April 1986, when numbers were unusually high. Lauer et al. (1996) provided details on some recent county records.

4197 ***P. rapae*** (L.)
Cabbage White
PARKS AND RESERVES: BEC, BLA, BLD, BMF, CAC, CBB, FAO, HHW, HWS, IRO, NBP, SWR. **COUNTIES:** ADA, ALL, AND, BAL, BAR, BAT, BEL, BOL, BOO, BRK, BUL, BUT, CAD, CAE, CAL, CAR, CHR, CLA, CLI, CLK, CRI, DAV, EDM, FAY, FLE, FRA, FUL, GAL, GAR, GRA, GRV, GRY, HAN, HEN, HIC, HLN, HND, HOP, HRD, HRT, HSN, JEF, JOH, LAR,

LAU, LAW, LET, LIN, LIV, LOG, LYO, MAD, MAR, MCC, MCY, MEA, MEN, MER, MRL, MTG, MUH, NEL, OHI, OLD, OWE, OWS, PEN, PER, POW, PUL, ROW, SCO, SHE, SIM, SPE, TOD, TRM, UNI, WAR, WAS, WAY, WOL, WOO. **DATES:** Feb. 29–Nov. 18. **REMARKS:** Common throughout the state, and often a pest of cole crops.

Genus ***Euchloe*** Hbn.

4202 ***E. olympia*** (Edw.)
Olympia Marble
PARKS AND RESERVES: BMF, JWS, MCV. **COUNTIES:** BUL, EDM, FLO, LAR, NEL, WAR. **DATES:** Apr. 12–29. Very local and uncommon.

Genus ***Anthocharis*** Bdv., Rambur and Graslin

4207 ***A. midea*** (Hbn.)
Falcate Orangetip
PARKS AND RESERVES: BMF, CAC, CBB, HWS, KGC, MCV, OCP, PCP, PNM, RRG, WVP. **COUNTIES:** ALL, BEL, BRE, BUL, CAM, CAR, CRI, EDM, EST, GRY, HLN, HND, HRD, HSN, JEF, JES, LAU, LEE, LYO, MAG, MCC, MCY, MEA, MEN, MER, MOR, NEL, OLD, OWE, OWS, POW, PUL, ROC, ROW, SPE, WAR. **DATES:** Mar. 26–May 17. **REMARKS:** Common in appropriate woodland habitat statewide.

Subfamily **Coliadinae**

Sulfurs

Genus ***Colias*** F.

4209 ***C. philodice*** Godt.
Clouded Sulphur
PARKS AND RESERVES: BBL, BBM, BEC, BLA, BMF, CAC, CBB, CBF, FAO, HHW, HWS, IRO, JCF, KGC, LCP, MCV, NBP, OCP, PYR, RRG, SGS, SWR. **COUNTIES:** ADA, ALL, AND, BAL, BAR, BAT, BEL, BOD, BOL, BOO, BRA, BRE, BRK, BUL, CAD, CAL, CAR, CAS, CHR, CLA, CLI, CLK, CRI, CRL, CUM, DAV, EDM, ELL, EST, FAY, FLE, FLO, FRA, FUL, GAL, GAR, GRA, GRE, GRY, GUP, HAN, HEN, HIC, HLN, HND, HOP, HRD, HRT, HSN, JAC, JEF, JES, JOH, KNO, KNX, LAR, LAU, LAW, LES, LET, LIN, LIV, LOG, LYO, MAD, MAG, MAR, MCC, MCL, MCY, MEA, MEN, MER, MET, MON, MOR, MRL, MTG, MUH, NEL, OHI, OLD, OWE, OWS, PEN, PER, POW, PUL, ROC, ROW, RUS, SCO, SHE, SIM, SPE, TAY, TOD, TRI, TRM, UNI, WAR, WAS, WAY, WEB, WHI, WOL, WOO. **DATES:** Mar. 17–Nov. 19, Dec. 16. **REMARKS:** Common and generally distributed throughout the state. Males often congregate in large numbers in "mud puddle clubs" at wet places. They are often seen on hot summer days where dirt roads meet paved highways.

4210 ***C. eurytheme*** Bdv.
Orange Sulphur; Alfalfa Butterfly
PARKS AND RESERVES: AJJ, BBL, BEC, BLA, BMF, CAC, CBB, FAO, GBT, HHW, HWS, IRO, JCF, KDV, KGC, LBL, LCP, MCV, NBP, OCP, PYR, RGR, SWR, WKM, YEL. **COUNTIES:** ADA, ALL, AND, BAL, BAR, BAT, BEL, BOD, BOL, BOO, BOU, BRA, BRE, BRK, BUL, BUT, CAD, CAE, CAL, CAM, CAR, CAS, CHR, CLI, CLK, CRI, CRL, CUM, DAV, EDM, ELL, EST, FAY, FLE, FLO, FRA, FUL, GAL, GAR, GRA, GRE, GRV, GRY, HAN, HEN, HIC, HLN, HND, HOP, HRD, HRT, HSN, JAC, JEF, JES, KNO, KNX, LAR, LAU, LAW, LEE, LET, LIN, LOG, LYO, MAD, MAG, MAR, MCC, MCL, MCY, MEA, MER, MET, MON, MOR, MRL, MTG, MUH, NEL, NIC, OHI, OLD, OWE, OWS, PEN, PIK, POW, PUL, ROC, ROW, RUS, SCO, SHE, SIM, SPE, TAY, TOD, TRI, TRM, UNI, WAR, WAS, WAY, WEB, WHI, WOL, WOO. **DATES:** Mar. 21–Nov. 16, Dec. 5, 29. **REMARKS:** L. D. Gibson took an aberrant female with entirely blackish wings in Bracken Co., on May 19, 1984. This is one of Kentucky's most common butterflies, and can be found in all counties. Intergrades between this and *C. philodice* occur frequently, suggesting widespread hybridization. The larva may be a pest of alfalfa crops. It is also a "mud puddler."

Genus ***Zerene*** Hbn.

4224 ***Z. cesonia*** (Stoll)
Southern Dogface
PARKS AND RESERVES: FAO, HHW, SWR, WVP. **COUNTIES:** BOO, FAY, FUL, GAL, GRV, HND, JEF, MEA, OLD, UNI. **DATES:** June 26–July 26, Aug. 17–Oct. 10. **REMARKS:** Apparently overwinters farther south and repopulates Kentucky during the summer as two broods. Generally uncommon, but sometimes frequently seen along the Ohio and Mississippi rivers in August and September.

Genus ***Phoebis*** Hbn.

4228 ***P. sennae*** (L.)
Cloudless Sulphur
PARKS AND RESERVES: BLD, BMF, FAO, HHW, KDV, LBL, NBP, PNM, SWR. **COUNTIES:** ADA, ALL, BAL, BAR, BEL, BOL, BOO, BUL, CAE, CAL, CHR, CLI, CRI, CUM, FUL, GRE, GRV, HIC, HND, HOP, HRT, JEF, KNO, LAR, LAU, LES, LIV, LOG, LYO, MAD, MCC, MCY, MEA, MRL, NEL, OHI, OWS, POW, PUL,

SPE, TOD, TRI, UNI, WAY, WHI, WOL. **DATES:** Apr. 25, 26, May 16, July 5–11, 23–27, Aug. 6, 8, 17–Nov. 15. **REMARKS:** Our subspecies is *eubule* Linnaeus. This butterfly can be very common in some summers (e.g., 1977, 1987, 1992, and 1998), especially in the western counties. It apparently does not survive harsh winters, but repopulates Kentucky each summer. Males have been seen in "mud puddle clubs" in Fulton Co., in early September.

4229 ***P. philea*** (Johansson)
Orange-barred Sulphur
One male of this subtropical species was sighted in Louisville on Apr. 30, 1981, by Alan Baron and Rosemary Baumann. Another was seen on Frankfort Avenue near the Louisville Water Co. The most recent was observed by R. A. Henderson on Sept. 5, 1997, along the Purchase Parkway in Graves Co. This strong flier has been reported straying as far north as New York, Ohio, and Nebraska, but rarely.

4231 ***P. agarithe*** (Bdv.)
Large Orange Sulphur
One male was sighted by both L. D. Gibson and R. A. Henderson along SR 94 about 1.6 km (1 mi) W of junction with SR 51, Fulton Co., Aug. 18, 1978. Henderson recorded another one 16.0 km (10 mi) W of SR 51 on SR 94 on Sept. 11, 1993. Presumably this was subspecies *maxima* (Neum.). It is a rare stray from the deep south.

Genus *Kricogonia* Reak.

4235 ***K. lyside*** (Godt.)
Lyside Sulphur
Ladd (1953) reported collecting a specimen of this butterfly in Hardin Co., Elizabethtown, in June, 1953. Its normal range limits are much farther south, but it is known to migrate long distances and in great numbers.

Genus *Eurema* Hbn.

4237 ***E. lisa*** Bdv. & Leconte
Little Yellow; Little Sulphur
PARKS AND RESERVES: BUC, FAO, FKN, HHW, HWS, IRO, KGC, LBL, MUR, NBP, OCP, PYR, RGR, SWR. **COUNTIES:** ADA, BAL, BAR, BEL, BRE, BRK, BUL, CAD, CAE, CAL, CAS, CHR, CLI, CRI, CUM, EDM, FUL, GRE, GRV, GRY, HIC, HLN, HND, HOP, HRD, HRT, JAC, JEF, KNO, LAR, LAU, LIN, LIV, LOG, LYO, MAD, MCC, MCY, MEA, MET, MRL, MUH, NEL, OHI, OLD, OWS, PER, POW, RUS, TRI, UNI, WAR, WAY, WHI. **DATES:** June 16–Oct. 30. **REMARKS:** Sometimes common, especially in the western counties and in late summer. R. A. Henderson collected a white male form "*alba*" in Meade Co., OCP, Sept. 22, 1973; and C.V. Covell, Jr., took another at Hickman, Fulton Co., Sept. 11, 1993.

4242 ***E. nicippe*** (Cram.)
Sleepy Orange
PARKS AND RESERVES: BBL, BMF, FAO, HWS, LBL, OCP, PYR, SWR. **COUNTIES:** ALL, BAL, BOO, BUL, BUT, CAD, CHR, CLI, CRI, CUM, FAY, FUL, GAL, GRE, GRV, HEN, HND, HSN, JEF, JES, LOG, MCY, MEA, MER, MET, MRL, MUH, NEL, OLD, SIM, TOD, TRI, UNI, WAY, WHI. **DATES:** Mar. 28, Apr. 15, May 21, June 1, 6, 19, July 2–11, 22–Oct. 16. **REMARKS:** Locally common, especially late in the season. Holland (1931) stated that it "fairly swarms at times in Kentucky...." He encountered "clouds of it near Jeffersonville, Indiana..." across the Ohio River from Louisville. I have never seen it that common anywhere in the United States. Ladd (1952) reported seeing what amounts to form "*flava*" in a Kentucky collection (no specific data).

4246 ***E. mexicanum*** (Bdv.)
Mexican Yellow
Ladd (1952) reported seeing a Kentucky specimen of "*Terias mexicana*" in another person's collection, but gave no details as to owner, collector, or locality of the capture. This is not an unlikely capture, as Opler and Krizek (1984) reported records of this species from Michigan and Illinois. A rare stray from the southwest.

Genus *Nathalis* Bdv.

4248 ***N. iole*** Bdv.
Dainty Sulphur
PARKS AND RESERVES: FAO, FKN. **COUNTIES:** BAL, BOO, CLI, FUL, HIC, HRD, JEF, MCC, MEA, UNI. **DATES:** July 9–Sept. 1, 11, 22, 29, Oct. 1, Nov. 2. **REMARKS:** Apparently this species does not overwinter in Kentucky, but repopulates some summers from farther south along the Mississippi and Ohio valleys. Locally common some years, but absent in others.

Family **LYCAENIDAE**

Harvester, Coppers, Hairstreaks and Blues

Subfamily **Liphyrinae**

Harvesters

Genus *Feniseca* Grt.

4249 ***F. tarquinius*** (F.)
Harvester
PARKS AND RESERVES: BBL, BMF, LBL, OCP, RRG, WVP. **COUNTIES:** ALL, BAL,

BOO, BUL, EST, FUL, GRV, HRT, JEF, KNO, KNX, LAR, LET, MAR, MCC, MEA, MEN, NEL, OWS, PUL, TRI, WAY, WHI. **DATES:** Apr. 6–Sept. 23. **REMARKS:** Uncommon, usually unexpectedly found as single individuals. Larvae eat aphids, particularly those on alder.

Subfamily **Lycaeninae**

Coppers

Genus ***Lycaena*** F.

4251 ***L. phlaeas*** (L.)
American Copper
PARKS AND RESERVES: HWS. **COUNTIES:** ADA, ALL, BAR, BOD, BOO, BUL, CAM, CAS, CLK, FAY, GRE, HLN, HND, HRD, JEF, MAS, MCC, MEA, NEL, OLD, SCO, SHE, SPE, TRI, WOO. **DATES:** Apr. 6–Oct. 28. **REMARKS:** Our subspecies is *americana* Harr. Uncommon and local.

Genus ***Hyllolycaena*** Miller & F. M. Brown

4256 ***H. hyllus*** (Cram.)
Bronze Copper
PARKS AND RESERVES: BBL, BMF, CBB, FAO, LBL, SWR. **COUNTIES:** ALL, BAL, BOO, BUL, CAM, CHR, FLE, FLO, FUL, HIC, HND, HRD, JEF, KEN, LAR, MCL, MEA, NEL, OLD, SHE, TOD, TRI, WAR. **DATES:** May 9–Oct. 15. **REMARKS:** May be common in very localized colonies in wet meadows.

Subfamily **Theclinae**

Hairstreaks

Genus ***Atlides*** Hbn.

4270 ***A. halesus*** (Cram.)
Great Purple Hairstreak
PARKS AND RESERVES: KGC, OKH. **COUNTIES:** FAY, FRA, HLN, JEF, JES, MCC, MER, NEL. **DATES:** June 7, 16, 26, July 12, 28–30, Aug. 2, 13–18, 31, Sept. 18, 19, 27, Oct. 6, Nov. 1. **REMARKS:** Rarely seen. W. R. Black, Jr., reported the only multiple sighting (14 individuals) of this species in two localities in Paducah, McCracken Co., on Sept. 18, 1997, along the Tennessee River in an area where mistletoe, the foodplant, was abundant. He found them common in Aug.–Oct., 1998, in other Paducah sites as well.

Genus ***Satyrium*** Scudder

4275a ***S. titus*** (F.)
Coral Hairstreak
PARKS AND RESERVES: BBL, BMF, HWS. **COUNTIES:** BOO, BUL, CAR, CAS, FLE, HLN, HOP, JEF, MCC, MEA, NEL, OLD, PEN, ROW, SCO, SPE, TRM, WAR, WEB. **DATES:** June 11–July 11. **REMARKS:** Our subspecies is *mopsus* (Hbn.). Locally common, especially on milkweed blossoms in old fields.

4278 ***S. acadicum*** (Edw.)
Acadian Hairstreak
Harlan Co., Cumberland, Aug. 30; Letcher Co., Jenkins, July 12; Whitley Co., SR 26, 10.5 km (6.5 mi) S of Corbin at Brummett, on *Asclepias syriaca* (common milkweed), June 29. Rare, and questionably a regular part of the Kentucky fauna.

4281 ***S. edwardsii*** (G. & R.)
Edwards' Hairstreak
PARKS AND RESERVES: BMF, GBT, HWS, PNM. **COUNTIES:** BEL, BUL, CHR, CRL, HLN, HRD, JEF, MEA, MET, NEL, OLD, PIK, SPE, TRM. **DATES:** May 17–July 27. **REMARKS:** Often common to abundant on milkweeds, daisy fleabane, and other summer flowers in fields near oaks, the larval foodplant.

4282 ***S. calanus*** (Hbn.)
Banded Hairstreak
PARKS AND RESERVES: BBM, BMF, GBT, HWS, LCP, LWO, PNM. **COUNTIES:** ALL, AND, BEL, BOD, BOO, BUL, CAR, CAS, CHR, CRL, FAY, GRA, HEN, HLN, HRD, JEF, LAR, LEE, LET, MAD, MAS, MCC, MCY, MEA, NEL, OLD, OWE, OWS, PEN, PIK, PUL, ROB, ROW, RUS, SCO, SPE, TRI, WAY, WHI, WOL. **DATES:** Apr. 12–July 24. **REMARKS:** Our subspecies is *falacer* (Godt.). Widespread throughout the state and often locally common. I found a small colony of them on the University of Louisville Belknap Campus.

4283 ***S. caryaevorum*** (McD.)
Hickory Hairstreak
Carroll Co., GBT, June 10; Fleming Co., June 29; Oldham Co., HWS, June 17; Rockcastle Co., Mt. Vernon, June 14; Rowan Co., June 29. More widespead than indicated by these records, and probably confused very often with the banded and Edwards' hairstreaks.

4285 ***S. liparops*** (LeConte)
Striped Hairstreak
PARKS AND RESERVES: AJJ, BBM, BMF, GBT, HWS. **COUNTIES:** BOO, BUL, CAM, CRL, FLE, HLN, JEF, NEL, OLD, SCO, TAY, WAY. **DATES:** June 4–July 4. **REMARKS:** Our subspecies is *strigosum* (Harr.). Usually seen one at a time, but occasionally frequent in the proper habitat.

4332 ***S. favonius*** J. E. Sm.
Northern Hairstreak
Fayette Co., Lexington, Mason Headley Rd. at Campbell House Restaurant, June 10, 11, 19, 23, C. J. Dempwolf; Jefferson Co., Louisville, taken by hand on a mayapple plant near IRO, May 31; McCracken Co., Ray Black farm, 9.7 km (6 mi) SW of Paducah, at lights, May 30; Oldham Co., near HWS, June 25. Uncommon to rare – or perhaps just difficult to find. Our subspecies is *ontario* (Edw.).

Genus *Calycopis* Scudder

4299 ***C. cecrops*** (F.)
Red-banded Hairstreak
PARKS AND RESERVES: BMF, CBB, CBF, DHW, JCF, KDV, KGC, MCV, OCP, PCP, PNM, SWR. **COUNTIES:** BEL, BUL, BUT, CLI, CRI, EDM, EST, GRY, HLN, HND, JAC, JEF, JES, LAR, LAU, LET, LIN, MAD, MCY, MEA, MOR, MRL, NEL, OLD, OWS, PIK, POW, PUL, WAY, WHI. **DATES:** Apr. 25–Sept. 11. Widespread and sometimes locally common, as it was in DBNF in 1996–97.

Genus *Callophrys* Billberg

NOTE: Some works use genus *Mitoura* Scudder for *grynea* and *Incisalia* Scudder for the four species beginning with *augustus*.

4318 ***C. grynea*** (Hbn.)
Juniper Hairstreak; Olive Hairstreak
PARKS AND RESERVES: BBL, BMF, CBB, HWS, LCP, MCV, OCP, PCP, PYR. **COUNTIES:** BEL, BOO, BRA, BUL, CAM, CHR, CLI, CRI, DAV, EDM, EST, HLN, JEF, JES, LAU, LOG, MAR, MCC, MEA, MRL, NEL, OLD, OWE, OWS, PEN, RUS, SPE, TRM. **DATES:** Apr. 11–Aug. 8, peaking in Apr. and July. **REMARKS:** Two broods. Common in red cedar stands, usually staying in medium to large trees. Sometimes found nectaring on flowers nearby.

4322 ***C. augustus*** (Kby.)
Brown Elfin
PARKS AND RESERVES: KGC, NBP. **COUNTIES:** HLN, JAC, LAU, MCY, POW, ROW, WHI. **DATES:** Apr. 7–21. **REMARKS:** Our subspecies is *croesioides* Scudder. Powell County specimen reared on *Kalmia latifolia* and emerged June 3. Local and uncommon.

4325 ***C. irus*** (Godt.)
Frosted Elfin
Bell Co., Apr. 5, 1938, 1 male in F. W. Case Collection, CMNHS; Harlan Co., Hi Lewis State Nature Preserve, Apr. 25, 1998, P.A. Florence; Mercer Co., near Shakertown, one female, Apr. 28, 1962, C. J. Dempwolf. Rare.

4326 ***C. henrici*** (G. & R.)
Henry's Elfin
PARKS AND RESERVES: AJJ, BMF, CAC, CBB, HWS, IRO, KGC, OCP, PCP, PYR, RRG. **COUNTIES:** ALL, BRA, BRE, BUL, CAM, CAR, CHR, FLO, GAL, HLN, HSN, JEF, JES, LAR, LAU, LAW, LEE, MCY, MEA, MEN, MOR, NEL, OLD, PEN, POW, ROC, SPE, TRM, WAR, WHI. **DATES:** Mar. 26–May 19. Widespread and locally common.

4328 ***C. niphon*** (Hbn.)
Eastern Pine Elfin
PARKS AND RESERVES: BMF, OCP, PCP, RRG. **COUNTIES:** BUL, JAC, JEF, LAU, MCY, MEA, MEN, NEL, POW, PUL, ROW, WHI, WOL. **DATES:** Mar. 16–June 12. Local and uncommon.

Genus *Parrhasius* Hbn.

4335 ***P. m-album*** (Bdv. & Leconte)
White M Hairstreak
PARKS AND RESERVES: MCV, OCP, PNM, PYR, RRG. **COUNTIES:** BEL, BOO, CHR, EDM, FAY, FUL, GAL, HLN, JEF, LAU, MAR, MCC, MEA, MEN, OLD, OWS, WAR, WAY, WHI. **DATES:** Apr. 4, 22–Oct. 4. **REMARKS:** Widespread but rarely seen. The only recorded multiple occurrence was in Paducah Sept. 12–13, 1998, where several were found nectaring on aster and *Eupatorium* sp.

Genus *Strymon* Hbn.

4336 ***S. melinus*** Hbn.
Gray Hairstreak
PARKS AND RESERVES: BBL, BEC, BMF, CBB, FAO, HHW, HWS, KGC, LBL, OCP, PYR, SWR, WVP. **COUNTIES:** AND, BAL, BEL, BOO, BUL, CAL, CAM, CAR, CAS, CHR, CLA, CLI, CRI, CUM, DAV, FAY, FUL, GRA, GRE, GRV, GRY, HIC, HLN, HND, HRD, HRT, JEF, JES, LAU, LIV, MAD, MAR, MAS, MCC, MCY, MEA, MEN, MER, MRL, NEL, OLD, OWE, OWS, PER, PIK, POW, PUL, SCO, SPE, TAY, TOD, TRI, UNI, WAY, WEB, WHI, WOL. **DATES:** Mar. 9–Oct. 21. **REMARKS:** Widespread and common, with the longest overall flight period among our hairstreaks.

Genus *Erora* Scudder

4349 ***E. laetus*** (Edw.)
Early Hairstreak
PARKS AND RESERVES: BBM, KGC, RRG. **COUNTIES:** GRE, HLN, LET, MEN. **DATES:**

Apr. 14, 24, May 1, 5, 6, July 2–17. **REMARKS:** Two broods. Cook (1948) recorded this species as a sighting in Green Co., near Crailhope, Apr. 1941, and he collected one specimen at that same spot on May 6, 1947. It had not been reported again until L. D. Gibson collected 1 male, in Harlan Co., BBM, July 17, 1975. Amos Showalter collected two on BBM on Apr. 24, 1976, and ten more were taken there on wild hydrangea and other flowers on July 10, 1976. In 1977, L. D. Gibson had a specimen under his net on Apr. 14, and at least 33 specimens were caught at this locality by numerous individuals from July 2 to July 10. Many were reported at that locality in early July 1978. A few have been reported there since then. J. S. Nordin was unable to capture one that perched on his net which lay on the ground in Menifee Co., RRG, May 5, 1985; Menifee Co., RRG, one female, Apr. 17, 1988, T.W. Carr. Generally considered rare, but I would rather call it "inaccessible" because of its reclusive habits and reluctance to fly up readily when approached.

Subfamily **Polyommatinae**

Blues

Genus ***Leptotes*** Scudder

4357 ***L. marina*** (Reak.)
Marine Blue
Jefferson Co., Louisville, N of SR 171, Mockingbird Valley Rd., one male, July 31, 1978, M. McInnis; Union Co., HHW, SR 56, Morganfield, Sept. 9, 1983, V.P. Lucas. A rare migrant from farther south and west.

Genus ***Everes*** Hbn.

4361 ***E. comyntas*** (Godt.)
Eastern Tailed Blue
PARKS AND RESERVES: BEC, BLA, BMF, CAC, CBB, FAO, HHW, HWS, IRO, JCF, LBL, MCV, OCP, PYR, RBF, RRG, SWR, WKM, YEL. **COUNTIES:** ADA, ALL, AND, BAL, BAR, BAT, BEL, BOD, BOL, BOO, BRE, BRK, BUL, BUT, CAD, CAE, CAL, CAM, CAR, CAS, CHR, CLI, CLK, CRI, CRL, CUM, DAV, EDM, ELL, EST, FAY, FLE, FLO, FRA, FUL, GAL, GAR, GRA, GRE, GRV, GRY, HAN, HIC, HLN, HND, HOP, HRD, HRT, HSN, JAC, JEF, JES, JOH, KNX, LAR, LAU, LAW, LEE, LES, LET, LIN, LIV, LOG, LYO, MAD, MAG, MAR, MCC, MCL, MCY, MEA, MEN, MER, MET, MON, MOR, MRL, MTG, MUH, NEL, OHI, OLD, OWE, OWS, PEN, POW, PUL, ROC, ROW, RUS, SCO, SHE, SIM, SPE, TAY, TOD, TRI, TRM, UNI, WAR, WAS, WAY, WEB, WHI, WOL, WOO. **DATES:** Mar. 24–Oct. 19. **REMARKS:** Common throughout the state in yards, gardens, fields, and other open habitats.

Genus ***Celastrina*** Tutt

4363 ***C. ladon*** (Cram.)
Spring Azure
PARKS AND RESERVES: BBL, BMF, CAC, CBB, CBF, FKN, IRO, MCV, NBP, OCP, RRG. **COUNTIES:** BOO, BUL, CAR, EDM, FAY, GRE, HLN, JEF, JOH, LAR, LAU, LAW, MCY, MEA, OLD, POW, SHE, TRI, WAR, WHI, WOL. **DATES:** Mar 5–May 16. **REMARKS:** Common throughout the state. This species is the earliest azure and has only one generation per year. It is followed by the Appalachian azure, then the summer azure, which has more than one generation and can be separated from the spring azure by wing scale and color differences (David M. Wright, *pers. comm.*).

4363.1 ***C. neglecta*** (Edw.)
Summer Azure
PARKS AND RESERVES: BBL, BBM, BMF, CBB, HWS, NBP, PYR. **COUNTIES:** BEL, BOO, BOY, BUL, BUT, CAL, CHR, CLA, FAY, FUL, HAR, HIC, HLN, JEF, LAU, LET, MAD, MEA, MCY, NEL, OLD, POW, RUS, SCO, SPE, TOD, TRI, WHI, WOO. **DATES:** May 26–Nov. 9. **REMARKS:** This is the multi-brooded azure which is common throughout the state after the other three single-brooded species have ceased to fly. All published records for *C. ladon* prior to 1998 and after mid–May, should be considered *C. neglecta*.

4363.2 ***C. neglectamajor*** (Tutt)
Appalachian Azure
PARKS AND RESERVES: JWS, RRG. **COUNTIES:** FLO, GRV, JAC, JEF, MCY, MEN, OWS, POW. **DATES:** Apr. 1, May 6, 7, June 2. **REMARKS:** This butterfly has just recently been elevated to species level, and is generally larger and paler blue than the spring and summer azures. According to Opler and Krizek (1984) it flies "from mid-May to early June." Uncommon.

4364 ***C. nigra*** (Forbes)
Dusky Azure
PARKS AND RESERVES: KGC, RBF, RRG, SGS. **COUNTIES:** BRE, BUL, FLO, HLN, MCY, MEN, OWS, POW. **DATES:** Apr. 1, 16, 27. Also a single specimen was taken and another sighted by R. A. Henderson at KGC, Harlan Co., Aug. 10, 1980, suggesting a partial second brood. **REMARKS:** This species, long confused with the spring azure, was described in 1972, and first discovered in Kentucky by Amos Showalter on Apr. 26, 1975, near Elkatawa, Breathitt Co. It was

later found common in RRG, Powell and Menifee counties, and the life cycle was described on the basis of material from there (Wagner and Mellichamp 1978). This butterfly has in the past been reported under the name *Celastrina ebenina* Clench, but I am convinced now that *nigra* is the correct choice.

Genus ***Glaucopsyche*** Scudder

4372 ***G. lygdamus*** (Doubleday)
Silvery Blue
PARKS AND RESERVES: CAC, NBP, RBF, RRG. **COUNTIES:** BRA, BRE, BUL, CAR, HLN, KNO, LAU, LAW, LES, LET, MAG, MCY, MEN, OWS, POW, PUL, ROW, WOL. **DATES:** Mar. 21, 22, Apr. 9–30, May 4–8, 18–22. **REMARKS:** Very local. May be common in small colonies where the foodplant, Carolina vetch, is well established and flowering.

Family **RIODINIDAE**

Metalmarks

Genus ***Calephelis*** G. & R.

4387 ***C. borealis*** (G. & R.)
Northern Metalmark
PARKS AND RESERVES: AJJ, BBL, HWS, KIN, RRG. **COUNTIES:** BOO, CAM, JEF, LES, MEN, OLD, OWS, PEN, TRM. **DATES:** June 17–July 11, 27. **REMARKS:** Uncommon. Only a few, very small local populations have been discovered thus far. Members of a colony in Oldham Co., were found nectaring on flowers of orange milkweed on July 8, 1995 by the author. One brood per year.

4391 ***C. mutica*** McAlpine
Swamp Metalmark
Christian Co., PYR, June 27, Aug. 24, 27, Sept. 11; Meade Co., OCP, near Camp Piomingo, one slightly worn male, Sept. 11, 1976, L. D. Gibson, and Battletown, Sept. 1996, J.A. Bess; Nelson Co., uncommon. Two broods per year.

Family **NYMPHALIDAE**

Brushfoots

Subfamily **Libytheinae**

Snouts

Genus ***Libytheana*** Michener

4411 ***L. carinenta*** (Cramer)
American Snout
PARKS AND RESERVES: BBL, BBM, BLA, BMF, CBB, COB, HWS, LBL, OCP, RGR, SWR, WVP. **COUNTIES:** BAL, BEL, BOO, BRA, BUL, CHR, CLI, CLK, CRI, EDM, FAY, FRA, FUL, GRE, GRV, GRY, HIC, HLN, HND, HOP, JEF, KNX, LAU, LIV, LOG, LYO, MAD, MAR, MCC, MCY, MEA, MEN, MON, MOR, MUH, NEL, OHI, OLD, PEN, PUL, SHE, TOD, TRI, TRM, UNI, WAR, WAS, WAY, WHI, WOO. **DATES:** Apr. 3–Oct. 13, Nov. 1. **REMARKS:** Our subspecies is *bachmanii* (Kirtland). Formerly known simply as *L. bachmanii* (Kirtland) (Hodges et al. 1983). Sometimes common, particularly in late summer and early fall in western Kentucky.

Subfamily **Heliconiinae**

Longwings

Genus ***Agraulis*** Bdv. & Leconte

4413 ***A. vanillae*** (L.)
Gulf Fritillary
PARKS AND RESERVES: LBL, LMP. **COUNTIES:** BOL, CAD, CAM, CHR, FAY, FLE, FUL, JEF, MCC, MUH, NEL, OLD, PIK, SCO, TRI, WAR, WHI. **DATES:** June 4–18, 28–Nov. 14, 26. **REMARKS:** A regular stray and sometimes summer breeder from farther south. R. A. Henderson collected a pupa at Crofton, Christian Co., Sept. 2, 1989, and an adult emerged Sept. 12. W. R. Black, Jr., (*pers. comm.*) counted 8 individuals nectaring on morning glory along the Tennessee River in Paducah, McCracken Co., Sept. 18–21, 1989. He, several Boy Scouts from Troop 1, and I visited the site on Oct. 7, and 6 adults were collected plus about 20 larvae and 2 pupae on *Passiflora*. Adults emerged from resulting pupae on Oct. 23–25, 1989, in Louisville. Black reported adults at the site as late as Nov. 6. One male from this group of larvae emerged on Oct. 25 and lived until Nov. 14 on sugar water. Black reported two sightings of adults in the same general area Sept. 9 and Oct 11, 1990, and in 1997, but no immature stages. He found no adults or immature stages in 1991 or 1992. In 1994, they were back, last seen flying on Dec. 7. Breeding was again observed in Paducah in 1995 and 1998, and in Scott County in 1998 (Sean Lube).

Genus ***Euptoieta*** Doubleday

4447 ***E. claudia*** (Cram.)
Variegated Fritillary
PARKS AND RESERVES: COB, FAO, FKN, HWS, PYR, SGS, SWR, WVP. **COUNTIES:** ADA, ALL, BAL, BAR, BOL, BUL, CAL, CHR, CLI, CLK, CRI, CUM, DAV, FAY, FLE, FUL, GRE, GRV, HAN, HIC, HLN, HND, HOP, HRD, HRT, JEF, LAR, LAU, LIV, MAD, MCC, MCY, MEA, MET, MRL, MUH, OHI, OLD, RUS, SIM, SPE, TAY, TOD, TRI, UNI, WAR, WAY, WHI,

WOO. **DATES:** Apr. 3–Nov. 9. **REMARKS:** Locally common, especially in open areas such as alfalfa and clover fields.

Genus *Speyeria* Scudder

4449 ***S. diana*** (Cram.)
Diana; Diana Fritillary
PARKS AND RESERVES: BBM, KGC, NBP, PNM. **COUNTIES:** BAL, BEL, CAR, HLN, JAC, KNX, LAU, LES, LET, LIN, MAG, MEN, PER, PIK, POW, PUL, WAY, WHI, WOL. **DATES:** June 12–Oct. 12. **REMARKS:** The one Ballard County record (East Cairo) was cited by Warner (1895), but all recent records are from eastern counties. May be locally common some years, uncommon in others. Males appear first, with females following a week or more later, and flying later in the season. A worn male and nearly fresh females were reported by D. Flaim on Sept. 2, 1985 at KGC; on Oct. 12 a female was found by Jeremy Fraser in the DBNF in a power transmission corridor. Because of its beauty and association with Appalachia, the male of this butterfly was chosen as the logo of the Society of Kentucky Lepidopterists, and both sexes are shown on the cover of this book.

4450 ***S. cybele*** (F.)
Great Spangled Fritillary
PARKS AND RESERVES: BBL, BBM, BEC, BMF, CBB, HHW, HLN, HWS, JCF, KGC, LBL, LCP, OCP, PYR, RRG, SGS, SWR. **COUNTIES:** ALL, BAT, BEL, BOD, BOL, BOO, BRA, BUL, BUT, CAD, CAL, CAR, CAS, CHR, CLI, CRL, CUM, ELL, EST, FAY, FLE, FUL, GRA, GRE, GRY, GUP, HLN, HND, HOP, HRD, JAC, JEF, KNX, LAR, LAU, LEE, LES, LET, LIV, LOG, MAD, MAR, MAS, MCY, MEA, MEN, MET, MON, MUH, NEL, OHI, OLD, OWE, OWS, PEN, PER, PIK, POW, PUL, ROB, ROC, ROW, RUS, SHE, SIM, SPE, TOD, TRI, TRM, UNI, WAY, WEB, WHI, WOL. **DATES:** May 4–Oct. 11. **REMARKS:** Common throughout the state.

4451 ***S. aphrodite*** (F.)
Aphrodite Fritillary
PARKS AND RESERVES: BBM, KGC, RRG. **COUNTIES:** BEL, CAR, FLE, GUP, HLN, LET, MEN, NEL, OWS, POW, ROW, WOL. **DATES:** May 23, June 5, 22, 24, July 3–12, 30, Aug. 26. **REMARKS:** Sometimes locally common in eastern counties.

4452 ***S. idalia*** (Drury)
Regal Fritillary
Clark Co., Winchester, worn male, July 15, 1967; Jefferson Co., Louisville, July 25, 1954; Knox Co., Scalf, Beech Creek Hollow, July 1, 2, 4, 1971; Laurel Co., LVJ; Nelson Co., Simpson Lake, July 1, 1973; Whitley Co., Johnnie Hollow, June, 1965. Apparently uncommon to rare, though possibly overlooked. Some unrecorded specimens have appeared in 4-H collections at the Kentucky State Fair in past years, but I haven't seen any since 1980. The Kentucky population is greatly reduced. There have been two unconfirmed sightings since 1994 in Kentucky, so it may still be present in the Commonwealth. Listed as historic (not verified in 20 or more years) in Kentucky (KSNPC 1996).

Genus *Boloria* Moore

4464 ***B. selene*** (D. & S.)
Silver-bordered Fritillary
One male was found in the late Siegfried Scholz Collection labelled "Valley Station, Jefferson Co., Ky., 9-12-1978." Since there is a railroad near this area, the record could be valid as a stray from farther north. Also, Scholz was not known to have visited a more likely habitat for this species. A very rare stray, if a valid record. Our subspecies is *myrina* (Cram.).

4465 ***B. bellona*** (F.)
Meadow Fritillary
PARKS AND RESERVES: BEC, BLA, BMF, CBF, FAO, HWS, KGC, RRG. **COUNTIES:** ALL, BAT, BEL, BOD, BOL, BOO, BOU, BRA, BUL, CAM, CAR, CLK, FAY, FLE, FRA, FUL, HLN, HRD, HSN, JEF, JES, LAR, LIN, MAD, MAR, MAS, MEN, MTG, OLD, OWE, PEN, ROB, ROW, SCO, SHE, SPE, TRM, WHI, WOL, WOO. **DATES:** Mar. 3, Apr. 13, 30–Nov. 16. **REMARKS:** Common. The occurrence of this species in Kentucky is very interesting, since it was unknown until L. C. Koehn's capture of it at CBF on May 29, 1970. A Boyd County record (Apr. 26, 1971) followed, but the species was taken in a number of counties from Jefferson eastward in 1973 (in Oct., 1973, I collected the first Jefferson Co. record by swatting a tired individual with a golf club on the Crescent Hill Golf Course). Since that time *B. bellona* has been common in the eastern half of the state, and is often seen in Louisville residential areas. It has spread to the south and west. A sight record by Roger Zebold along SR 94, Willingham Bottoms, 4.0 km (2.5 mi) E of Cayce, Fulton Co., Sept 12, 1993, indicates its presence all across the state. I believe it colonized Kentucky and spread westward via the interstate highway system, utilizing park-like medians and roadsides.

Subfamily **Nymphalinae**

Brushfoots

Genus ***Polygonia*** Hbn.

4420 ***P. interrogationis*** (F.)
Question Mark
PARKS AND RESERVES: BBL, BEC, BLD, BMF, CBB, FAO, HWS, JCF, KGC, LBL, LCP, NBP, RRG, SGS, SWR. **COUNTIES:** ALL, BAL, BAR, BAT, BOL, BOO, BUL, CAE, CAL, CHR, CLI, CLK, CRI, CRL, CUM, EDM, EST, FAY, FLE, FRA, FUL, GAL, GRA, GRE, GRV, GRY, HIC, HLN, HND, HOP, HRD, HRT, HSN, JEF, JES, KEN, LAR, LAU, LET, LIN, LOG, MAD, MAR, MCY, MEA, MEN, MRL, MTG, MUH, NEL, OHI, OLD, OWE, OWS, PEN, POW, PUL, ROW, RUS, SCO, SHE, SIM, SPE, TOD, TRI, TRM, UNI, WAR, WAY, WEB, WHI, WOL, WOO. **DATES:** Mar. 20–Nov. 20. **REMARKS:** Common throughout the state.

4421 ***P. comma*** (Harr.)
Eastern Comma
PARKS AND RESERVES: BBL, BEC, BLD, BMF, CAC, CBB, COB, FAO, GBT, HHW, HWS, IRO, JCF, KGC, OCP, RRG. **COUNTIES:** ALL, BAL, BOL, BOO, BRA, BUL, CAE, CAL, CAR, CHR, CLI, CRI, CRL, CUM, EST, FAY, FLE, FRA, FUL, GAL, GRA, GRE, HIC, HLN, HND, HRD, HRT, JAC, JEF, JES, KEN, LAU, LAW, LOG, MAD, MAG, MAR, MCC, MCY, MEA, MEN, MER, MON, NEL, OHI, OLD, OWS, PEN, PER, POW, ROW, SCO, SPE, TOD, TRM, UNI, WAY. **DATES:** Mar. 1, 20–Nov. 9. **REMARKS:** Common throughout the state.

4423 ***P. faunus*** (Edw.)
Green Comma
Bell Co., "Cumberland Gap, Ky.," 3 males in CMNHS, July 26, 1937, A. H. Clark; Knox Co., near Scalf, Beech Creek Hollow, Aug. 15; Meade Co., OCP, Sept., 20, 21, Oct. 2; Whitley Co., Butcher Hollow, near Woodbine, Aug. 4. Very local and apparently rare. Our subspecies is *smythi* A. H. Clark.

4429 ***P. progne*** (Cram.)
Gray Comma
Harlan Co., July 30; Knox Co., Beech Creek Hollow, near Scalf, Aug. 15; Martin Co., Inez (sight record), Aug. 23; Whitley Co., Johnnie Hollow, near Rockholds, Sept. 3. Rare.

Genus ***Nymphalis*** Kluk

4430 ***N. vau-album*** (D. & S.)
Compton Tortoiseshell
Meade Co., fall, 1939, the late David A. Young, collector. Possibly a mislabeled specimen from Michigan, but Dr. Young (*pers. comm.*) attested to the authenticity of this record. A rare stray from northward.

4432 ***N. antiopa*** (L.)
Mourning Cloak
PARKS AND RESERVES: BBM, BMF, CAC, CBB, FAO, HWS, OCP, PNM, RNR. **COUNTIES:** BEL, BUL, CAR, FUL, HLN, JEF, KEN, LAW, MAD, MCY, MEA, OLD, POW, PUL, SCO, WAR, WHI. **DATES:** Feb. 19, Mar. 3–May 5, 15–31, June 20–24, July 14, Aug. 7, 12, Sept. 2–16, 29, 30, Oct. 1–18. **REMARKS:** Widespread but uncommon.

4433 ***N. milberti*** (Godt.)
Milbert's Tortoiseshell
Campbell Co., Ft. Thomas, 2 specimens in a local collection reported by B. L. Monroe, Jr., in the 1940's; Harlan Co., BBM, July 30, Aug. 26; Meade Co., OCP, Sept. 21; Nelson Co., Leslie Farm, SR 62, Boston, Apr. 16, and May 24. A rare stray from farther north.

Genus ***Vanessa*** F.

4434 ***V. virginiensis*** (Drury)
American Lady; American Painted Lady
PARKS AND RESERVES: BBL, BLA, BMF, CAC, DHW, HHW, HWS, KGC, LBL, LCP, SWR, WKM. **COUNTIES:** ALL, BAL, BAT, BEL, BOO, BUL, BUT, CAD, CAE, CAM, CAR, CLA, CLI, CRI, CUM, FAY, FLO, FUL, GAL, GRE, GRV, HAN, HEN, HIC, HLN, HND, HOP, HRD, HRT, JAC, JEF, KNO, LAR, LAU, LEE, LET, LOG, MAD, MCC, MCL, MCY, MEA, MEN, MON, NEL, OHI, OLD, OWE, OWS, PER, POW, PUL, ROW, RUS, SHE, SIM, TAY, TOD, TRI, UNI, WAR, WAY, WHI. **DATES:** Mar. 19–Nov. 9. **REMARKS:** Common throughout the state.

4435 ***V. cardui*** (L.)
Painted Lady
PARKS AND RESERVES: BBL, BEC, BMF, CBB, FAO, HHW, HWS, IRO, LBL, OCP, SWR. **COUNTIES:** BAL, BOO, BUL, CAL, CLA, CLI, CLK, CRI, DAV, FAY, FUL, GAR, GRA, GRE, GRV, GUP, HLN, HND, HRD, HRT, HSN, JAC, JEF, LET, MAD, MAR, MCC, MEA, MEN, MTG, MUH, NEL, OLD, OWS, POW, SHE, SIM, SPE, TOD, TRI, UNI, WAR, WAS, WOL. **DATES:** Mar. 19, Apr. 6–Nov. 9. **REMARKS:** Throughout Kentucky the abundance of this species is highly variable; common one year, but nearly absent another.

4437 ***V. atalanta*** (L.)
Red Admiral
PARKS AND RESERVES: BEC, BLA, BLD, BMF, CBB, FAO, HHW, HWS, IRO, KGC, LBL, MUR, OCP, PYR, SWR. **COUNTIES:** ALL, BAL, BAT, BOL, BOO, BRA, BRK, BUL, CAD, CAE, CAL, CHR, CLA, CLI, CLK, CRI, CUM, DAV, EDM, EST, FLE, FUL, GAL, GRV, GUP, HAN, HIC, HLN, HND, HOP, HRD, HRT, JEF, JES, KNO, LAR, LES, LET, LOG, MAD, MCC, MCY, MEA, MEN, MET, NEL, OHI, OLD, OWS, PEN, PER, PUL, ROC, ROW, RUS, SIM, SPE, TOD, TRI, TRM, UNI, WAR. **DATES:** Mar. 19–Oct. 10. **REMARKS:** Common throughout the state. Unusually abundant in May, 1990. Among the first hibernators to appear in late winter and early spring.

Genus *Junonia* Hbn.

4440 ***J. coenia*** (Hbn.)
Common Buckeye
PARKS AND RESERVES: BLD, BMF, FAO, HHW, HWS, LBL, LCP, PYR, SWR, YEL. **COUNTIES:** AND, BAL, BAR, BEL, BOL, BOO, BRK, BUL, BUT, CAD, CAE, CAL, CAS, CHR, CLI, CRI, CRL, CUM, FAY, FRA, FUL, GRE, GRV, GRY, HIC, HLN, HND, HOP, HRD, HRT, JEF, KNX, LAR, LAU, LIV, LOG, LYO, MAD, MCC, MCL, MCY, MEA, MER, MET, MRL, MUH, NEL, OHI, OLD, OWS, PUL, RUS, SCO, SPE, TOD, TRI, TRM, UNI, WAR, WAY, WEB, WHI, WOO. **DATES:** June 11–Nov. 25, Dec. 1. **REMARKS:** Locally common, especially late in the season in western counties.

Genus *Anartia* Hbn.

4443 ***A. jatrophae*** (Johansson)
White Peacock
One male with right wing only slightly frayed taken by James Harrod of Jeffersontown, Ky., in the Hurstbourne area of Jefferson Co., probably in Aug. 1969. A very rare stray from the Gulf states. The North American subspecies is *guantanamo* Mun.

Genus *Phyciodes* Hbn.

4481 ***P. tharos*** (Drury)
Pearl Crescent
PARKS AND RESERVES: BBL, BBM, BEC, BLA, BMF, CBB, COB, DAV, FAO, HHW, HWS, IRO, JCF, KGC, LBL, NBP, OCP, PYR, RGR, RRG, SWR, YEL. **COUNTIES:** ADA, ALL, AND, BAL, BAR, BAT, BOL, BOO, BOY, BRA, BRE, BRK, BUL, BUT, CAD, CAE, CAL, CAR, CAS, CHR, CLA, CLI, CLK, CRI, CRL, CUM, DAV, EDM, EST, FAY, FLE, FLO, FRA, FUL, GAL, GRA, GRE, GRV, GRY, HEN, HIC, HLN, HND, HOP, HRD, HRT, HSN, JAC, JEF, JES, JOH, KNO, KNX, LAR, LAU, LAW, LEE, LES, LET, LIN, LIV, LOG, LYO, MAG, MAR, MCC, MCL, MCY, MEA, MEN, MER, MET, MOR, MRL, MTG, MUH, NEL, OHI, OLD, OWE, OWS, PEN, POW, PUL, ROC, ROW, RUS, SCO, SHE, SPE, TAY, TOD, TRI, TRM, UNI, WAR, WAS, WAY, WHI, WOL. **DATES:** Mar. 26–Nov. 15. **REMARKS:** Common to abundant throughout the state. I observed a "puddle club" aggregation of males at a spring seepage in Fulton County in early September which included over 100 individuals.

4482 ***P. batesii*** (Reak.)
Tawny Crescent
Harlan Co., Cumberland, 1 female, May 10, 1977, Gregory Florence. Rare. Listed as a threatened species in Kentucky (KSNPC 1996).

Genus *Chlosyne* Btlr.

4489 ***C. gorgone*** (Hbn.)
Gorgone Checkerspot
Edmonson Co., MCV, May 2, 1874, F.G. Sanborn (specimen in MCZ); Jefferson Co., June 29, 1946, D. K. Weniger (Weniger 1946) (since Weniger misidentified *Speyeria cybele* as *S. aphrodite*, this record could be in error); Whitley Co., US 25, 18.5 km (11.5 mi) S of Corbin, May 30, 1973, L.C. Koehn, who indicated this site was destroyed by developers in 1975. Very rare, if still present in Kentucky.

4490 ***C. nycteis*** (Doubleday)
Silvery Checkerspot
PARKS AND RESERVES: BBL, BEC, BMF, CBB, FAO, HWS, KGC, LBL, MUR, PNM, PYR, RGR, TIO. **COUNTIES:** ALL, AND, BEL, BOD, BOO, BOU, BRE, BUL, CHR, CLK, CRI, FAY, FUL, GAR, GRA, GRE, GRY, HIC, HLN, HND, HRD, HRT, HSN, JAC, JEF, JES, JOH, LAR, LAU, LES, LET, LIV, LOG, LYO, MAD, MAR, MCC, MCY, MEA, MRL, NEL, NIC, OLD, OWE, OWS, PEN, PER, PIK, ROW, SCO, SIM, SPE, TOD, TRI, TRM, WAS, WHI. **DATES:** Apr. 23–Oct. 1. **REMARKS:** Two broods. Common.

Genus *Euphydryas* Scudder

4516 ***E. phaeton*** (Drury)
Baltimore Checkerspot
PARKS AND RESERVES: BBM, NBP, OCP, PNM, WVP. **COUNTIES:** BEL, CLA, CLI, GRY, HLN, JEF, KNX, LAR, MEA, MEN, NEL, OWS, PER, POW, PUL, ROW, TRI, WHI. **DATES:** May 17–July 23. **REMARKS:** Local where

colonies of turtlehead, its foodplant, occur. Generally uncommon.

Subfamily **Limenitidinae**

Genus ***Limenitis*** F.

4522 ***L. arthemis arthemis*** (Drury)
White Admiral
This northern subspecies is represented by one specimen in the ULK Collection collected by the late D. A. Young in Meade Co., during Aug. of the period 1936–1939. It may be a mislabeled specimen from farther north, but Young (*pers. comm.*) attested to its authenticity.

4522 ***L. arthemis*** (Drury)
Red-spotted Purple
PARKS AND RESERVES: BBL, BEC, BLD, BMF, CBB, FAO, HHW, HWS, IRO, JCF, KGC, LBL, OCP, PNM, PYR, RBF, RGR, SWR, TIO. **COUNTIES:** ALL, BAL, BEL, BOD, BOL, BOO, BRA, BRE, BUL, BUT, CAD, CAE, CAL, CAS, CHR, CLI, CRI, CRL, CUM, EST, FAY, FLE, FRA, FUL, GAL, GRE, GRV, GRY, HIC, HLN, HND, HOP, HRD, HRT, HSN, JAC, JEF, KNO, LAR, LAU, LES, LET, LIV, LOG, MAD, MAR, MCC, MCY, MEA, MEN, MRL, MUH, NEL, OHI, OLD, OWE, OWS, PEN, POW, PUL, ROW, RUS, SIM, SPE, TOD, TRI, TRM, UNI, WAR, WAY, WEB, WHI, WOL. **DATES:** Apr. 30–Oct. 19. **REMARKS:** Common throughout the state. Our subspecies is *astyanax* (F.). See reference to hybridization with *L. archippus* in next account.

4523 ***L. archippus*** (Cram.)
Viceroy
PARKS AND RESERVES: BBL, BLD, BMF, CBB, FAO, HHW, HWS, KDV, LBL, NBP, PYR, SGS, SWR. **COUNTIES:** ALL, AND, BAL, BAT, BOO, BRA, BUL, BUT, CAD, CAE, CAL, CAR, CHR, CLI, CRI, CRL, FAY, FUL, GRE, GRV, HIC, HND, HOP, HRD, HSN, JAC, JEF, KNO, LAR, LES, LIV, LOG, LYO, MAD, MCC, MCL, MEA, MRL, MUH, NEL, OHI, OLD, OWE, OWS, POW, ROW, SHE, SPE, TOD, TRI, TRM, UNI, WAR, WAS, WEB, WHI, WOL. **DATES:** Apr. 27–Oct. 19. **REMARKS:** Common throughout the state. In 1990 the Kentucky General Assembly named the viceroy as Kentucky's State Insect on a bill sponsored by the Honorable John "Eck" Rose and initiated by Mrs. Lillian Pace of Bowling Green. Monroe (1953) reported a hybrid male *L. archippus* x *L. arthemis astyanax* (*"rubidus"*) taken in Jefferson Co., Sept., 1948, and a second was taken by S. Scholz at Valley Station, Jefferson Co., Oct. 21, 1978. I collected another on the bank of the Mississippi River at Hickman, Fulton Co., Sept. 14, 1980. On Oct. 7, 1989, I also discovered a female *L. archippus* mating with a male *L. arthemis astyanax* only a few hundred yards from the site of the 1980 hybrid capture (Covell 1994). The same phenomenon was observed by Taylor Paris and others at the same spot on Sept. 6, 1997. A fourth hybrid was reported by B. Palmer-Ball, Jr., west of London, Laurel Co., June 28, 1991 (sight record), and R. V. Gregg and N. Zufelt took one each in Fulton Co. (1993 and 1995). W. R. Black, Jr., took the most recent at Massac Creek Bottoms, Paducah, on Sept. 13, 1998 (Covell 1999).

Subfamily **Charaxinae**

Leaf Butterflies

Genus ***Anaea*** Hbn.

4554 ***A. andria*** Scudder
Goatweed Leafwing
PARKS AND RESERVES: BEC, BMF, CBB, COB, FAO, FKN, HWS, IRO, LBL, OCP, PCP, PYR. **COUNTIES:** ALL, BOL, BOO, BUL, CAD, CAE, CHR, CLI, DAV, FAY, FUL, GRE, HIC, HLN, HRD, JEF, JES, KEN, LAR, LEW, LOG, MAD, MAR, MCC, MCY, MEA, MER, NEL, OLD, POW, PUL, ROC, SIM, TOD, TRI, TRM, WAR, WHI. **DATES:** Mar. 3–Nov. 15. **REMARKS:** Uncommon. Rare in the mountains. Overwinters as an adult.

Subfamily **Apaturinae**

Hackberry Butterflies

Genus ***Asterocampa*** Rober

4557 ***A. celtis*** (Bdv. & LeConte)
Hackberry Emperor
PARKS AND RESERVES: BBL, BEC, BLD, BMF, CBB, HHW, JCF, LBL, LCP, OCP, RGR, SWR, WVP. **COUNTIES:** ALL, BAL, BAT, BOO, BRA, BRK, BUL, CAD, CAE, CAM, CAR, CHR, CLK, CRI, CRL, EDM, FAY, FLE, FUL, GRA, GRE, GRV, GRY, HIC, HLN, HND, HOP, HRD, HRT, HSN, JEF, JES, LAR, LIV, LOG, MAR, MCC, MCY, MEA, MEN, MER, MON, MRL, MUH, NEL, OHI, OLD, OWE, PEN, POW, ROW, RUS, SCO, SPE, TOD, TRI, TRM, UNI, WEB, WOO. **DATES:** May 10–Nov. 8. Common.

4562.1 ***A. clyton*** (Bdv. & LeConte)
Tawny Emperor
PARKS AND RESERVES: BBL, BEC, BLA, BMF, CBB, HWS, LCP, SGS, SWR, YEL. **COUNTIES:** AND, BAL, BOO, BRA, BRK, BUL, BUT, CAD, CAE, CHR, CLK, CRI, CRL, CUM, EST, FAY, FLE, FRA, FUL, GRE, HIC, HLN, HND, HOP, HRD, JEF, JES, LAR, LAU, LOG, LYO, MAD, MAR, MCY, MEA, MET,

MRL, MUH, NEL, OLD, PEN, POW, RUS, SPE, TRM, UNI, WOL, WOO. **DATES:** June 13–Nov. 2. **REMARKS:** Not as common as *A. celtis*.

Subfamily **Satyrinae**

Satyrs and Woodnymphs

Genus ***Enodia*** Hbn.

4568 ***E. portlandia*** (F.)
Southern Pearly-eye
Christian Co., PYR, Sept. 5; Fulton Co., SR 94, Willingham Bottoms, 4.0 km (2.5 mi) E of Cayce, May 27, Aug. 22, 27, Sept. 4–16; Graves Co., Sept. 11; McCracken Co., Paducah, Sept. 6, 14; Union Co., HHW, Sept. 9. This warm, brown subspecies with washed out, golden brown spots on the hindwing underside is sometimes common in the swamps of Fulton and nearby counties in stands of its foodplant, *Arundinaria gigantea*. At Willingham Bottoms, along Little Bayou de Chien, this species and the following three sometimes occur together from late August into early September. Our subspecies is *missarkae* Heitzman & dos Passos.

4568.1 ***E. anthedon*** Clark
Northern Pearly-eye
PARKS AND RESERVES: BBL, BEC, BLD, BMF, CBB, HWS, KGC, LBL, LCP, OCP, PYR, RRG, SWR, WVP. **COUNTIES:** BAL, BOO, BRA, BUL, BUT, CAR, CHR, EST, FUL, GRA, GRE, GRV, GUP, HIC, HLN, HND, HOP, HRT, HSN, JEF, JES, KEN, LAR, LES, LOG, MAD, MCC, MCY, MEA, MEN, MER, MUH, NEL, OHI, OLD, OWE, OWS, PEN, POW, ROW, RUS, TRI, TRM, WAY, WHI. **DATES:** May 10, 31, June 2, 16, July 16, Aug. 10–18, Sept. 4, 13, 19. **REMARKS:** Widespread and sometimes locally common to abundant.

4568.2 ***E. creola*** (Skin.)
Creole Pearly-eye
PARKS AND RESERVES: NBP, RRG. **COUNTIES:** BEL, FUL, GRV, LOG, MEN, NEL, POW, SPE. **DATES:** May 31, June 2, Aug. 10–18, 27, Sept. 6–13. **REMARKS:** Sometimes locally common, but few colonies have been found so far. The eastern records in the mountains are quite interesting as this species is typically found in swampy, lowland habitats.

Genus ***Satyrodes*** Scudder

4569 ***S. appalachia*** (R.L. Chermock)
Appalachian Brown
PARKS AND RESERVES: MUR, SWR. **COUNTIES:** BUL, FUL, GRV, HIC, HND, JEF, MCY, WHI. **DATES:** May 27–29, June 17–26, Aug. 8–22, Sept. 6–18. **REMARKS:** Local, but sometimes common to abundant where found.

Genus ***Cyllopsis*** Felder

4573 ***C. gemma*** (Hbn.)
Gemmed Satyr
PARKS AND RESERVES: BBM, BMF, CBF, DHW, JWS, KGC, MCV, MUR, NBP, PCP, PNM, PYR, RBF, RRG. **COUNTIES:** BEL, BRE, BUL, BUT, CAD, CHR, CLI, CRI, CUM, EDM, FLO, FUL, GRE, HIC, HLN, JAC, JEF, JES, LAR, LAU, LES, LET, LIV, MAD, MAG, MAR, MCY, MEA, MEN, NEL, OWS, POW, PUL, ROW, WAY, WHI, WOL. **DATES:** Apr. 13–Oct. 9. **REMARKS:** Colonies seem widespread, but are very local. Gemmed satyrs may be moderately common when found.

Genus ***Hermeuptychia*** Forster

4575 ***H. sosybius*** (F.)
Carolina Satyr
PARKS AND RESERVES: CBF, DHW, MUR, PNM, PYR, RBF, RRG. **COUNTIES:** BAL, BEL, BRE, CHR, CLI, CUM, EDM, EST, FUL, GRE, GUP, HIC, HLN, LAU, LES, LEW, LOG, MAD, MCC, MCY, MEN, MUH, OWS, POW, PUL, ROW, TOD, WAR, WAY, WHI, WOL. **DATES:** Apr. 23–Aug. 24, Sept. 5, 6, 15, 29, Oct. 9. **REMARKS:** Local and sometimes common where found.

Genus ***Megisto*** Hbn.

4578 ***M. cymela*** (Cram.)
Little Wood Satyr
PARKS AND RESERVES: BBL, BLD, BMF, CBB, HWS, LBL, NBP, OCP, RRG. **COUNTIES:** ALL, BAL, BEL, BOO, BRK, BUL, BUT, CAE, CAR, CAS, DAV, EDM, FLE, GRA, GRE, GRY, HLN, HND, HOP, HRD, HRT, JAC, JEF, JES, KNX, LAU, LAR, LET, MAR, MCC, MCY, MEA, MEN, MER, MET, MUH, NEL, OHI, OLD, OWE, OWS, PEN, PER, POW, PUL, ROW, RUS, SCO, SIM, SPE, TRI, TRM, WAR, WAY, WEB, WHI. **DATES:** May 11–Aug. 23. **REMARKS:** Local and occasionally very common.

Genus ***Cercyonis*** Scudder

4587 ***C. pegala*** (F.)
Common Wood Nymph
PARKS AND RESERVES: AJJ, BBL, BBM, BLA, HWS, KGC, NBP, OCP, PNM. **COUNTIES:** BAR, BAT, BEL, BOO, BRA, BUL, CAM, FLE, GRA, HEN, HLN, HRD, JAC, JEF, LAU, LEE, MAD, MCY, MEA, NEL, OLD, OWE, PEN, POW, PUL, SCO, SHE, SPE, TRM, WAY, WHI.

DATES: May 25, June 29–Sept. 13. Locally common.

Subfamily **Danainae**

Monarchs

Genus ***Danaus*** Kluk

4614 ***D. plexippus*** (L.)
Monarch
PARKS AND RESERVES: BBM, BEC, BLA, BMF, CBB, COB, DHW, FAO, FKN, HHW, HWS, IRO, JCF, KGC, LBL, PYR, SGS, SWR, WVP. **COUNTIES:** ADA, ALL, AND, BAL, BAR, BAT, BEL, BOL, BOO, BUL, CAD, CAE, CAM, CAR, CAS, CHR, CLA, CLI, CLK, CRI, CRL, CUM, DAV, EDM, EST, FAY, FRA, FUL, GAL, GRA, GRE, GRV, GRY, GUP, HAN, HEN, HIC, HLN, HND, HOP, HRD, HRT, JAC, JEF, JOH, KNO, KNX, LAR, LAU, LEE, LET, LIV, LOG, LYO, MAD, MAR, MCC, MCY, MEA, MET, MRL, MTG, MUH, NEL, OHI, OLD, OWE, OWS, PEN, PER, PIK, POW, PUL, ROC, ROW, RUS, SCO, SHE, SIM, SPE, TAY, TOD, TRI, TRM, UNI, WAR, WAY, WEB, WHI, WOL, WOO. **DATES:** Apr. 3–Nov. 20. **REMARKS:** Common throughout the state with unusually high numbers noted in 1997. Migration through the Louisville area is usually during the last few days of September into early October, stragglers remain longer. In the fall of 1992 several lepidopterists in Jefferson, Oldham, and McCracken counties tagged over 300 monarchs to help Dr. Orley R. Taylor (University of Kansas "Monarch Watch") study the unique migration pattern of this species.

4615 ***D. gilippus*** (Cram.)
Queen
B. Palmer-Ball, Jr., reported sighting this deep southern species along a roadside in Ballard Co., BLD, on June 20, 1980. The butterfly was nectaring on butterweed, *Senecio glabellus*. A mating pair was found at the Louisville Zoo June 16, 1996, but they had been imported from Florida and released as part of an opening ceremony for a new exhibit there ten days earlier. Our subspecies is *berenice* (Cram.).

Superfamily **ZYGAENOIDEA**

Family **ZYGAENIDAE**

Smoky Moths

Genus ***Harrisina*** Pack.

4624 ***H. americana*** (Guer.)
Grapeleaf Skeletonizer Moth
PARKS AND RESERVES: BAN, BMF, CBB, CBF, HWS, OCP, WVP. **COUNTIES:** BAR, BUL, FAY, FUL, HIC, JEF, JES, MCC, MEA, OLD, ROW, WAR, WHI. **DATES:** Apr. 8, 30–May 15, 31–July 10, 19, 31, Aug. 12–20. Uncommon.

Genus ***Acoloithus*** Clem.

4629 ***A. falsarius*** Clem.
Bullitt Co., BMF, June 19; Jefferson Co., BEC, July 16; Owsley Co., 4.8 km (3 mi) NE of Booneville, June 16. Uncommon to rare.

Genus ***Pyromorpha*** H.-S.

4639 ***P. dimidiata*** H.-S.
Orange-patched Smoky Moth
PARKS AND RESERVES: BMF, CBF, IRO, NBP, OCP, PNM, TIO. **COUNTIES:** BEL, BUL, HRD, JEF, LAU, MEA, POW, WHI, WOL **DATES:** May 20–June 8, 30, Sept. 1. Uncommon.

Family **MEGALOPYGIDAE**

Flannel Moths

Genus ***Lagoa*** Harr.

4644 ***L. crispata*** (Pack.)
Black-waved Flannel Moth
PARKS AND RESERVES: BBM, BLA, BMF, BUC, JWS, KRC, LCP, LWO, NBP, PNM, PYR, RBF, WVP. **COUNTIES:** BEL, BRE, BUL, CAD, FLO, HLN, JAC, JEF, LET, MEA, NEL, PER, PIK, POW, ROW, RUS, TAY, TRI, WOL. **DATES:** Apr. 6, 20, 21, May 7, June 5–Aug. 18, 31, Sept. 22. Common.

Genus ***Norape*** Wlk.

4650 ***N. ovina*** (Sepp)
White Flannel Moth
PARKS AND RESERVES: BLA, BMF, LBL, MUR, OCP, RRG, SGS, WVP. **COUNTIES:** BUL, FAY, GAL, HIC, HOP, JEF, MCC, MEA, MEN, ROC, TRI, WOO. **DATES:** Apr. 12, 18, 29, May 3, 15, 16, 27–31, June 4, 26–30, July 6–30, Aug. 2–21, Sept. 1, 13, 18. Locally common.

Family **LIMACODIDAE**

Slug Caterpillar Moths

Genus ***Tortricidia*** Pack.

4652 ***T. testacea*** Pack.
PARKS AND RESERVES: BBM, CBB, KGC, LCP, LMP, NBP, PNM. **COUNTIES:** BEL, BRA, BUL, HLN, HRD, JEF, MUH, OLD, OWS, POW, ROW, RUS. **DATES:** May 2–28, June 8, July 1–15, Aug. 8–19, Sept. 8. Common.

4654 ***T. flexuosa*** (Grt.)
PARKS AND RESERVES: BBM, LBL, LCP, LWO, NBP, PNM, RBF, WKM. **COUNTIES:** BEL, BRE, JEF, HLN, LET, MCC, MER, POW, ROW, RUS, TRI, WOL. **DATES:** June 27–Aug. 18. Common.

Genus ***Packardia*** G. & R.

4659 ***P. geminata*** (Pack.)
PARKS AND RESERVES: CBB, LMP, NBP, PNM, RRG. **COUNTIES:** BEL, JEF, MEN, MUH, OLD, POW, WHI. **DATES:** Apr. 29, May 14, 28, June 28–July 15. Common.

4661 ***P. elegans*** (Pack.)
Menifee Co., RRG, June 27, 28, July 13. Rare.

Genus ***Lithacodes*** Pack.

4665 ***L. fasciola*** (H.-S.)
PARKS AND RESERVES: BBM, BMF, BUC, HWS, LBL, LVJ, NBP, OCP, PMS, PYR, RRG. **COUNTIES:** BRE, BUL, CAD, HLN, JEF, LAU, MEA, MEN, MET, OLD, OWS, PER, POW, ROW, TRI, WHI. **DATES:** Apr. 21, May 26–Aug. 10, Sept. 3. Common.

Genus ***Apoda*** Haw.

4667 ***A. y-inversum*** (Pack.)
PARKS AND RESERVES: BAN, BBL, BBM, BMF, HWS, KLK, LBK, LBL, LCP, LMP, NBP, OCP, PMS, PYR, PNM, WVP. **COUNTIES:** BAR, BEL, BOO, BRA, BUL, CAD, HLN, JEF, MCC, MEA, MRL, MUH, OLD, POW, ROW, RUS, TRI. **DATES:** May 15–July 23, Aug. 14. Common.

4669 ***A. biguttata*** (Pack.)
PARKS AND RESERVES: BBM, BMF, CBB, HWS, KGC, LBL, LCP, NBP, OCP, PMS, PNM, PYR, RBF, WVP. **COUNTIES:** BEL, BRE, BUL, CAD, HLN, HOP, JEF, MCC, MEA, NEL, OLD, OWS, POW, ROW, RUS, TRI, WHI. **DATES:** Apr. 16, May 26–Aug. 8. Common.

Genus ***Prolimacodes*** Schaus

4671 ***P. badia*** (Hbn.)
Skiff Moth
PARKS AND RESERVES: BBL, BBM, BEC, BMF, BUC, CBB, HWS, KLK, LCP, LBL, NBP, OCP, PNM, PYR, RBF. **COUNTIES:** BEL, BOO, BRE, BUL, CAD, HLN, JEF, MCC, MEA, MRL, OLD, OWS, PER, POW, ROW, RUS, TRI. **DATES:** May 29–Sept. 6, 16–22. Common.

Genus ***Isochaetes*** Dyar

4675 ***I. beutenmuelleri*** (Hy. Edw.)
PARKS AND RESERVES: BBL, BMF, HWS, LBL, LCP, MUR, NBP, PNM, RBF. **COUNTIES:** BEL, BOO, BRA, BRE, BUL, HIC, JEF, OLD, OWS, POW, ROW, RUS, TRI. **DATES:** May 8–Aug. 8, 18. Common.

Genus ***Phobetron*** Hbn.

4677 ***P. pithecium*** (J. E. Smith)
Hag Moth
PARKS AND RESERVES: AJJ, LBL, WVP. **COUNTIES:** CAM, HEN, JEF, MEA, OWS, RUS, TRI. **DATES:** May 16–Aug. 4, 14, Sept. 8. **REMARKS:** Uncommon. Found as larva (monkey slug) in Henry Co., on Sept. 11.

Genus ***Natada*** Wlk.

4679 ***N. nasoni*** (Grt.)
Nason's Slug Moth
PARKS AND RESERVES: BMF, BUC, CBF, HWS, IRO, KLK, LBL, LCP, NBP, PYR, RBF. **COUNTIES:** BAR, BRE, BUL, CAD, JEF, MEA, MRL, NEL, OLD, OWS, PER, POW, ROW, RUS, TRI, WHI. **DATES:** May 30, June 11–July 18. Common.

Genus ***Isa*** Pack.

4681 ***I. textula*** (H.-S.)
PARKS AND RESERVES: BBL, BBM, BMF, KRC, LBL, MUR, NBP, RBF. **COUNTIES:** BOO, BRA, BRE, BUL, HIC, HLN, JEF, MEA, NEL, POW, TRI, WOL. **DATES:** Apr. 26, 29, May 14, June 10–Aug. 20. Common.

Genus ***Adoneta*** Clem.

4684 ***A. bicaudata*** Dyar
Owsley Co., 4.8 km (3 mi) NE Booneville, July 20, 1991, L. D. Gibson. Rare.

4685 ***A. spinuloides*** (H.-S.)
PARKS AND RESERVES: BBL, BMF, NBP, RRG. **COUNTIES:** BAR, BOO, BUL, JEF, MCC, MEA, MEN, OLD, POW, PUL, ROW, TAY. **DATES:** May 7–Aug. 17, Sept. 8. Common.

Genus ***Monoleuca*** G. & R.

4691 ***M. semifascia*** (Wlk.)
Jefferson Co., Valley Station, June 8, 20, 21, July 7, 15; Owsley Co., 4.8 km (3 mi) NE of Booneville, June 14, July 19. Uncommon.

Genus ***Euclea*** Hbn.

4697 ***E. delphinii*** (Bdv.)
Spiny Oakslug Moth
PARKS AND RESERVES: BBL, BMF, BUC, HWS, LBL, LWO, NBP, RBF, WVP. **COUNTIES:** BOO, BRE, BUL, HOP, JEF, LET, MCC, MEA, MEN, OWS, OLD, PER, POW, ROW, TRI, WHI. **DATES:** Apr. 20, May 26–Aug. 26. Locally common.

Genus ***Parasa*** Moore

4698 ***P. chloris*** (H.-S.)
Smaller Parasa
PARKS AND RESERVES: BBL, BMF, KGC, LCP, NBP, OCP, RBF, WVP. **COUNTIES:** BOO, BRA, BRE, BUL, HLN, JEF, MEA, PIK, POW, ROW, RUS. **DATES:** Apr. 14, May 5, 20, June 7–Aug. 10, 22, Sept. 8. **REMARKS:** Common. S. Scholz collected a bilateral gynandromorph of this species in Jefferson Co., Valley Station, Sept. 5, 1974.

4699 ***P. indetermina*** (Bdv.)
Stinging Rose Caterpillar Moth
Fleming Co., Goddard, Apr. 10; Jefferson Co., WVP, June 11–20, July 4–14, Aug. 10, 16; Owsley Co., 4.8 km (3 mi) NE of Booneville, June 22, July 30; Rowan Co., Morehead, June 26–30, July 17; Trigg Co., LBL, July 10–15. Uncommon.

Genus ***Sibine*** H.-S.

4700 ***S. stimulea*** (Clem.)
Saddleback Caterpillar Moth
PARKS AND RESERVES: BMF, CBB, CBF, CMI, HWS, LBL, NBP, PYR, WVP. **COUNTIES:** BOO, BRA, BUL, CAD, JEF, MCC, OLD, POW, ROW, TRI, WHI. **DATES:** June 5–July 28. Common.

Family **EPIPYROPIDAE**

Planthopper Parasite Moths

Genus ***Fulgoraecia*** Newman

4701 ***F. exigua*** (Hy. Edw.)
Planthopper Parasite Moth
Boone Co., BBL, Oct. 26; Fayette Co., Lexington, in Malaise trap, Aug. 22–25; Jefferson Co., BEC, July 16; McCracken Co., Ray Black farm, 9.7 km (6 mi) SW of Paducah, July 28, Sept. 4, 6; Oldham Co., Goshen, July 30, Aug. 31, Sept. 2, 17. Moderately common.

Superfamily **PYRALOIDEA**

Family **CRAMBIDAE**

NOTE: With the publication of Solis (1997) the former Pyralidae has been divided into Crambidae and Pyralidae as presented here. The subfamily arrangement is still the same as in Hodges et al. (1983).

Subfamily **Scopariinae**

Genus ***Scoparia*** Haw.

4716 ***S. biplagialis*** Wlk.
PARKS AND RESERVES: BBL, BBM, BMF, CBF, PNM. **COUNTIES:** BEL, BOO, BUL, HLN, JEF, WHI. **DATES:** Apr. 30, May 4–8, 18–29, June 4, 13, July 6–18, Aug. 4. Common.

4717 ***S. penumbralis*** Dyar
Letcher Co., Pine Mountain, May 24, A. F. Braun.

4718 ***S. cinereomedia*** Dyar
Harlan Co., BBM, July 14, 1979, L. D. Gibson.

4719 ***S. basalis*** Wlk.
McCreary Co., Yahoo Ridge, Sept. 6; Russell Co., LCP, Aug. 11.

Genus ***Eudonia*** Billberg

4722 ***E. rectilinea*** (Zell.)
Letcher Co., Big Black Mountain, June 30, July 7, A. F. Braun.

4738 ***E. strigalis*** (Dyar)
Fulton Co., SR 94, 3.2 km (2 mi) E of Cayce, Sept. 11, 1993, R. A. Watkins.

4739 ***E. heterosalis*** (McD.)
Fayette Co., Lexington, in black-light trap, June 6; McCracken Co., 14.5 km (9 mi) SW of Paducah, June 29; Powell Co., NBP, May 13.

Subfamily **Nymphulinae**

Genus ***Chrysendeton*** Grt.

4744 ***C. medicinalis*** Grt.
PARKS AND RESERVES: AJJ, BBL, BMF, JWS, LBL, OCP. **COUNTIES:** BEL, BOO, BUL, CAM, FLO, JEF, MCY, MEA, MEN, NEL, OWS, ROW, TRI. **DATES:** June 14, 22–July15, Aug. 14, 30, Sept. 6, 7. Common.

4746 ***C. imitabilis*** (Dyar)
Bullitt Co., BMF, Aug. 7; Rowan Co., Morehead, July 13.

Genus ***Munroessa*** Lange

4748 ***M. icciusalis*** (Wlk.)
Fulton Co., SR 94, 4.0 km (2.5 mi) E of Cayce in Willingham Bottoms swamp, Sept. 8, 13; Hopkins Co., SR 502, 16.0 km (10 mi) W of Madisonville, in light trap, June 26. Local and uncommon.

4751 ***M. gyralis*** (Hulst)
McCracken Co., Ray Black farm, 9.7 km (6 mi) SW of Paducah, June 29, July 24, and "Aug.–Oct."; Menifee Co., RRG, Leatherwood Fork, Indian Creek, May 31; Trigg Co., LBL, July 10.

Genus ***Synclita*** Led.

4755 ***S. obliteralis*** (Wlk.)
PARKS AND RESERVES: BBL, BMF, CER, HWS, MUR, OCP. **COUNTIES:** BOO, BUL, FAY, FUL, HIC, JEF, MCC, MEA, OLD. **DATES:** May 13–18, June 1–14, 20, 29–Sept. 15,

Oct. 2. **REMARKS:** Common to abundant in appropriate wetlands habitat.

Genus ***Parapoynx*** Hbn.

4760 ***P. obscuralis*** (Grt.)
Owsley Co., 4.8 km (3 mi) NE of Booneville, June 14, 1980, L. D. Gibson.

Genus ***Petrophila*** Guilding

4774 ***P. bifascialis*** Rob.
Monroe Co., East Fork of Barren River, Aug. 21, A. F. Braun.

4777 ***P. fulicalis*** (Clem.)
PARKS AND RESERVES: BBL, CBB, OCP. **COUNTIES:** BOO, JEF, MCY, MEA, MEN, OLD, SPE. **DATES:** May 30, 31, June 12, 24, July 30, Aug. 30, Sept. 11. Common.

4779 ***P. canadensis*** (Mun.)
Meade Co., OCP, June 13, 1980, R. A. Henderson.

Subfamily **Odontiinae**

Genus ***Metrea*** Grt.

4789 ***M. ostreonalis*** Grt.
Menifee Co., RRG, at light, July 6, 13–20. These captures represent a significant range extension southward from southeastern Michigan.

Genus ***Eustixia*** Hbn.

4794 ***E. pupula*** Hbn.
PARKS AND RESERVES: BBL, BMF, HWS, OCP. **COUNTIES:** BOO, BUL, FAY, JEF, MCC, MEA, NEL, OLD, OWS, ROC, ROW. **DATES:** June 14–July 11, Aug. 6–14, 30. Common.

Genus ***Frechinia*** Mun.

4798 ***F. helianthiales*** (Murt.)
Barren Co., along SR 70, 8.0 km (5 mi) W of Cave City, May 5, 24; Bracken Co., SR 1159 at Locust Creek, May 8; Campbell Co., AJJ, July 20; Rowan Co., Clack Mtn., July 20–30. Uncommon.

Subfamily **Glaphyriinae**

Genus ***Hellula*** Gn.

4846 ***H. rogatalis*** (Hulst)
Rockcastle Co., Sept. 11, 1955, Ralph Beebe. Munroe (1973) stated that this species "ranges north at least to Kentucky and Arkansas, and Nova Scotia."

Genus ***Nephrogramma*** Mun.

4857 ***N. reniculalis*** (Zell.)
Bullitt Co., BMF, June 12; Owsley Co., 4.8 km (3 mi) NE of Booneville, June 14, July 5; Rockcastle Co., June 20, 30. Uncommon.

Genus ***Glaphyria*** Hbn.

4869 ***G. glaphyralis*** (Gn.)
Harlan Co., KGC, July 11; Jefferson Co., Louisville, July 4, Valley Station, July 6, Sept. 15; McCracken, 14.5 km (9 mi) SW Paducah, June 29; Owsley Co., 4.8 km (3 mi) NE of Booneville, July 5. Uncommon.

4870 ***G. sequistrialis*** Hbn.
PARKS AND RESERVES: BBL, PCP. **COUNTIES:** BOO, BRA, BUL, JEF, MCC, MCY, OLD, ROC. **DATES:** June 4, 5, 15–30, July 1–17, 25–Sept. 20. **REMARKS:** Locally common and very common at Goshen, Oldham Co.

4872 ***G. peremptalis*** (Grt.)
Boone Co., BCN, July 5; Menifee Co., RRG, in black-light trap, Sept. 1, 2; Owsley Co., 4.8 km (3 mi) NE of Booneville, July 5.

4873 ***G. fulminalis*** (Led.)
McCracken Co., 14.5 km (9 mi) SW of Paducah, June 29, 1981, L. D. Gibson.

Genus ***Aethiophysa*** Mun.

4877 ***A. lentiflualis*** (Zell.)
Jefferson Co., Louisville, Sept. 19; Oldham Co., Goshen, July 10, 1927; Owsley Co., 4.8 km (3 mi) NE of Booneville, July 14.

4878 ***A. consimilis*** Mun.
Jefferson Co., WVP, Valley Station, July 31; McCracken Co., Ray Black farm, 9.7 km (6 mi) SW of Paducah, July 24.

Genus ***Xanthophysa*** Mun.

4879 ***X. psychialis*** (Hulst)
Grayson Co., RGR, Aug. 11; Bullitt Co., BMF, July 20–21; Jefferson Co., Valley Station, Aug. 1; McCracken Co., Ray Black farm, 9.7 km (6 mi) SW of Paducah, Sept. 6.

Genus ***Lipocosma*** Led.

4881 ***L. sicalis*** (Wlk.)
PARKS AND RESERVES: LBL, PYR. **COUNTIES:** CAL, JEF, MCC, MCY, OWS, ROW, TRI. **DATES:** June 12–July 15. Common.

4883 ***L. adelalis*** (Kft.)
Boone Co., BBL, July 17; Jefferson Co., Jeffersontown, July 16, and Valley Station, June 29; McCracken Co., 14.5 km (9 mi) SW of Paducah, June 29; Meade Co., OCP, July 1; Oldham Co, Goshen, June 7, 8, 23–30, July 3–29. Locally common.

Genus *Lipocosmodes* Mun.

4888 ***L. fuliginosalis*** (Fern.)
PARKS AND RESERVES: BBM, CER, RRG. **COUNTIES:** BOO, FAY, FUL, HLN, HRD, JEF, MEN, OLD. **DATES:** May 26, 29, June 16, 14–17, 28–30, July 2, 14, 15, 23–Sept. 19. Common.

Genus *Dicymolomia* Zell.

4889 ***D. julianalis*** (Wlk.)
PARKS AND RESERVES: BAN, BBL, GBT, HWS, LBK, LBL. **COUNTIES:** BAR, BOD, BOO, BUL, CRL, FUL, HRD, JEF, MCC, OLD, OWS, TRI, WAR. **DATES:** May 1–15, 25, 26, June 7–18, July 5, 6, 16, 28–31, Aug. 6–14, 22, Sept. 1, 8, Oct. 12. Common.

Genus *Chalcoela* Zell.

4895 ***C. iphitalis*** (Wlk.)
Fayette Co., Lexington, found as a pupa in a wasp nest (*Polistes* sp.), July 13 (an account of rearings from *Polistes fuscatus* nests in Fayette Co. is given by Neary and Dahlman (1989), moths emerged on May 26, 1988); Gallatin Co., Markland Dam, June 15; Jefferson Co., Louisville, May 14, Prospect, Sept. 16, and Valley Station, June 10; Kenton Co., Taylor Mill, July 26, Aug. 28; McCracken Co., Ray Black farm, 9.7 km (6 mi) SW of Paducah, May 29; Oldham Co., Goshen, May 12, 23–31, June 2–12, July 13, 27, Aug. 1, 12–16, Sept. 6–13, 29, Oct. 2. Locally common.

4896 ***C. pegasalis*** (Wlk.)
Jefferson Co., Valley Station, May 27, Aug. 22, 24.

Subfamily **Evergestinae**

Genus *Evergestis* Hbn.

4897 ***E. pallidata*** (Hufn.)
Purplebacked Cabbageworm Moth
Boone Co., Florence, July 20; Jefferson Co., Louisville, May 25, Sept. 6, 11, and Caperton Swamp, Sept. 19. Uncommon.

4898 ***E. rimosalis*** (Gn.)
Fayette Co., Lexington, Aug. 13, Nov. 21; Jefferson Co., Valley Station, Aug. 15; McCracken Co., Ray Black farm, 9.7 km (6 mi) SW of Paducah, Sept. 26; Owsley Co., 4.8 km (3 mi) NE of Booneville, May 23; Rowan Co., DBNF, E side of SR 1274, 3.2 km (2 mi) W of SR 519, Aug. 18. Uncommon.

4901 ***E. unimacula*** (G. & R.)
Bullitt Co., BMF, Aug. 31; Floyd Co., JWS, June 7; Jefferson Co., Valley Station, May 20, June 23, July 9; Meade Co., OCP, May 3, June 7, Aug. 10; Powell Co., NBP, May 7. Uncommon.

Subfamily **Pyraustinae**

Genus *Saucrobotys* Mun.

4935 ***S. fumoferalis*** (Hulst)
Owsley Co., 4.8 km (3 mi) NE of Booneville, May 10, 1980, L. D. Gibson.

4936 ***S. futilalis*** (Led.)
Henderson Co., Henderson, July 18; McCreary Co., along SR 90, June 10; Oldham Co., Goshen, June 21, July 13–31, Sept. 14, and HWS, Aug. 5; Rowan Co., near Morehead, reared on *Asclepias tuberosa*, emerged June 29–July 3.

Genus *Crocidophora* Led.

4943 ***C. pustuliferalis*** Led.
Jefferson Co., Valley Station, July 31; Madison Co., Richmond, May 9; Oldham Co., CBB, May 14.

4944 ***C. serratissimalis*** Zell.
Letcher Co., Pine Mountain, June 21; Rowan Co., Morehead, June 25.

4945 ***C. tuberculalis*** Led.
PARKS AND RESERVES: BBL, CBB, HWS, OCP, PMS, RGR. **COUNTIES:** BOO, BRK, GRY, HLN, JEF, LET, MEA, OLD, ROW. **DATES:** May 19, June 3–11, July 8, 23–31, Aug. 3–11. Common.

Genus *Ostrinia* Hbn.

4946 ***O. penitalis*** (Grt.)
Jefferson Co., Louisville, Caperton Swamp, May 7; McCracken Co., Paducah, July 24; Trigg Co., LBL, July 13, 14.

4947 ***O. obumbratalis*** (Led.)
Rockcastle Co., Renfro Valley, Aug. 13, 1955, Ralph Beebe.

4949 ***O. nubilalis*** (Hbn.)
European Corn Borer
PARKS AND RESERVES: BAN, BMF, BUC, CBB, HNK, HWS, JWS, LBL, LCP, LMP, OCP, PNM, PYR, RGR, SWR. **COUNTIES:** BAR, BEL, BRE, BRK, BUL, CAD, CAL, FLO, FUL, GAL, GRV, GRY, HND, HRD, JEF, MCC, MCY, MEA, MRL, MUH, NEL, OLD, OWS, PER, ROW, RUS, TRI. **DATES:** Apr. 26–Sept. 18. **REMARKS:** A common pest of corn throughout the Commonwealth.

Genus *Fumibotys* Mun.

4950 ***F. fumalis*** (Gn.)
Campbell Co., AJJ, Sept. 13; Oldham Co., Goshen, Aug. 26, 31, Sept. 2, 12. Rare.

Genus ***Perispasta*** Zell.

4951 ***P. caeculalis*** Zell.
PARKS AND RESERVES: BBL, BMF, HWS, LBL, NBP, OCP, PMS, PNM, PYR. **COUNTIES:** BEL, BOO, BUL, CHR, HLN, JEF, MCC, MEA, NEL, OLD, OWS, POW, ROC, TRI. **DATES:** Apr. 9–24, May 1, 12, 17, June 1, 2, 12, 19, July 8–16, Aug. 4, 8, 27, Sept. 4, 5. Common.

Genus ***Phlyctaenia*** Hbn.

4953 ***P. coronata*** (Hufn.)
Bullitt Co., Mt. Washington, July 28; Fulton Co., SR 94, Willingham Bottoms, 4.0 km (2.5 mi) E of Cayce, Aug. 3, Sept. 8; Hardin Co., Glendale, in Malaise trap, July 22–31, Aug. 1–14; Harlan Co., BBM, July 8, 11; Hart Co., Horse Cave, May 13. Our subspecies is *tertialis* (Gn.). Uncommon.

Genus ***Anania*** Hbn.

4958 ***A. funebris*** (Strom)
Boone Co., no date; Bracken Co., SR 1159 at Locust Creek, May 23, June 9; Harlan Co., BBM, June 3; all L. D. Gibson. Our subspecies is *glomeralis* (Wlk.).

Genus ***Hahncappsia*** Mun.

4962 ***H. marculenta*** (G. & R.)
PARKS AND RESERVES: BBM, PNM, WVP. **COUNTIES:** BEL, BUL, FAY, HLN, JEF, WHI. **DATES:** May 20, June 5, 9, July 28, Aug. 6–10, Oct. 22. Common.

4967 ***H. mancalis*** (Led.)
Fayette Co., Lexington, Aug. 8–12; Jefferson Co., Anchorage, July 15; Jefferson Co., Louisville, Nov. 14, and Valley Station, Aug. 3; McCracken Co., Ray Black farm, 9.7 km (6 mi) SW of Paducah, July 24; Meade Co., OCP, June 7; Russell Co., LCP, Aug. 14. Common.

Genus ***Achyra*** Gn.

4975 ***A. rantalis*** (Gn.)
Garden Webworm Moth
PARKS AND RESERVES: BOW, FAO, LMP, NBP. **COUNTIES:** BOO, BUL, FAY, FUL, HRD, JEF, MUH, POW, ROC, ROW. **DATES:** Apr. 20, July 16–31, Aug. 1–16, 24–27, Sept. 7, 8, 20, Oct. 2, 3, Nov. 1. A very common vegetable pest.

Genus ***Helvibotys*** Mun.

4980 ***H. helvialis*** (Wlk.)
Hardin Co., Glendale, in Malaise trap, Aug. 22–31, Sept. 13; Jefferson Co., Louisville, July 16; McCracken Co., Ray Black farm, 9.7 km (6 mi) SW of Paducah, July 24, "Aug.–Oct." (label); Russell Co., LCP, Sept. 13. Common.

Genus ***Sitochroa*** Hbn.

4986 ***S. dasconalis*** (Wlk.)
Harlan Co., BBM, July 12, 1972, C. V. Covell, Jr.

4987 ***S. chortalis*** (Grt.)
Oldham Co., Goshen, June 1, 10, July 1, Sept. 11; Trigg Co., LBL, July 16. Uncommon.

Genus ***Sericoplaga*** Warr.

4991 ***S. externalis*** Warr.
Jefferson Co., Valley Station, May 17, Aug. 9; Oldham Co., CBB, May 14, HWS, Aug. 8, and Goshen, May 23, July 12, 26–30, Aug. 1–15, Sept. 4–6; Trigg Co., LBL, July 12.

Genus ***Uresiphita*** Hbn.

4992 ***U. reversalis*** (Gn.)
Fayette Co., Lexington, July 15–16, Aug. 2, Aug. 11, 1928 "in electric light trap," 18, 20, and Lexington, on *Viburnum*, larva collected Aug. 1, emerged Aug. 20; Gallatin Co., Warsaw, larva on *Cytisus*, emerged July 20, 1908, all H. Garman.

Genus ***Loxostege*** Hbn.

5017 ***L. cereralis*** (Zell.)
Alfalfa Webworm Moth
Fayette Co., "at lighted window," Sept. 12; Meade Co., OCP, Aug. 27. A vegetable pest.

Genus ***Pyrausta*** Schr.

5018 ***P. demantrialis*** (Druce)
Hardin Co., FKN, May 13; Jefferson Co., Valley Station, July 3. Uncommon.

5036 ***P. inveterascalis*** B. & McD.
Boone Co., BCN, June 15; Bullitt Co., BMF, June 5; Harlan Co., BBM, July 11.

5040 ***P. bicoloralis*** (Gn.)
PARKS AND RESERVES: AJJ, BBL, NBP, OCP, RRG, SWR. **COUNTIES:** BOO, CAM, FAY, FUL, HLN, HND, HOP, JEF, MEA, POW, ROC, ROW, WHI. **DATES:** Apr. 26, May 5, 7, 27–30, June 8, 10, 24, 26, July 3–13, Aug. 3–27, Sept. 8, Oct. 7. Common.

5049 ***P. phoenicealis*** (Wlk.)
Letcher Co., Big Black Mountain, July 8; Oldham Co., HWS, June 22; Trigg Co., LBL, July 10; Wayne Co., Beaver Creek, May 16. Uncommon.

5051 ***P. rubricalis*** (Hbn.)
Jefferson Co., Valley Station, Aug. 8; McCracken Co., Ray Black farm, 9.7 km (6 mi) SW of Paducah, July 26; Oldham Co., HWS, June 22, and Goshen, Apr. 28, Sept. 1, 2, 14; Trigg Co., LBL, July 10; Whitley Co., Goldbug, Aug. 2. Common.

5055 ***P. homonymalis*** Mun.
PARKS AND RESERVES: LCP, PCP, RRG. **COUNTIES:** BUL, LAU, LEE, MCY, MEN, RUS. **DATES:** May 9, 28, June 4, 10, Aug. 19. Moderately common.

5056 ***P. generosa*** (G. & R.)
Boone Co., BBL, July 31, Aug. 25; Oldham Co., HWS, Aug. 8, and Goshen May 20, 23, 30, June 11, 22, Aug. 18. Uncommon.

5058 ***P. orphisalis*** Wlk.
PARKS AND RESERVES: BBL, BMF, HWS. **COUNTIES:** BOO, BUL, FAY, JEF, OLD, OWS. **DATES:** June 4–6, 30–July 5, Aug. 4–12, 31. Common.

5060 ***P. subsequalis*** (Gn.)
PARKS AND RESERVES: BBL, BMF, BUC, HWS. **COUNTIES:** BOO, BRA, BRE, BUL, CAM, FAY, JEF, MCY, MEA, OLD, PER. **DATES:** Apr. 3, May 23, June 6–15, July 16–Aug. 3, 25, Sept. 12. Common.

5069 ***P. tyralis*** (Gn.)
PARKS AND RESERVES: BBL, CCR, LBL, OCP, PCP, PYR, RRG. **COUNTIES:** BOO, BUL, CHR, JEF, MCC, MEA, MEN, ROC, ROW, TRI. **DATES:** June 1, 10, 25, July 9–22, Aug. 9, 23–Sept. 12, Oct. 2. Common.

5070 ***P. laticlavia*** (G. & R.)
Bell Co., PNM, June 9; Jefferson Co., Valley Station, June 20, 21. Uncommon.

5071 ***P. acrionalis*** (Wlk.)
PARKS AND RESERVES: BBL, BMF, CBB, HWS, LCP, OCP, RRG. **COUNTIES:** BOO, BRA, BUL, FAY, JEF, MCC, MEA, MEN, MRL, NEL, OLD, OWS, ROW, RUS, WHI. **DATES:** Mar. 20–Aug. 27. Very common.

5073 ***P. niveicilialis*** (Grt.)
PARKS AND RESERVES: BBM, BUC, CBF, PNM. **COUNTIES:** BEL, BRE, HIC, HLN, JEF, MCC, PER, WHI. **DATES:** Apr. 9–19, 30, May 1, 9–17, June 1, 2, 25, 26, July 9, 14, 31, Aug. 31, Sept. 5, 6, 23, Oct. 10. Common.

Genus ***Udea*** Gn.

5079 ***U. rubigalis*** (Gn.)
Celery Leaftier Moth
PARKS AND RESERVES: BBL, BLA, HWS, KLK, LBK, LCP, OCP, PNM, PYR, RGR. **COUNTIES:** BEL, BOO, BRK, CHR, FAY, GRV, GRY, HRD, JEF, MCC, MEA, MRL, OLD, OWS, PEN, ROC, ROW, RUS, SPE, TAY, TRI. **DATES:** Feb. 15–Oct. 27. A common garden pest.

Genus ***Lineodes*** Gn.

5107 ***L. integra*** (Zell.)
Henderson Co., SWR, Sept. 10, 11; Menifee Co., RRG, FS 9A NE of Slade, Indian Creek, 229 m (750 ft) elevation, Oct. 12, 13. Uncommon.

Genus ***Choristostigma*** Warr.

5130 ***C. roseopennalis*** (Hulst)
PARKS AND RESERVES: ASC, CAV, MCV, RRG. **COUNTIES:** BAT, EDM, GRY, LEW, MEN, OWS, ROC. **DATES:** July 8, 19, Aug. 9, 23, 28, Sept. 5, 7, 16–26. Locally common.

Genus ***Diacme*** Warr.

5142 ***D. elealis*** (Wlk.)
PARKS AND RESERVES: BBL, BMF, HWS, KLK, LBL, LCP, SWR. **COUNTIES:** BOO, BUL, HND, JEF, MCC, MRL, NEL, OLD, OWS, ROC, RUS, TRI. **DATES:** Apr. 26, June 9, July 5–22, Aug. 19, 29, Sept. 7–23. Common, especially in late summer.

5143 ***D. adipaloides*** (G. & R.)
Meade Co., OCP, Apr. 11, Russell Co., LCP, Mar. 19. Uncommon.

Genus ***Epipagis*** Hbn.

5147 ***E. huronalis*** (Gn.)
PARKS AND RESERVES: BAN, LBL, LCP. **COUNTIES:** BAR, BUL, FUL, GRA, JEF, MCC, MRL, RUS, TRI. **DATES:** May 31, June 29, July 14, 15, 30, Aug., Sept. 4–13. Locally common in some years.

Genus ***Samea*** Gn.

5150 ***S. ecclesialis*** Gn.
Russell Co., LCP, Mar. 19, 1976, C. Cornett. Rare.

Genus ***Nomophila*** Hbn.

5156 ***N. nearctica*** Mun.
PARKS AND RESERVES: BAN, CBB, GBT, HWS, LCP, LMP, NBP, PYR, RGR. **COUNTIES:** BAR, BRK, CHR, CLK, CRL, FLO, FUL, GRV, GRY, HLN, HND, JEF, MCC, MEA, MUH, OLD, POW, ROW, RUS, SHE, SPE, WHI. **DATES:** Apr. 12–Oct. 31, Nov. 24. Very common.

Genus ***Desmia*** Westwood

5159 ***D. funeralis*** (Hbn.)
Grape Leaffolder Moth
PARKS AND RESERVES: BAN, BBL, BEC, BMF, BUC, CBF, HWS, KLK, LCP, MCV, NBP, PNM, PYR, RGR. **COUNTIES:** BAR, BEL, BOO, BRA, BRE, BRK, BUL, CAR, CHR, EDM, FAY, FUL, GRV, GRY, HRT, JEF, MCC, MCY,

MEA, MOR, MRL, NEL, OLD, PER, POW, PUL, RUS, SHE, WAR, WHI. **DATES:** Apr. 19, May 1, 2, 11–Aug. 25, Sept. 1, Oct. 13. Common.

5160 ***D. maculalis*** Westwood
PARKS AND RESERVES: BMF, CBB, HWS, KLK, LBK, OCP, PYR, RGR. **COUNTIES:** BRK, BUL, CHR, FAY, GRY, HRD, HRT, JEF, MCC, MEA, MRL, OLD, TRI. **DATES:** May 11, June 3, 14–18, July 8, 21–31, Aug. 1–18. **REMARKS:** Common, but difficult to separate from *D. funeralis*. All of the records above were identified by E. G. Munroe.

Genus *Hymenia* Hbn.

5169 ***H. perspectalis*** (Hbn.)
Spotted Beet Webworm Moth
PARKS AND RESERVES: BAN, BBL, BIP, BLA, BMF, BUC, CAV, FAO, GBT, LCP, NBP, PNM, PYR. **COUNTIES:** BAR, BAT, BEL, BOO, BRE, BUL, CAD, CRL, FAY, FUL, GAL, HRT, JEF, MCC, NEL, PER, PIK, POW, ROB, ROW, RUS. **DATES:** May 15, 16, June 11, 18, 26, July 1–Nov. 23. Generally common and locally abundant.

Genus *Spoladea* Gn.

5170 ***S. recurvalis*** (F.)
Hawaiian Beet Webworm Moth
Boone Co., BOW, Oct. 2; Jefferson Co., Valley Station, Aug. 14, Oct. 3, 26; Russell Co., LCP, Sept. 13; Whitley Co., CBF, May 16. Uncommon.

Genus *Diasemiopsis* Mun.

5171 ***D. leodocusalis*** (Wlk.)
McCracken Co., Mayfield Creek bottoms, Oct. 6, 1989, at light, C. V. Covell, Jr.

Genus *Diasemiodes* Mun.

5172 ***D. janassialis*** (Wlk.)
Bell Co., PNM, Pound Gap, Aug. 20; Menifee Co., RRG, NE of Slade, Indian Creek, Apr. 30; Owsley Co., 4.8 km (3 mi) NE of Booneville, July 18; Rowan Co., W side of SR 1274, 4.3 km (3.3 mi) E of SR 519. Uncommon.

Genus *Diathrausta* Led.

5174 ***D. reconditalis*** (Wlk.)
PARKS AND RESERVES: BMF, CAV. **COUNTIES:** BAT, BUL, JEF, OLD, OWS, ROW. **DATES:** June 12, 14, July 24, 25, Aug. 25, 28, Sept. 9. Uncommon.

Genus *Anageshna* Mun.

5176 ***A. primordialis*** (Dyar)
PARKS AND RESERVES: BMF, NBP, OCP, RRG. **COUNTIES:** BUL, JEF, LAR, LET, MEA, MEN, OWS, POW. **DATES:** May 10, 19, 27, June 7, 14–22, July 1, Aug. 7, 21, 24, Sept. 17. Common.

Genus *Apogeshna* Mun.

5177 ***A. stenialis*** (Gn.)
PARKS AND RESERVES: BMF, KLK, LBK, OCP. **COUNTIES:** BUL, JEF, MCC, MEA, MRL, NEL, OWS, TRI, WHI. **DATES:** Apr. 28, May 28, June 4, 16–July 3. Common.

Genus *Blepharomastix* Led.

5182 ***B. ranalis*** (Gn.)
PARKS AND RESERVES: BAN, BMF, BUC, HWS, JWS, KLK, KRC, LBK, LCP, LMP, NBP, PNM, PYR, RBF. **COUNTIES:** BAR, BEL, BRE, BUL, CHR, FLO, JEF, MCC, MCY, MRL, MUH, NEL, OLD, OWS, PER, POW, ROW, RUS, TRI, WAR, WOL. **DATES:** Apr. 28, May 15–July 15, Aug. 14. Very common.

5188 ***B. aplicalis*** (Gn.)
Bullitt Co., Pine Creek Forest Rd., 0.8 km (0.5 mi) N of SR 480, July 22, 1989, D. J. Wright. Rare.

Genus *Colomychus* Mun.

5200 ***C. talis*** (Grt.)
McCracken Co., Ray Black farm, 9.7 km (6 mi) SW of Paducah, Aug. 26, 1978, W. R. Black, Jr. Rare.

Genus *Diaphania* Hbn.

5202 ***D. nitidalis*** (Stoll)
Pickleworm Moth
Bell Co., PNM, Oct. 18; Fayette Co., Lexington, Aug. 3, 25, 28, Sept. 12, 21, Oct. 2, 14; Jefferson Co., Louisville, Sept. 11, 28, Oct. 1, and Valley Station, Sept. 4, 17, 27, Oct. 3, 13; McCracken Co., Ray Black farm, 9.7 km (6 mi) SW of Paducah, Aug. 24, Sept. 26, Oct. 27. Uncommon.

5204 ***D. hyalinata*** (L.)
Melonworm Moth
Fayette Co., Lexington, Oct. 25; Jefferson Co., Louisville, Oct. 7, Nov. 1; Jessamine Co., Oct. 4; McCracken Co., Ray Black farm, 9.7 km (6 mi) SW of Paducah, Oct. 1; Montgomery Co., Mt. Sterling, Oct. 5. Uncommon.

Genus *Omiodes* Gn.

5212 ***O. indicata*** (F.)
Owsley Co., 4.8 km (3 mi) NE of Booneville, 1 male at light, Oct. 8, 1979, L. D. Gibson.

Genus *Palpita* Hbn.

5218 ***P. quadristigmalis*** (Gn.)
Bath Co., CAV, FS 918, Aug. 28; Warren Co., Bowling Green, June 25. Uncommon.

5221 ***P. cincinnatalis*** Mun.
PARKS AND RESERVES: BUC, CAC, CBB, CBF, HWS, LBL, LCP, OCP. **COUNTIES:** BRE, CAR, JEF, MEA, OLD, PER, RUS, TRI, WHI. **DATES:** Apr. 6, 14–27, May 14, 19, June 4, 10, July 16, 25, Aug. 6–16, Nov. 14. Common.

5222 ***P. arsaltealis*** (Wlk.)
Oldham Co., CBB, May 14, 1970, C. V. Covell, Jr. Rare.

5223 ***P. illibalis*** Hbn.
McCreary Co., South Fork Cumberland River, emerged July 13, A. F. Braun.

5226 ***P. magniferalis*** (Wlk.)
PARKS AND RESERVES: BAN, BEC, BIP, BMF, BUC, CAC, CBB, CBF, HWS, JWS, LBL, LCP, NBP, OCP, PNM, PYR, RNR, RRG. **COUNTIES:** BAR, BEL, BRE, BUL, CAR, CHR, FLO, FUL, HRT, JEF, MCC, MEA, MEN, OLD, PER, PIK, POW, ROC, ROW, RUS, SPE, TRI, WAR, WHI. **DATES:** Apr. 18–Oct. 10. Very common.

5227 ***P. aenescentalis*** Mun.
Carroll Co., GBT, Apr. 11; Harlan Co., BBM near Lynch, Apr. 21, and KGC, Apr. 20; Meade Co., OCP, Apr. 11.

Genus *Polygrammodes* Gn.

5228 ***P. flavidalis*** (Gn.)
PARKS AND RESERVES: BEC, BMF, CBB, HWS, JWS, MCV, NBP, OCP, PMS, RGR. **COUNTIES:** BRA, BRK, BUL, EDM, FAY, FLO, GRY, HLN, HND, JEF, MEA, NEL, OLD, OWS, POW, ROC, ROW. **DATES:** May 10, June 3, 15–Sept. 6. Common.

5229 ***P. langdonalis*** (Grt.)
PARKS AND RESERVES: HWS. **COUNTIES:** BOO, BRA, BUL, GAL, JEF, OLD, ROW. **DATES:** June 2, July 4, 15–Aug. 10, 18. Common.

Genus *Compacta* Amsel

5233 ***C. capitalis*** (Grt.)
PARKS AND RESERVES: BBL, BMF, MCV. **COUNTIES:** BOO, BUL, CAM, EDM, GAL, OLD. **DATES:** June 12-26, July 13. Local and rare.

Genus *Pantographa* Led.

5241 ***P. limata*** (G. & R.)
Basswood Leafroller Moth
PARKS AND RESERVES: BAN, BBM, HWS, JCF, KGK, LCP, LWO, MCV, NBP, OCP, PMS, PNM. **COUNTIES:** BAR, BEL, EDM, HLN, JEF, LET, MCC, MEA, MOR, OLD, OWS, POW, RUS, SPE, WAR. **DATES:** Apr. 26, May 23–Aug. 8. Common.

Genus *Pleuroptya* Meyr.

5243 ***P. silicalis*** (Gn.)
Henderson Co., SWR, Sept. 11; Jefferson Co., Valley Station, Oct. 14; Russell Co., LCP, Oct. 12. Uncommon.

Genus *Lygropia* Led.

5250 ***L. rivulalis*** Hamp.
PARKS AND RESERVES: BBL, CMI, RRG. **COUNTIES:** BOO, GRV, HRD, JEF, MCC, MEN, OLD. **DATES:** May 15, 31–Aug. 21. Uncommon.

Genus *Diastictis* Hbn.

5255 ***D. ventralis*** (G. & R.)
PARKS AND RESERVES: BBL, LBL. **COUNTIES:** BOO, GAL, HAR, JEF, ROW, TRI, WHI. **DATES:** June 28, July 10, Aug. 2–22, Oct. 10. Uncommon.

Genus *Psara* Snell.

5268 ***P. obscuralis*** (Led.)
Trigg Co., LBL, July 10, 1965, C. V. Covell, Jr. Rare.

Genus *Herpetogramma* Led.

5272 ***H. bipunctalis*** (F.)
Jefferson Co., Louisville, Oct. 27, and Valley Station, July 8; McCracken Co., Paducah, Oct. 1; Russell Co., LCP, Oct. 11.

5273 ***H. ipomoealis*** (Capps)
Fulton Co., SR 94, Willingham Bottoms, 4.0 km (2.5 mi) E of Cayce, Sept. 13; Jefferson Co., Valley Station, Aug. 28.

5275 ***H. pertextalis*** (Led.)
Harlan Co., BBM, June 5; Letcher Co., LWO, July 11.

5275.1 ***H.*** undescribed species near ***pertextalis***
PARKS AND RESERVES: CBB, CMI, HWS, OCP, PNM, PYR. **COUNTIES:** BEL, BOO, CHR, FUL, JEF, MEA, MCC, OLD, SPE, WHI. **DATES:** May 14, June 2, 4, 8, 14–20, July 14, 23–26, Aug. 1–4 (Malaise trap), 6–10, 18, Sept. 8, 24. **REMARKS:** Common. E. G. Munroe identified this series in the ULK Collection as an undescribed *Herpetogramma* species.

5277 ***H. thestealis*** (Wlk.)
PARKS AND RESERVES: BBM, CAC, HWS, MCV, PNM. **COUNTIES:** BEL, CAR, EDM, HLN, JEF, OLD, OWS. **DATES:** May 2, 3, 23, 28, June 13, 24, July 2–18.

5279 ***H. theseusalis*** (Wlk.)
Elliott Co., Little Sandy River, emerged July 4, 15; Rockcastle Co., Renfro Valley, May 30, and Rockcastle Co., Aug. 3, 14.

5280 ***H. aeglealis*** (Wlk.)
Grayson Co., RGR, Aug. 17; Jefferson Co., Valley Station, June 27, 30, July 7, 24, Aug. 6, 16; Russell Co., LCP, May 28, July 18, Sept. 13.

Genus ***Pilocrosis*** Led.

5281 ***P. ramentalis*** Led.
Caldwell Co., PYR, Sept. 12; Jefferson Co., WVP, Oct. 6, Nov. 1; Perry Co., BUC, Sept. 14; Pike Co., BIP, Oct. 15, 16; Robertson Co., Blue Licks, Oct. 27.

Genus ***Conchylodes*** Gn.

5292 ***C. ovulalis*** (Gn.)
PARKS AND RESERVES: BBL, BCN, BMF, HWS, LBL, OCP, PYR. **COUNTIES:** BOO, BUL, CHR, FAY, FUL, HRT, JEF, LAU, MCC, MEA, NEL, OLD, OWS, ROC, TRI. **DATES:** May 2–Aug. 30. Locally common

Subfamily **Schoenobiinae**

Genus ***Donacaula*** Meyr.

5315 ***D. tripunctella*** (Rob.)
This species is part of a complex needing revision, so identification is uncertain. Owsley Co., 4.8 km (3 mi) NE of Booneville, Sept. 2, 1977, L. D. Gibson.

5317 ***D. aquilella*** (Clem.)
Fulton Co., SR 94, Willingham Bottoms, 4.0 km (2.5 mi) E of Cayce, Sept. 8; Meade Co., FKN, week of May 28; Oldham Co., Goshen, May 17–30, June 3–20, Aug. 23. Very local.

5324 ***D. maximella*** (Fern.)
Menifee Co., RRG, Leatherwood Fork of Indian Creek, May 31, 1991, L. D. Gibson.

Subfamily **Ancylolomiinae**

Genus ***Mesolia*** Rag.

5331 ***M. incertella*** (Zinck.)
A specimen in the NMNH bears the label "Kentucky Sanborn."

Subfamily **Crambinae**

Genus ***Crambus*** F.

5355 ***C. praefectellus*** (Zinck.)
Fayette Co., Lexington, July 11, 12; "Ky. near Cincinnati, Ohio," Aug. 9.

5357 ***C. leachellus*** (Zinck.)
Barren Co., Diamond Caverns, Sept. 11; Perry Co., BUC, May 1; Butler Co., Oct. 10; Jefferson Co., Louisville, May 27, Sept. 29, and Valley Station, May 18; Warren Co., Sept. 26. Common.

5361 ***C. albellus*** Clem.
Bell Co., PNM, June 21; Harlan Co., BBM, June 30; Jefferson Co., BEC, June 13; Menifee Co., RRG, FS 9b, Indian East Fork, Kelly Branch, black-light trap, June 7–8; Whitley Co., CBF, June 25. Common.

5362 ***C. agitatellus*** Clem.
PARKS AND RESERVES: BBL, BMF, HWS, KLK, LBK, LBL, OCP, PMS, PYR. **COUNTIES:** BOO, BUL, CHR, HLN, JEF, LET, MEA, MRL, NEL, OLD, TRI. **DATES:** June 10–29, July 7–28, Aug. 5, 6, 18. Common.

5365 ***C. girardellus*** Clem.
Menifee Co., RRG, FS 6b, Indian East Fork, Kelley Branch, 220 m (720 ft) elevation, June 8–9, July 19–20, 1985, J. S. Nordin.

5378 ***C. laqueatellus*** Clem.
PARKS AND RESERVES: BAN, BBL, CBB, RRG. **COUNTIES:** BAR, BOO, BUL, FAY, JEF, MCC, OLD, POW, WAR. **DATES:** Apr. 19–30, May 7–31, Aug. 1. Common.

5379 ***C. luteolellus*** Clem.
Fayette Co., July 3, 15–26; Pendleton Co., Falmouth, July 12; Rowan Co., June 20; Trigg Co., LBL, July 10.

5381 ***C. caliginosellus*** Clem.
Boone Co., BBL, June 28; Grayson Co., RGR, Aug. 10; Fayette Co., Lexington, June 19; Jefferson Co., BEC, June 13, and Valley Station, Aug. 29; Trigg Co, LBL, July 10. Common.

5382 ***C. murellus*** Dyar
Bullitt Co., SR 480, 11.0 km (6.9 mi) E of SR 61, May 28, 1989, L. D. Gibson and D. J. Wright.

Genus ***Chrysoteuchia*** Hbn.

5391 ***C. topiaria*** (Zell.)
Edmonson Co., MCV, July 1; Fayette Co., Lexington, June 12, July 12, 18, Sept. 8; Gallatin Co., Markland Dam, July 15; Harlan Co., BBM, July 14; Rowan Co., DBNF, E side of SR 1274, 3.2 km (2 mi) W of SR 519, July 1.

Genus ***Raphiptera*** Hamp.

5393 ***R. argillaceella*** (Pack.)
Laurel Co., London, Sept. 5, A. F. Braun.

Genus *Agriphila* Hbn.

5399 ***A. ruricolella*** (Zell.)
PARKS AND RESERVES: BBL, RRG. **COUNTIES:** BOO, FAY, FUL, JEF, LAU, MEN, NEL, OLD. **DATES:** June 23, July 1, 9, 14, Aug. 29, Sept. 5, 19, Oct. 6. Common.

5403 ***A. vulgivagella*** (Clem.)
Vagabond Crambus
PARKS AND RESERVES: CBF, LCP. **COUNTIES:** FAY, JEF, MCC, OLD, RUS, WHI. **DATES:** Sept. 8–Oct. 6. Common.

Genus *Pediasia* Hbn.

5413 ***P. trisecta*** (Wlk.)
Sod Webworm Moth
PARKS AND RESERVES: BAN, BUC, GBT, LBL, LCP. **COUNTIES:** BAR, BRE, BUL, CRL, FAY, HND, JEF, NIC, PER, RUS, TRI. **DATES:** May 23–Oct. 14. **REMARKS:** A common to abundant pest of lawns and other turf.

Genus *Microcrambus* Bleszynski

5419 ***M. biguttellus*** (Fbs.)
Bell Co., PNM, July 18; Harlan Co., BBM, no date; Jefferson Co., Louisville, Sept. 28, and Valley Station, June 4, 20, 29, Aug. 6; Oldham Co., HWS, Aug. 25, Sept. 6; Owsley Co., 4.8 km (3 mi) NE of Booneville, Sept. 1.

5420 ***M. elegans*** (Clem.)
PARKS AND RESERVES: BEC, BMF, CAH, KRC, NBP, OCP. **COUNTIES:** BUL, FAY, FUL, HRT, JEF, MCC, MEA, OLD, OWS, POW, WOL, WOO. **DATES:** Apr. 19, May 27–July 16, Aug. 1, 2, 10–25, Sept. 5, 7. Common.

Genus *Fissicrambus* Bleszynski

5435 ***F. mutabilis*** (Clem.)
Perry Co., BUC, June 2; Fayette Co., Lexington, at sugar bait, Aug. 22, 1891 on bluegrass, June 5, 1904 at light, May 10, June 19, Aug. 10 (UK); Jefferson Co., Valley Station, Aug. 9; "Morgan, Ky." (county uncertain), larvae from corn, emerged May 11.

Genus *Thaumatopsis* Morr.

5438 ***T. edonis*** (Grt.)
Bullitt Co., PCP, Sept. 6; Jefferson Co., Valley Station, Sept. 29. Rare.

Genus *Parapediasia* Bleszynski

5451 ***P. teterrella*** (Zinck.)
Bluegrass Webworm Moth
PARKS AND RESERVES: BAN, BBL, BMF, HWS, RGR. **COUNTIES:** BAR, BOO, BRK, BUL, FAY, GRY, HND, JEF, MRL, OLD. **DATES:** May 10–26, June 10–28, July 3, 6, 24–31, Aug. 4–9, 25, Sept. 2–5, 10. **REMARKS:** Common. Prey of the bolas spider, *Mastophora hutchinsoni*, in Fayette Co. (Yeargan 1988), responding to pheromones emitted by the spider.

Genus *Euchromius* Gn.

5454 ***E. ocelleus*** (Haw.)
Boone Co., BBL, Sept. 5, and Florence, Nov. 20; Harlan Co., KGC, Aug. 9; Oldham Co., Goshen, Apr. 14. Rare.

Genus *Argyria* Hbn.

5460 ***A. nummulalis*** Hbn.
McCreary Co., May 15, A. F. Braun.

Genus *Urola* Wlk.

5464 ***U. nivalis*** (Drury)
PARKS AND RESERVES: BAN, BLA, FKN, HWS, KLK, LCP, LMP, OCP, PYR, RGR. **COUNTIES:** BAR, BRK, BUL, CAD, FAY, GAL, GRY, HND, HRD, JEF, MCC, MEA, MRL, MUH, NEL, OLD, OWS, RUS, WAR. **DATES:** Apr. 28–Aug. 30, Sept. 2, 13, Nov. 14. Locally common.

Genus *Vaxi* Bleszynski

5465 ***V. auratella*** (Clem.)
PARKS AND RESERVES: BBL, HWS, KGC, SWR. **COUNTIES:** BEL, BOO, HLN, HND, JEF, OLD. **DATES:** June 4, 18–28, July 10, 11, Aug. 12. Common.

5466 ***V. critica*** (Fbs.)
Fulton Co., SR 94, Willingham Bottoms, 4.0 km (2.5 mi) E of Cayce, in light trap, Sept. 5; Menifee Co., RRG, Indian East Fork, Kelley Branch, July 18–19, in light trap. Uncommon.

Genus *Diatraea* Guilding

5479 ***D. grandiosella*** Dyar
Southwestern Corn Borer Moth
PARKS AND RESERVES: PYR. **COUNTIES:** CHR, FUL, HIC, JEF, MAR, MCC, MON. **DATES:** Apr. 2, May 26, July 24, Aug. 14, 22, 30, Sept. 7–11. **REMARKS:** Recorded damaging corn in the western two-fifths of Kentucky (*Kentucky Cooperative Economic Insect Report* 1974).

Genus *Haimbachia* Dyar

5482 ***H. squamulella*** (Zell.)
Meade Co., OCP, June 24, 1995, L. D. Gibson.

Genus *Eoruma* Ely

5492 ***E. densella*** (Zell.)
Laurel Co., DBNF, Marsh Branch Road at powerline, June 27, 1997, and DBNF, powerline E side

of S end of FS 775, July 10 and 25, and DBNF powerline W side of Craigs Creek Road, Sept. 19; McCracken Co., 9 miles SW of Paducah, June 29; all collected by L.D. Gibson.

Genus *Xubida* Schaus

5500 ***X. panalope*** (Dyar)
Oldham Co., Goshen, July 6, 11, 26, 28, 31, Aug. 1, 5, 6, Sept. 27. All collected by R.V. Gregg. Locally common.

Family **PYRALIDAE**

Subfamily **Pyralinae**

Genus *Pyralis* L.

5510 ***P. farinalis*** L.
Meal Moth
Fayette Co., Lexington, May 14, 19, June 9, July 18, Aug. 22, Sept. 21; Hart Co., May 13; Jefferson Co., Louisville, Apr. 3, June 2, July 25, Sept. 14, 25, Oct. 20, and Valley Station, July 3l; Owsley Co., 4.8 km (3 mi) NE of Booneville, May 23. A pest species that may be found infesting grains and grain products in any place and season.

5512 ***P. disciferalis*** Dyar
Jefferson Co., Valley Station, June 20, Aug. 12; Oldham Co., HWS, June 16, and Goshen, May 26, June 5–13, 25, 27, July 22, Aug. 21.

Genus *Aglossa* Latr.

5517 ***A. caprealis*** (Hbn.)
Murky Meal Moth
Type locality of synonym, *Tetralopha enthealis* Hulst, cited by Rindge (1955) as "Kentucky." Kenton Co., Covington, Oct., 1979 (larvae feeding around wine casks in small winery); Jefferson Co., Louisville, Dec. 27, 1979, larvae boring in corks of wine bottles (*Kentucky Cooperative Plant Pest Report* 1980).

5518 ***A. cuprina*** Zell.
Grease Moth
PARKS AND RESERVES: BBL, HWS, OCP, PNM, RBF, RRG. **COUNTIES:** BEL, BOO, BRE, JEF, MCC, MEA, OLD, OWS, POW, ROW. **DATES:** May 18–Sept. 2. Common.

Genus *Hypsopygia* Hbn.

5524 ***H. costalis*** (F.)
Clover Hayworm Moth
PARKS AND RESERVES: CER, HWS, LMP, OCP, PYR. **COUNTIES:** BOO, CHR, FAY, JEF, MEA, MOR, MUH, OLD, SHE, WAR. **DATES:** May 11, 23, June 3, 12–17, 25–30, July 12, 15, Aug. 11–25, Sept. 11, Oct. 21, Nov. 5. **REMARKS:** Common. Occasionally reported infesting hay in various parts of the Commonwealth (Lee Townsend, Sr., *pers. comm.*).

Genus *Herculia* Wlk.

5526 ***H. intermedialis*** (Wlk.)
Oldham Co., Goshen, June 10–22; Russell Co., LCP, in light trap, June 9. Uncommon.

5532 ***H. infimbrialis*** Dyar
PARKS AND RESERVES: BBM, BUC, LBL, LCP, NBP, PNM, PYR, RRG. **COUNTIES:** BEL, BOO, BRE, CHR, FAY, HLN, JEF, PER, POW, RUS, TRI. **DATES:** June 7–13, 30–July 14, Aug. 8, 14, Sept. 12, 18. Common.

5533 ***H. olinalis*** (Gn.)
PARKS AND RESERVES: BAN, BMF, BUC, CBB, FKN, HWS, KLK, KRC, LBL, LCP, LMP, LWO, PNM, PYR, RRG, SWR. **COUNTIES:** BAR, BEL, BRE, BUL, CHR, GAL, HEN, HLN, HND, JEF, LET, MCY, MEA, MRL, MUH, NEL, OLD, OWS, PER, ROW, RUS, TRI, WOL. **DATES:** Mar. 18–Sept. 18. Common.

Subfamily **Chrysauginae**

Genus *Parachma* Wlk.

5538 ***P. ochracealis*** Wlk.
Nelson Co., Leslie farm, June 23, 1971, G. Florence.

Genus *Galasa* Wlk.

5552 ***G. nigrinodis*** (Zell.)
PARKS AND RESERVES: BBL, BMF, CAV, KGC, OCP, PMS. **COUNTIES:** BAT, BOO, BUL, FAY, HLN, HRT, JEF, MEA, NEL, ROW. **DATES:** Apr. 28, May 27, June 2, 11, 19–30, July 7–16, Aug. 17, 25, Sept. 6, 9. Common.

Genus *Tosale* Wlk.

5556 ***T. oviplagalis*** (Wlk.)
PARKS AND RESERVES: BEC, CBB, LCP, OCP, PYR. **COUNTIES:** CHR, JEF, MCC, MEA, OLD, OWS, RUS. **DATES:** May 14–July 16, Aug. 9, 10, 28, 29, Sept. 1, Oct. 10. Common.

Genus *Clydonopteron* Riley

5563 ***C. sacculana*** (Bosc)
PARKS AND RESERVES: BAN, BUC, LBL, LCP, SWR. **COUNTIES:** BAR, BRE, BUL, HND, JEF, MCC, OWS, PER, RUS, TRI, WAR. **DATES:** May 22, 23, June 8, 29, July 4, 16–27, Aug. 10, 14, Sept. 2, 11. **REMARKS:** Locally common. This species was known as *C. tecomae* Riley until the earlier name was rediscovered by Miller and Becker (1989).

Genus *Arta* Grt.

5566 ***A. statalis*** Grt.
Bullitt Co., BMF, July 7; Fayette Co., Lexington, in Malaise trap, July 7–11; Hopkins Co., 16.0 km (10 mi) W of Madisonville, in black-light trap, June 26; Menifee Co., RRG, June 27–28; Taylor Co., July 5.

5568 ***A. olivalis*** Grt.
Bell Co., PNM, July 19, 1975, A. J. Brownell.

Genus ***Condylolomia*** Grt.

5571 ***C. participialis*** Grt.
PARKS AND RESERVES: BBM, BMF, LCP, PMS, PNM, RRG, WVP. **COUNTIES:** BEL, BUL, HLN, JEF, MEA, MEN, NEL, OLD, OWS, PIK, ROW, RUS, WHI. **DATES:** June 3–July 22, Aug. 5. Common.

Subfamily **Epipaschiinae**

Genus ***Epipaschia*** Clem.

5577 ***E. superatalis*** Clem.
PARKS AND RESERVES: BAN, BMF, HWS, LBK, LBL, LCP, OCP. **COUNTIES:** BAR, BUL, JEF, MEA, OLD, ROC, RUS, TRI. **DATES:** May 25–31, June 10–29, July 4, 6, 14, Aug. 4, 10. Common.

5579 ***E. zelleri*** (Grt.)
PARKS AND RESERVES: BBL, HWS, KGC, OCP. **COUNTIES:** BOO, BRA, JEF, HLN, MCC, MEA, OLD, WHI. **DATES:** May 23, June 14, 18, 27, July 7, 14, 25–31, Aug. 2, 13–17. Common.

Genus ***Oneida*** Hulst

5588 ***O. lunulalis*** (Hulst)
PARKS AND RESERVES: BBM, BUC, LCP, MCV, NBP, OCP, PNM, WVP. **COUNTIES:** BAR, BEL, BRE, HLN, JEF, MCC, MEA, PER, POW, ROW, RUS, WHI. **DATES:** Apr. 24–31, May 11, June 2–15, July 14–Aug. 6, 14. Common.

Genus ***Pococera*** Zell.

NOTE: *Pococera* was reinstated by Solis (1993) for species formerly in *Tetralopha* Zell.

5595 ***P. robustella*** (Zell.)
PARKS AND RESERVES: OCP. **COUNTIES:** HOP, MEA, NEL, OWS, POW, ROW, WHI. **DATES:** June 23–July 3, 16, 18, Aug. 4, 7.

5603 ***P. maritimalis*** (McD.)
PARKS AND RESERVES: BBL, BMF, LCP, OCP. **COUNTIES:** BOO, BUL, GAL, JEF, MEA, NEL, OWS, RUS. **DATES:** May 15–25, June 8–23, July 18, Aug. 10. Common.

5604 ***P. militella*** (Zell.)
PARKS AND RESERVES: LBL, OCP, PYR, RRG. **COUNTIES:** CAD, FUL, GAL, JEF, MEA, MEN, OLD, OWS, POW, TRI. **DATES:** May 27–June 13, 21, 29, July 14, 28, Aug. 9–10, 19, 24, Sept. 8, 16. Common.

5606 ***P. asperatella*** (Clem.)
PARKS AND RESERVES: BAN, BBM, BMF, CER, LBL, LCP, OCP, PMS, PNM, RRG. **COUNTIES:** BAR, BEL, BOO, BUL, FUL, HLN, JEF, MEA, MEN, ROW, RUS, TRI, WHI. **DATES:** May 15, 31, June 10, 13, 21–29, July 5–14, 24–Aug. 19, Sept. 5. Common.

5608 ***P. expandens*** (Wlk.)
PARKS AND RESERVES: BAN, BBL, BBM, BIP, KGC, LBL, LCP, NBP, OCP, PMS, PNM, PYR, RRG. **COUNTIES:** BAR, BEL, BOO, BUL, CAD, HLN, JEF, MEA, MEN, OWS, PIK, POW, RUS, TRI, WHI. **DATES:** Apr. 30–July 31. Common.

5617 ***P. humerella*** (Rag.)
PARKS AND RESERVES: BBL, BMF, OCP, RGR, WVP. **COUNTIES:** BOO, BRE, BUL, GAL, GRY, JEF, MEA. **DATES:** May 2, 18, 24, June 3, 9, 19, Aug. 9, 10.

5619 ***P. baptisiella*** (Fern.)
Boone Co., BBL, May 26, and BOW, June 9; Bullitt Co., N side of SR 480, 11.0 km (6.9 mi) E of SR 61, Sept. 6; Gallatin Co., Markland Dam, May 25, 29; Oldham Co., Goshen, July 25.

Subfamily **Galleriinae**

Genus ***Galleria*** F.

5622 ***G. mellonella*** (L.)
Greater Wax Moth
PARKS AND RESERVES: GBT, LBL, LCP, OCP. **COUNTIES:** CRL, FAY, JEF, MCC, MEA, OWS, RUS, TRI. **DATES:** Mar. 12, Apr. 2, June 7, 15, 25, July 8, 30, Aug. 15–20, Sept. 9, 13, 21, Oct. 7, 12, 24, 29, Dec. 15. **REMARKS:** A pest in bee hives.

Genus ***Achroia*** Hbn.

5623 ***A. grisella*** (F.)
Lesser Wax Moth
Oldham Co., HWS, one female in black-light trap, Aug. 25, 1977, C. V. Covell, Jr.

Genus ***Omphalocera*** Led.

5627 ***O. munroei*** Martin
PARKS AND RESERVES: LBL, OCP. **COUNTIES:** BOO, BUL, CAM, JEF, MEA, OLD, TRI. **DATES:** July 1–Aug. 18. **REMARKS:** A larva

found feeding in Louisville on pawpaw leaves pupated Sept. 20 and emerged on Oct. 16.

Genus *Aphomia* Hbn.

5630 ***A. terrenella*** Zell.
Oldham Co., Goshen, at lights, June 3 and July 23, R.V. Gregg. Rare.

Genus *Paralipsa* Butler

5633.1 ***P. fulminalis*** (Zell.)
Oldham Co., HWS, June 14, 1979, C. V. Covell, Jr. Rare.

Subfamily **Phycitinae**

Genus *Acrobasis* Zell.

5651 ***A. indiginella*** (Zell.)
Boone Co., BBL, June 18, July 7, Aug. 21; Fayette Co., Lexington, July 3; Hardin Co., Glendale, in Malaise trap, July 15–22; Menifee Co., RRG, Indian Creek, July 8; Owsley Co., 4.8 km (3 mi) NE of Booneville, June 25, July 3, 5.

5654 ***A. amplexella*** Rag.
Owsley Co., 4.8 km (3 mi) NE of Booneville, June 14, July 5, 14; Laurel Co., DBNF, FS 121 near Bolton Branch, June 11; all L. D. Gibson.

5659 ***A. palliolella*** Rag.
Oldham Co., Goshen, June 18, July 3, 5, 7, Aug. 12; Meade Co., OCP, June 21, July 10; Powell Co., NBP, July 4, and RRG, Tunnel Ridge, June 19.

5660 ***A. caryalbella*** Ely
Gallatin Co., SR 42, 11.2 km (7 mi) E of Warsaw, June 23; Meade Co., OCP, June 7, 21; Owsley Co., 4.8 km (3 mi) NE of Booneville, June 25, July 14; Rowan Co., DBNF, SR 1274, 3.2 km (2 mi) W of SR 519, July 16.

5661 ***A. juglandis*** (LeBaron)
Boone Co., BBL, July 11; Jefferson Co., WVP, July 3 and 6.

5663 ***A. kearfottella*** Dyar
Boone Co., BBL, July 11, and BCN, July 5; Laurel Co., DBNF, FS 121 near Bolton Branch, June 17; Menifee Co., RRG, Salt Fork of Gladie Creek, 366 m (1,200 ft) elevation, July 3–8; Powell Co., NBP, July 13, and RRG, Tunnel Ridge, July 4. Common.

5664 ***A. caryae*** Grt.
Harlan Co., BBM, July 14; Meade Co., OCP, July 10; Oldham Co., HWS, Aug. 5; Owsley Co., 4.8 km (3 mi) NE of Booneville, June 25; Russell Co., LCP, Aug. 11, 14. Uncommon.

5665 ***A. carpinivorella*** Neunzig
Rowan Co., DBNF, west end of Clack Mtn. Rd., from larva on *Carpinus caroliniana*, collected May 8, 1994, female emerged indoors June 8, 1994, L. D. Gibson.

5666 ***A. elyi*** Neunzig
Boone Co., CER, Aug. 17; Owsley Co., 4.8 km (3 mi) NE of Booneville, June 25; both L. D. Gibson.

5669 ***A. stigmella*** Dyar
Menifee Co., RRG, July 13; Powell Co., NBP, July 9, Aug. 17, and RRG, Tunnel Ridge, June 19; Rowan Co., DBNF, SR 1274, 3.2 km (2 mi) W of SR 519, July 16. Uncommon.

5670 ***A. aurorella*** Ely
Bullitt Co., Leigh's Prairie, 1.6 km (1 mi) N of SR 480, July 22; Jefferson Co., Valley Station, Sept. 1, 8; Powell Co., RRG, Tunnel Ridge, June 28; Rowan Co., DBNF, SR 1274, 3.2 km (2 mi) W of SR 519, July 16; Russell Co., LCP, Aug. 16.

5672 ***A. exsulella*** (Zell.)
Bell Co., PNM, June 8, 9; Laurel Co., DBNF, FS 121 near Bolton Branch, June 11; Oldham Co., Goshen, May 11–29, June 3–11, July 25; Powell Co., RRG, Tunnel Ridge, June 28; Whitley Co., Goldbug, June 5.

5673 ***A. angusella*** Grt.
Bell Co., PNM, July 19; Powell Co., NBP, July 8, Aug. 17, 19; Russell Co., LCP, Aug. 14. Uncommon.

5674 ***A. demotella*** Grt.
Bell Co., PNM, June 9; Owsley Co., 4.8 km (3 mi) NE of Booneville, May 28; Powell Co., RRG, Tunnel Ridge, May 21.

5675 ***A. latifasciella*** Dyar
PARKS AND RESERVES: BBL, BMF, NBP, OCP, PNM, RRG. **COUNTIES:** BEL, BOO, BUL, GAL, MEA, MEN, POW. **DATES:** June 19, 23, July 7–19, Aug. 9–17. Common.

5677 ***A. normella*** Dyar
Menifee Co., RRG, reared from *Corylus americana*, larva found May 29, adult June 23; Owsley Co., 4.8 km (3 mi) NE of Booneville, from larvae on *Corylus americana*, collected May 14, 1994, emerged indoors June 12 and 19, 1994; Rowan Co., DBNF, 8.0 km (5 mi) NE of Morehead, from larva on *Corylus americana*, collected May 8, 1994, male emerged indoors June 12, 1994; Trigg Co., LBL, at lights, July 15. Common.

5680 ***A. ostryella*** Ely
Edmonson Co., MCV, June, 1924. A specimen in the NMNH is labelled "Mammoth C., Ky. VI.1924" and "Collection G. P. Engelhardt."

5685 ***A. cunulae*** Dyar & Heinr.
Bell Co., PNM, June 9, July 18; Laurel Co., Bolton Branch, June 11; Meade Co., OCP, Aug. 8; Powell Co., RRG, Tunnel Ridge, June 19.

5686 ***A. caryivorella*** Rag.
Rowan Co., Morehead, June 20 and 25, 1962, T. N. Freeman. Rare.

5689 ***A. betulivorella*** Neunzig
Menifee Co., RRG along Edwards Branch, from larva on *Betula nigra*, collected May 29, 1994, one male emerged June 20, 1994, L. D. Gibson.

5690 ***A. rubrifasciella*** Pack.
Harlan Co., KGC, collected as larvae on *Alnus* on May 17, 1993, adults June 17, 1993; Laurel Co., FS 121, larva on *Alnus serrulata*, collected May 22, 1994, one male emerged June 20, 1994; both L. D. Gibson.

Genus ***Myelopsis*** Heinr.

5718 ***M. subtetricella*** (Rag.)
Jefferson Co., Valley Station, July 3, 6, A. J. Brownell. Rare.

Genus ***Amyelois*** Amsel

5724 ***A. transitella*** (Wlk.)
A specimen in the NMNH under this name is labelled "*Ex* codling moth bands, Henderson, Ky., 1931, D. O. Eddy, collector."

Genus ***Monoptilota*** Hulst

5736 ***M. pergratialis*** (Hulst)
Harlan Co., KGC, 1 female, Aug. 9, 1980, L. D. Gibson.

Genus ***Glyptocera*** Rag.

5745 ***G. consobrinella*** (Zell.)
Powell Co., NBP, July 9, 1981, C. Cornett.

Genus ***Oreana*** Hulst

5767 ***O. unicolorella*** (Hulst)
Fayette Co., Lexington, collected skeletonizing leaf of *Crataegus* on July 9, 1917, emerged July 31, 1917, H. Garman; Powell Co., RRG, Tunnel Ridge, June 19, 1993, D. J. Wright.

Genus ***Salebriaria*** Heinr.

5773 ***S. engeli*** (Dyar)
Bell Co., PNM, June 9; Russell Co., LCP, Aug. 14; Trigg Co., LBL, July 15 (Neunzig 1988).

5773.1 ***S. fasciata*** Neunzig
The following are listed as paratypes in the original description (Neunzig 1988): Harlan Co., PMS, 1 female, July 8, 1977, C. V. Covell, Jr.; Jefferson Co., Valley Station, 1 male, 1 female, May 9, June 8, 1974, Siegfried Scholz; Whitley Co., CBF, 1 female, Apr. 30, 1974, C. Cornett, and Goldbug, 1 female at lights, July 7, 1981, W. B. Early.

5773.2 ***S. rufimaculatella*** Neunzig
Neunzig (1988) named the following as paratypes: Bell Co., PNM, 2 females, June 8, 1981, C. Cornett; Bullitt Co., BMF, 1 female, June 17, 1976, A. J. Brownell; Jefferson Co., Valley Station, 2 females, May 15, June 13, 1974, S. Scholz; Muhlenberg Co., LMP, 1 male, May 15, 1977, C. Cornett.

Genus ***Meroptera*** Grt.

5786 ***M. cviatella*** Dyar
Gallatin Co., Markland Dam, May 27, June 2, Aug. 3, L. D. Gibson and D. J. Wright.

Genus ***Nephopterix*** Hbn.

5796 ***N. subcaesiella*** (Clem.)
PARKS AND RESERVES: BBL, BBM, BMF, CER, HWS, LCP, OCP, PNM. **COUNTIES:** BEL, BOO, BUL, FAY, HLN, JEF, MEA, OLD, OWS, RUS. **DATES:** Apr. 8, 22, 28, May 11, 28–June 20, July 12, 14, Aug. 1–8, 16–21, Sept. 6. Common.

5797 ***N. virgatella*** (Clem.)
Bell Co., PNM, June 9; Fayette Co., Lexington, May 9; Jefferson Co., Louisville, Aug. 12, and WVP, June 20; Whitley Co., Goldbug, June 5.

5802 ***N. uvinella*** (Rag.)
Fayette Co., Lexington, reared from sweetgum, emerged Apr. 26 and July 31; Henderson Co., SWR, Sept. 10; Jefferson Co., Valley Station, May 18, June 13, 20, Sept. 1; Meade Co., OCP, Aug. 10.

5803 ***N. celtidella*** (Hulst)
Boone Co., BBL, May 18, 29, June 18; Bullitt Co., BMF, in Malaise trap, Apr. 17–20, also reared from *Celtis laevigata*, emerged Mar. 5–17 and June 1–3 (A. J. Brownell); Oldham Co., CBB, May 14, and HWS, June 15. Common.

Genus ***Tlascala*** Hulst

5808 ***T. reductella*** (Wlk.)
PARKS AND RESERVES: BAN, HWS, NBP, OCP, PNM. **COUNTIES:** BAR, BEL, GAL, MEA, OLD, POW. **DATES:** Apr. 27, May 11–June 18, 28–30, July 22, 29, Aug. 3, 13–19. Common.

Genus ***Actrix*** Heinr.

5818 ***A. nyssaecolella*** (Dyar)
Laurel Co., FS 4176, June 11; Powell Co., RRG, Tunnel Ridge, May 16; both D. J. Wright. Uncommon.

Genus *Dioryctria* Zell.

5847 ***D. disclusa*** Heinr.
Bullitt Co., BMF, May 22; Owsley Co., 4.8 km (3 mi) NE of Booneville, June 14, 15; Powell Co., RRG, Tunnel Ridge, July 8.

Genus *Elasmopalpus* Blanchard

5896 ***E. lignosellus*** (Zell.)
Jefferson Co., Jeffersontown, Sept. 2, 1983, R. A. Henderson. Rare.

Genus *Canarsia* Hulst

5926 ***C. ulmiarrosorella*** (Clem.)
Fayette Co., Lexington, "From trunk of live elm bored at Vivar [*sic*] (Viva, Laurel Co. ?), Sept., 1899, emerged June 2, 1900, H. Garman." Jefferson Co., WVP, May 27; Menifee Co., RRG, at black-light, June 23.

Genus *Psorosina* Dyar

5931 ***P. hammondi*** (Riley)
Fayette Co., Lexington, emerged July 21, 23; Hickman Co., Clinton, emerged Sept. 3, 13, H. Garman (UK). These specimens were erroneously labelled "*Mineola indiginella*."

Genus *Homoeosoma* Curt.

5936 ***H. stypticellum*** Grt.
Jefferson Co., Anchorage, May 22; Meade Co., OCP, Aug. 10.

Genus *Phycitodes* Hamp.

5946 ***P. albatella*** (Rag.)
Harlan Co., BBM, July 14; Jefferson Co., May 16. The previous records represent subspecies *mucidella* (Rag.). Uncommon.

5946.1 ***P. reliquellum*** (Dyar)
Jefferson Co., Anchorage, May 22, 1951, B. L. Monroe, Jr. (determined by H. H. Neunzig as a good species.)

Genus *Laetilia* Rag.

5949 ***L. coccidivora*** (J. H. Comstock)
Jefferson Co., CAH, Aug. 2, 1976, C. V. Covell, Jr.

5951 ***L. myersella*** Dyar
Bell Co., PNM, July 19; Owsley Co., 4.8 km (3 mi) NE of Booneville, June 16.

Genus *Euzophera* Zell.

5995 ***E. semifuneralis*** (Wlk.)
Fayette Co., Lexington, in Malaise trap, June 27–30; Harlan Co., BBM, July 9–14, and PMS, July 8; Menifee Co., RRG, July 13; Powell Co., NBP, May 13, and RRG, Apr. 28; Whitley Co., Goldbug, Apr. 16. Common.

5995.2 ***E. habrella*** Neunzig
One paratype female was from Harlan Co., BBM, July 14, 1979, C. V. Covell, Jr. (Neunzig 1990).

5997 ***E. ostricolorella*** Hulst
Bath Co., CAV, Sept. 5; Bell Co., PNM, June 6–10, Oct. 23; Jefferson Co., Valley Station July 16, 24, Sept. 2–19, 29, Oct. 9–15, 24, Nov. 4; Russell Co., LCP, June 9, Oct. 13; Whitley Co., CBF, Oct. 4, and Goldbug, June 5. Common.

Genus *Eulogia* Heinr.

5999 ***E. ochrifrontella*** (Zell.)
PARKS AND RESERVES: BBL, BMF, OCP, RRG. **COUNTIES:** BOO, BUL, MCC, MEA, MEN, OLD. **DATES:** May 31–June 10, 28, July 11, Aug. 10–13, Sept. 1, 7, Oct. 6. Common.

Genus *Ephestiodes* Rag.

6001 ***E. infimella*** Rag.
"Kentucky (Aug.)" (Heinrich 1956); Jefferson Co., WVP, July 11.

Genus *Moodna* Hulst

6005 ***M. ostrinella*** (Clem.)
Powell Co., RRG, Apr. 28; Rockcastle Co., Aug. 10, 25, 31.

Genus *Vitula* Rag.

6007 ***V. edmandsii*** (Pack.)
Fayette Co., Lexington, in *Bombus* nest, emerged Feb. 6 and 24, and Lexington, Mar. (Heinrich 1956); Grayson Co., ASC, Sept. 11; Harlan Co., BBM, July 14; Oldham Co., Goshen, May 18–31, June 7, 9, July 2, 19, 26, Aug. 2–19, 27, Sept. 2–27, Oct. 6, 28; Russell Co., LCP, Oct. 6. Locally common.

Genus *Caudellia* Dyar

6013 ***C. albovittella*** Dyar
Meade Co., OCP, at light, Oct. 11, 1986, B. S. Nichols. Rare.

Genus *Plodia* Gn.

6019 ***P. interpunctella*** (Hbn.)
Indianmeal Moth
Daviess Co., adults active in home (Feb. 1, 1978); Jefferson Co., Louisville, Jan. 19, Feb. 4, 7 (in Mother's Oats), Mar. 29, Aug. 17, Sept., Nov. 24, and Valley Station, June 20; Madison Co., Apr. 27; Pulaski Co., Apr. 27; Todd Co., infesting grain bin, (*Kentucky Cooperative Plant Pest Report* 1977). A common pest that may be found infesting grain products anywhere and in all months.

Genus ***Anagasta*** Heinr.

6020 ***A. kuehniella*** (Zell.)
Mediterranean Flour Moth
Fayette Co., Lexington, Dec. 5; Mercer Co., Harrodsburg, reared from flour, Oct. 27, 28, Nov. 4; Rockcastle Co., Sept. 20. A grain products pest.

Genus ***Varneria*** Dyar

6029 ***V. postremella*** Dyar
"Kentucky (July)" (Heinrich 1956); Bell Co., PNM, July 18–19; Menifee Co., RRG, July 19–20.

Subfamily **Peoriinae**

Genus ***Peoria*** Rag.

6042 ***P. longipalpella*** (Rag.)
PARKS AND RESERVES: CAV, WVP. **COUNTIES:** BAT, BRA, BUL, FUL, JEF, LAU, LIV, MCC, OWS. **DATES:** July 3, 6, Sept. 5–25. **REMARKS:** Previously known only from Virginia and North Carolina in August (Shaffer 1968).

6044 ***P. tetradella*** (Zell.)
Bullitt Co., 1.6 km (1 mi) N of SR 480 on Pine Creek Forest Rd., at light, July 22, 1989, D. J. Wright. Uncommon.

6053 ***P. approximella*** (Wlk.)
PARKS AND RESERVES: AJJ, LBL, PYR. **COUNTIES:** BUL, CAM, CHR, JEF, OWS, ROC, TAY, TRI, WHI. **DATES:** June 5, 6, 24, 27, July 5–22. Uncommon.

6054 ***P. luteicostella*** (Rag.)
Fleming Co., Aug. 21, A. F. Braun.

Genus ***Atascosa*** Hulst

6087 ***A. glareosella*** (Zell.)
Christian Co., PYR, Aug. 26; Whitley Co., DBNF, powerline cut on S side of SR 90, 2.5 miles W of US 25W, Aug. 29; both L.D. Gibson. A wet prairie species (A. Solis, *pers. comm.*).

Family **THYRIDIDAE**

NOTE: Our records for adults of this family are from daytime captures or observations, as thyridids rarely come to lights.

Window-winged Moths

Genus ***Thyris*** Laspeyres

6076 ***T. maculata*** Harr.
Spotted Thyris
Bullitt Co., BMF, in Malaise trap, Apr. 8–20, May 14, June 1–15; Fayette Co., Lexington, May 24 on timothy, July 24, 28, on flowers of *Houstonia tenuifolia*; Meade Co., OCP, May 29; Owsley Co., 4.8 km (3 mi) NE of Booneville, June 4.

6077 ***T. sepulchralis*** Guér.
Mournful Thyris
PARKS AND RESERVES: BMF, CBF, MCV, OCP, RRG. **COUNTIES:** BUL, EDM, FAY, JEF, KEN, LET, MAD, MCY, MEA, MEN, OWS, WHI. **DATES:** Apr. 8–30, May 2–20, June 4–9, 18, Sept. 12–19. Common.

Genus ***Dysodia*** Clem.

6078 ***D. oculatana*** Clem.
Eyed Dysodia
Breathitt Co., RBF, July 13–19, Aug. 16–23, in Malaise trap; Harlan Co., KGC, July 13; Jefferson Co., Valley Station, May 22, Aug. 4, 16; Owsley Co., 4.8 km (3 mi) NE of Booneville, June 14.

Superfamily **PTEROPHOROIDEA**

Family **PTEROPHORIDAE**

Plume Moths

Genus ***Sphenarches*** Meyr.

6090 ***S. ontario*** (McD.)
Lexington, Fayette Co., in Malaise trap, June 12–16, 1975, P. H. Freytag.

Genus ***Geina*** Tutt

6091 ***G. periscelidactyla*** (Fitch)
Fayette Co., Lexington, reared on grape, June 1, 4; Jefferson Co., Valley Station, June 6, 27; Letcher Co., Pine Mountain, June 10; Rowan Co., Clack Mtn., June 20.

6091.1 ***G. sheppardi*** B. Landry
Fayette Co, Lexington, "emerged June 4, 1901"; Jefferson Co., Valley Station, May 27.

6092 ***G. tenuidactyla*** (Fitch)
Bullitt Co., BMF, June 10–15; Fayette Co., Lexington, in Malaise trap, June 16–20; Jefferson Co., Valley Station, June 4; McCreary Co., "Natural Bridge," June 22, 1932.

6093 ***G. buscki*** (McD.)
Bullitt Co., BMF, June 19; Fayette Co., Lexington, in Malaise trap, June 12–27; Hart Co., Green River Bluffs, May 20; Letcher Co., Pine Mountain, June 16.

Genus ***Oxyptilus*** Zell.

6097 ***O. delawaricus*** Zell.
Letcher Co., Big Black Mountain, June 23, Joe Day Branch, July 8, and Pine Mountain, June 11; Powell Co., date unknown; all A. F. Braun.

Genus *Platyptilia* Hbn.

6109 ***P. carduidactyla*** (Riley)
Fayette Co., Lexington, in Malaise trap, June 9–12, July 7–11; Jefferson Co., Louisville, Apr. 10, 30. Common.

6112 ***P. williamsii*** Grinnell
Floyd Co., JWS, Apr. 22, 1974, C. Cornett.

Genus *Amblyptilia* Hbn.

6118 ***A. pica*** (Wlsm.)
Letcher Co., Big Black Mountain, June 8, July 9, A. F. Braun.

Genus *Stenoptilodes* Zimmerman

6125 ***S. carolina*** (Kft.)
Letcher Co., Gilley, June 13, and Pine Mountain, June 24, July 7; McCreary Co., Yahoo Ridge, June 18; Perry Co., Lynn Fork, July 11; Rowan Co., Clack Mtn., June 20; all A. F. Braun.

6127 ***S. auriga*** (B. & L.)
Letcher Co., Big Black Mountain, May 4, 24, Joe Day Branch, emerged May 19, 23, and Pine Mountain, May 24, June 11; McCreary Co., Pine Knot, emerged May 29, June 14; Metcalfe Co., May 21; Powell Co., NBP, May 18, 23; Rowan Co., Morehead, emerged May 18 and 22; all A. F. Braun.

6128 ***S. edwardsii*** (Fish)
Harlan Co., BBM, May 23, A. F. Braun.

Genus *Adaina* Tutt

6155 ***A. bipunctata*** (Mösch.)
Fulton Co., SR 94, 3.2 km (2 mi) E of Cayce, Sept. 11, 1993, R. A. Watkins.

6160 ***A. ambrosiae*** (Murt.)
Fayette Co., Lexington, in Malaise trap, Aug. 25–29; Hardin Co., Glendale, in Malaise trap, July 1–8; Jefferson Co., Louisville, May 10, Sept. 3, 9, 10.

Genus *Oidaematophorus* Wallgr.

6186 ***O. inquinatus*** Zell.
"Ky." "Aug." August Busck, 5 specimens in the NMNH; McCracken Co., Paducah, Sept. 25.

6203 ***O. homodactylus*** (Wlk.)
Bullitt Co., BMF, June 19; Jefferson Co., Valley Station, May 29; Letcher Co., head of Line Fork, June 28.

6204 ***O. elliotti*** (Fern.)
McCreary Co., South Fork of Cumberland River, June 18; Powell Co., June 12, 17; all A. F. Braun.

6207 ***O. paleaceus*** (Zell.)
Two specimens are in the NMNH labelled "Ky.," "Aug," and "Aug. Busck Collection"; Wolfe Co., KRC, July 15.

6212 ***O. kellicottii*** (Fish)
Jefferson Co., WVP, July 6; Letcher Co., Big Black Mountain, July 17, 29.

6213 ***O. lacteodactylus*** (Cham.)
Kentucky is the type locality; Chambers (1873) wrote "Kentucky, in June."

Genus *Emmelina* Tutt

6234 ***E. monodactyla*** (L.)
Bullitt Co., BMF, June 19; Fayette Co., Lexington, June 16–23, July 7–11, Aug. 11–15; Hardin Co., Glendale, July 15–22; Jefferson Co., Louisville, Mar. 18, July 4, 9, Oct. 11; Oldham Co., CBB, July 10, and HWS, June 24. Common.

Family **THYATIRIDAE**

Thyatirid Moths

Genus *Habrosyne* Hbn.

6235 ***H. scripta*** (Gosse)
Lettered Habrosyne
Harlan Co., BBM, May 17, July 9, 11, 14, Aug. 8. Sometimes common at this site.

6236 ***H. gloriosa*** (Gn.)
Bullitt Co., PCP, July 22; Meade Co., OCP, Aug. 9; Oldham Co., CBB, May 14. Rare.

Genus *Pseudothyatira* Grt.

6237 ***P. cymatophoroides*** (Gn).
Tufted Thyatirid
PARKS AND RESERVES: BBM, BUC, CBF, NBP. **COUNTIES:** BOO, BRE, FAY, GAL, HLN, JEF, OLD, OWS, PER, POW, ROW, WHI. **DATES:** Apr. 20–Aug. 31, Oct. 13. Common.

Genus *Euthyatira* Smith

6240 ***E. pudens*** (Gn.)
Dogwood Thyatirid
PARKS AND RESERVES: BBM, BIP, BMF, CAC, CBF, JWS, NBP, PMS, PNM, RRG, WVP. **COUNTIES:** BEL, BUL, CAR, FLO, HLN, JEF, LAU, MEN, NEL, PIK, POW, WHI. **DATES:** Apr. 4–May 27. Common.

Superfamily **DREPANOIDEA**

Family **DREPANIDAE**

Hooktip Moths

Genus ***Drepana*** Schr.

6251 ***D. arcuata*** Wlk.
Arched Hooktip
PARKS AND RESERVES: BUC, CAV, CBF, JWS, KGC, NBP, OCP, PNM, RRG. **COUNTIES:** BAT, BEL, BRE, FLO, HLN, JEF, LEE, LET, MEA, MEN, OWS, PER, POW, ROW, WHI. **DATES:** Apr. 4, 11, 24–June 4, 16, 30, July 9, 21, Aug. 22–Sept. 21. Common.

Genus ***Eudeilinia*** Pack.

6253 ***E. herminiata*** (Gn.)
Northern Eudeilinia
McCreary Co., near Bell farm, SR 1363, Apr. 18; Menifee Co., RRG, along East Fork of Indian Creek, June 22; Rowan Co., Morehead, July 25; Trigg Co., LBL, July 13. Uncommon. Sometimes seen at rest during the day on leaves of plants in dense forests.

Genus ***Oreta*** Wlk.

6255 ***O. rosea*** (Wlk.)
Rose Hooktip
PARKS AND RESERVES: ASC, BBL, BMF, CAC, KRC, PMS, SWR, WVP. **COUNTIES:** BOO, BRA, BUL, CAD, CAR, GAL, GRY, HLN, HND, JEF, LEE, NEL, ROW, WHI, WOL. **DATES:** May 4–June 18, July 6–25, Aug. 7–16, 27, 30, Sept. 4–11, 23. Common.

Superfamily **GEOMETROIDEA**

Family **GEOMETRIDAE**

Inchworm or Geometer Moths

Subfamily **Oenochrominae**

Genus ***Alsophila*** Hbn.

6258 ***A. pometaria*** (Harr.)
Fall Cankerworm Moth
Franklin Co., Game Farm, Frankfort, Dec. 15 (male); Harlan Co., Lynch, larvae collected on red maple, June 15; Jefferson Co., Louisville, Nov. 27. Uncommon.

Subfamily **Ennominae**

Genus ***Heliomata*** G. & R.

6261 ***H. cycladata*** G. & R.
Common Spring Moth
PARKS AND RESERVES: BBL, BMF, COB, HWS, JWS, LCP, LMP, OCP, PNM. **COUNTIES:** BEL, BOO, BRA, BUL, FAY, FLO, HIC, JEF, MCC, MEA, MUH, NEL, OLD, RUS, WHI. **DATES:** Apr. 19–July 15. Common.

6263 ***H. infulata*** (Grt.)
Rare Spring Moth
Knott Co., Vest, 11.3 km (7 mi) NW of Hindman, June 19, 1987, Roger A. Zebold. Rare.

Genus ***Protitame*** McD.

6270 ***P. virginalis*** (Hulst)
Virgin Moth
PARKS AND RESERVES: BCN, LBL, OCP. **COUNTIES:** BOO, BUL, GAL, JEF, MEA, OLD, TRI. **DATES:** May 25, June 29–July 24, Aug. 10–29. Uncommon.

Genus ***Eumacaria*** Pack.

6272 ***E. latiferrugata*** (Wlk.)
Brown-bordered Geometer
Boone Co., BBL, June 1; Jefferson Co., Louisville, July 9, and Valley Station, May 25; McCracken Co., Ray Black farm, 9.7 km (6 mi) SW of Paducah, May 20; Rowan Co., Morehead, Sept. 1. Uncommon.

Genus ***Itame*** Hbn.

6273 ***I. pustularia*** (Gn.)
Lesser Maple Spanworm Moth
PARKS AND RESERVES: BBM, BEC, BMF, CBF, JWS, KGC, KRC, LCP, LWO, NBP, PNM, RRG. **COUNTIES:** BEL, BUL, FAY, FLO, GUP, HLN, HRD, JEF, LEE, LET, MAD, MCC, OWS, PIK, POW, ROW, RUS, TAY, WHI, WOL. **DATES:** Apr. 20–Oct. 1. Very common and sometimes abundant.

6303 ***I. subcessaria*** (Wlk.)
Barred Itame
Harlan Co., BBM, July 9; Letcher Co., KGC, July 9; Mercer Co., Shaker's Landing, June 14–15; Russell Co., LCP, May 29. Rare.

Genus ***Mellilla*** Grt.

6322 ***M. xanthometata*** (Wlk.)
Orange Wing
PARKS AND RESERVES: AJJ, BAN, BBL, BMF, CBB, HWS, LBL, OCP, RGR. **COUNTIES:** BAR, BOO, BRA, BUL, CAM, CRI, FAY, GAL, GRY, JEF, MEA, MEN, NEL, OLD, TRI, WOD. **DATES:** Apr. 8–Aug. 19, 27, Oct. 22. Locally common; active day and night.

Genus ***Semiothisa*** Hbn.

6326 ***S. aemulataria*** (Wlk.)
Common Angle
PARKS AND RESERVES: BAN, BBL, BEC, BIP, BMF, BUC, HWS, LBL, LCP, LWO, NBP, OCP, PNM, RBF, RGR. **COUNTIES:** BAR, BEL, BOO, BRE, BRK, BUL, EST, GRE, GRY,

HND, HRT, JEF, LAU, LET, MEA, NEL, OLD, PER, PIK, POW, RUS, TRI, WHI. **DATES:** Apr. 13–Oct. 11. Common.

6331 ***S. promiscuata*** Fgn.
Promiscuous Angle
PARKS AND RESERVES: BBL, BBM, BEC, BUC, CAC, HWS, JCF, LBK, LCP, LMP, LWO, NBP, OCP, PNM, PYR, RBF, RGR, RRG. **COUNTIES:** BEL, BOD, BOO, BRE, BRK, CAR, CHR, FAY, GRY, HLN, HRT, JEF, LET, MCC, MEA, MEN, MUH, OLD, PER, POW, ROC, ROW, RUS, TRI, WHI. **DATES:** Apr. 19–Sept. 8, 20. **REMARKS:** Common. Paratypes of this species include specimens from HWS, Morehead, Renfro Valley, and Valley Station.

6332 ***S. punctolineata*** (Pack.)
McCracken Co., 14.5 km (9 mi) SW of Paducah, July 24, 1976, C. V. Covell, Jr. A rare stray into Kentucky from the deep south.

6335 ***S. aequiferaria*** (Wlk.)
Woody Angle
PARKS AND RESERVES: BBM, BEC, CAC, GBT, HWS, KGC, KLK, LCP, LMP, MUR, OCP. **COUNTIES:** CAR, CRL, FAY, FUL, HIC, HLN, JEF, MCC, MEA, MRL, MUH, OLD, RUS, SPE. **DATES:** Apr. 18–27, May 5–24, June 13–18, July 13–24, Aug. 6, 8, 24, 31, Sept. 2–16, Oct. 2. Common.

6339 ***S. transitaria*** (Wlk.)
Blurry Chocolate Angle
PARKS AND RESERVES: BAN, CBF, GBO, JWS, KGC, NBP, PNM, PYR, RGR, RRG. **COUNTIES:** BAR, BEL, BRK, CHR, FLO, GRY, GUP, HLN, JEF, MCC, MEN, OWS, POW, ROW, WHI. **DATES:** Apr. 15–May 6, 18, 31–June 13, July 7, 13, Aug. 1–10, Sept. 1, 10. Locally common.

6340 ***S. minorata*** (Pack.)
Minor Angle
Floyd Co., JWS, June 4; Powell Co., RRG, Tunnel Ridge, at light, Apr. 28, May 21. Apparently rare in Kentucky.

6341 ***S. bicolorata*** (F.)
Bicolored Angle
PARKS AND RESERVES: BAN, BMF, BUC, CBB, CBF, JWS, LCP, NBP, PMS, PNM, PYR, RBF, RGR, RRG. **COUNTIES:** BAR, BEL, BRE, BRK, BUL, CHR, FLO, GRV, GRY, HLN, JEF, MCC, MEN, OLD, OWS, PER, POW, ROW, RUS, WHI. **DATES:** Apr. 28–Oct 9, Nov. 14. Common.

6342 ***S. bisignata*** (Wlk.)
Red-headed Inchworm Moth
PARKS AND RESERVES: BBL, KRC, LMP, NBP, PMS, PNM, PYR, RBF, RRG. **COUNTIES:** BEL, BOO, BRE, CAD, CHR, HLN, MOR, MUH, NEL, OWS, POW, ROW, WOL. **DATES:** May 12, 13, 27, June 7–12, July 8–20, 30, Aug. 11, 23, Sept. 5, 8, Oct 11. Common.

6348 ***S. fissinotata*** (Wlk.)
Hemlock Angle
PARKS AND RESERVES: BIP, CAC, KRC, NBP, PMS, PNM, RBF, RRG. **COUNTIES:** BEL, BRE, CAR, HLN, JEF, LEE, MEN, PIK, POW, WOL. **DATES:** Apr. 14–May 3, 11–24, June 8–16, July 7–18, Aug. 18, 26. Locally common.

6352 ***S. granitata*** (Gn.)
Granite Moth
PARKS AND RESERVES: BAN, BBM, BIP, BMF, CBF, KGC, LCP, LMP, NBP, PMS, PNM, PYR, RBF, RRG. **COUNTIES:** BAR, BEL, BOD, BRE, BUL, CAD, CHR, EST, GRV, GUP, HLN, JEF, MCC, MEN, MUH, NEL, OWS, PIK, POW, ROW, RUS, WHI. **DATES:** Apr. 15–Sept. 29. Common.

6353 ***S. multilineata*** (Pack.)
Many-lined Angle
PARKS AND RESERVES: BBL, BMF, CBB, FKN, HWS, LBK, LCP, LMP, OCP, PYR, RGR. **COUNTIES:** BRK, BOO, BUL, CHR, GRE, GRY, HRD, HRT, JEF, LEE, MCC, MEA, MEN, MUH, NEL, OLD, RUS, TRI, WOO. **DATES:** May 1–Aug. 18. Common.

6357 ***S. eremiata*** (Gn.)
Three-lined Angle
Jefferson Co., Valley Station, May 17, 1975, S. Scholz. Rare.

6360 ***S. quadrinotaria*** (H.-S.)
Four-spotted Angle
PARKS AND RESERVES: BBL, BBM, BMF, CAC, CBB, HWS, JWS, KGC, LCP, NBP, OCP, PMS, RRG. **COUNTIES:** BOO, BUL, CAR, EST, FLO, GAL, HLN, JEF, MEA, MEN, NEL, OLD, POW, RUS, SPE, TRM. **DATES:** Apr. 3–July 9, 27, Aug. 8. **REMARKS:** Common to abundant in woods in springtime. Active in the daytime as well as at night.

6362 ***S. continuata*** (Wlk.)
Curved-lined Angle
PARKS AND RESERVES: BBL, BMF, HWS, LBK, LCP, LMP, OCP, PNM, PYR, RGR. **COUNTIES:** BAR, BEL, BOO, BRK, BUL, CAD, FAY, GRE, GRY, JEF, MCC, MEA, MUH,

NEL, OLD, RUS, TRI. **DATES:** Mar. 20–Sept. 12. Common.

6386 ***S. ocellinata*** (Gn.)
Faint-spotted Angle
PARKS AND RESERVES: BAN, BBL, BBM, BEC, BMF, BUC, CBB, FKN, GBT, HWS, JCF, KGC, LBK, LBL, LCP, OCP, PMS, PNM, PYR, RBF, RGR. **COUNTIES:** BAR, BEL, BOO, BRA, BRE, BRK, BUL, CAD, CHR, CRL, EST, FAY, GAL, GRE, GRV, GRY, HLN, HND, HRD, JEF, MAD, MCC, MEA, MER, NEL, OLD, OWS, PER, ROW, RUS, TRI, WOL. **DATES:** Apr. 2–Oct 22. Very common.

6397 ***S. mellistrigata*** (Grt.)
Jefferson Co., Louisville, J. R. Merritt. A rare stray from the west.

6399 ***S. subminiata*** (Pack.)
Fulton Co., RNR, May 28; Jefferson Co., FAO, Aug. 21, Sept. 20, and Valley Station, July 6, Sept. 12; McCracken Co., Paducah, July 24; Meade Co., OCP, Apr. 11. Uncommon.

6405 ***S. gnophosaria*** (Gn.)
Hollow-spotted Angle
PARKS AND RESERVES: HWS, LCP, NBP, PYR, RGR. **COUNTIES:** BRK, CAD, CHR, FAY, FUL, GRY, JEF, MCC, MRL, NEL, OLD, POW, ROW, RUS, SPE. **DATES:** Apr. 17–Sept. 17. **REMARKS:** Locally common, especially late in the season.

Genus *Enconista* Led.

6419 ***E. dislocaria*** (Pack.)
Pale-veined Enconista
Boone Co., BBL, Apr. 25; Fayette Co., Lexington, Apr. 20–30, May 20, June 13, July 1, 25, Aug. 2; Gallatin Co., 16.0 km (10 mi) E of Warsaw, June 2; Jefferson Co., Louisville, Apr. 19, May 1, 10, 14, 23–29; Oldham Co., Anchorage, May 5, Crestwood, June 23, and HWS, May 17. Uncommon.

Genus *Orthofidonia* Pack.

6428 ***O. tinctaria*** (Wlk.)
Powell Co., RRG, Tunnel Ridge, 1 female at light, Apr. 28, 1989, D. J. Wright. Local and rare.

6430 ***O. flavivenata*** (Hulst)
Laurel Co., DBNF, FS 615A/B, Apr. 18, 19, and ridge top SW of Ned Branch, Apr. 24; Owsley Co., 4.8 km (3 mi) NE of Booneville, Apr. 18. Rare.

Genus *Hesperumia* Pack.

6431 ***H. sulphuraria*** Pack.
Sulphur Moth
Harlan Co., KGC, July 7, 9, 13, 1977, L. D. Gibson. Local and uncommon in eastern mountains.

Genus *Hypomecis* Hbn.

6439 ***H. umbrosaria*** (Hbn.)
Umber Moth
PARKS AND RESERVES: BUC, HNK, HWS, LBL, NBP, PYR. **COUNTIES:** BRE, CAD, CAL, CHR, LAU, MCC, OLD, PER, POW, TRI, WOO. **DATES:** May 4, 26, June 13, 20, 30, July 3, 13, 16. Common.

Genus *Glenoides* McD.

6443 ***G. texanaria*** (Hulst)
Texas Gray
PARKS AND RESERVES: BLA, BMF, BUC, CBF, FKN, GBT, HWS, LBL, LCP, LWO, MUR, PNM. **COUNTIES:** BEL, BRE, BUL, CRL, EST, HIC, JEF, LET, MCC, MEA, MER, OLD, PER, ROW, RUS, TRI, WHI, WOL, WOO. **DATES:** May 25–Oct. 15. **REMARKS:** Common, especially late in the season.

Genus *Glena* Hulst

6449 ***G. cribrataria*** (Gn.)
Dotted Gray
PARKS AND RESERVES: BAN, BBL, BUC, CAC, LBL, NBP, OCP, PNM, RRG. **COUNTIES:** BAR, BEL, BOO, BRE, CAR, JEF, LAR, MCC, MEA, MEN, OWS, PER, POW, TRI. **DATES:** Apr. 17–June 2, 14, July 15–19, 27, Aug. 30. Moderately common.

6450 ***G. cognataria*** (Hbn.)
Blueberry Gray
Powell Co., RRG, Tunnel Ridge, 1 female, May 21, 1988, C. V. Covell, Jr. Rare.

6452 ***G. plumosaria*** (Pack.)
Plumose Gray
PARKS AND RESERVES: AJJ, BBL, BMF, BUC, GBS, HWS, LBL, LCP, OCP, PCP. **COUNTIES:** BOO, BRE, BUL, CAM, HOP, MEA, NEL, OLD, PER, PUL, RUS, TRI, WHI, WOO. **DATES:** May 13–Oct 12. Moderately common.

Genus *Stenoporpia* McD.

6459 ***S. polygrammaria*** (Pack.)
McCracken Co., Ray Black farm, 9.7 km (6 mi) SW of Paducah, July 24, 1976, C. V. Covell, Jr. Rare.

Genus *Exelis* Gn.

6478 ***E. pyrolaria*** Gn.
Fine-lined Gray
PARKS AND RESERVES: BAN, BMF, HWS, LBL, LCP, OCP, PNM, PCP, PYR. **COUNTIES:** BAR, BEL, BUL, CAD, HRD, JEF, MCC, MEA, NEL, OLD, OWS, ROW, RUS, TRI, WHI. **DATES:** May 16, 28–Aug. 31. **REMARKS:** The

type of a synonym, *approximaria* Packard (1876), is labelled "Kentucky Sanborn." Common.

Genus ***Tornos*** Morr.

6486 ***T. scolopacinarius*** (Gn.)
Dimorphic Gray
PARKS AND RESERVES: BBL, BMF, HWS, LCP, LMP, OCP. **COUNTIES:** BOO, BUL, FAY, GRE, JEF, MAD, MCC, MEA, MUH, OLD, RUS. **DATES:** Mar. 14–Aug. 23. Moderately common.

Genus ***Aethalura*** McD.

6570 ***A. intertexta*** (Wlk.)
Four-barred Gray
PARKS AND RESERVES: CAC, LMP, PMS, PNM, RRG. **COUNTIES:** BEL, CAR, FUL, HLN, JEF, MEN, MUH. **DATES:** Mar. 28–Apr. 12, 21, 26, "June," July 8, Sept. 7. Uncommon.

Genus ***Anacamptodes*** McD.

6582 ***A. vellivolata*** (Hulst)
Large Purplish Gray
PARKS AND RESERVES: CAC, CBF, PNM, RRG. **COUNTIES:** BEL, CAR, NEL, POW, ROW, WHI. **DATES:** Apr. 18, 20, June 23, 29, July 1, 8, 18, 25. Uncommon.

6583 ***A. ephyraria*** (Wlk.)
Pale-winged Gray
PARKS AND RESERVES: PMS, PYR. **COUNTIES:** CHR, FAY, HLN, HND, JEF, MEN, WOL. **DATES:** June 13, 25, 26, July 8, 15, Sept. 4, 10–11. Moderately common.

6584 ***A. humaria*** (Gn.)
Small Purplish Gray
PARKS AND RESERVES: BBL, BIP, HWS, KGC, OCP, RGR. **COUNTIES:** BOO, BRK, BUL, FAY, FUL, GRE, GRY, HLN, HRD, JEF, MAD, MEA, MEN, NEL, OLD, PIK, ROW, WHI. **DATES:** Mar. 6, Apr. 3–6, 13–29, June 22–30, Aug. 8–25, Sept. 1–7, 18. Common.

6586 ***A. defectaria*** (Gn.)
Brown-shaded Gray
PARKS AND RESERVES: ASC, BAN, BBL, BLA, BMF, BUC, CBB, GBT, HWS, JWS, LBK, LCP, NBP, OCP, PYR, RGR. **COUNTIES:** BAR, BOO, BRE, BRK, BUL, CAD, CHR, CRL, FAY, FLO, FUL, GRE, GRY, JEF, LEE, MCC, MEA, MEN, NEL, OLD, PER, POW, RUS, TRI, WOO. **DATES:** Mar. 19–Nov. 11. **REMARKS:** Common, especially late in the season.

Genus ***Iridopsis*** Warr.

6588 ***I. larvaria*** (Gn.)
Bent-line Gray
PARKS AND RESERVES: AJJ, BBM, BIP, BMF, CBF, HWS, JWS, KGC, KRC, LCP, LWO, MCV, NBP, OCP, PMS, PNM, RBF, RGR. **COUNTIES:** BEL, BRA, BRE, BUL, CAM, EDM, EST, FAY, FLO, FUL, GRY, HLN, JEF, LAU, LET, MEA, MEN, NEL, OLD, OWS, PIK, POW, RUS, WHI, WOL, WOO. **DATES:** Apr. 19–Sept. 2. **REMARKS:** Common. No melanic individuals have been observed in Kentucky (they are common in Ohio).

Genus ***Anavitrinella*** McD.

6590 ***A. pampinaria*** (Gn.)
Common Gray
PARKS AND RESERVES: BBL, BMF, BUC, CBF, GBO, HWS, KLK, LBL, LMP, MCV, NBP, PMS, PNM, PYR, RGR. **COUNTIES:** BEL, BOO, BRE, BRK, BUL, CAD, EDM, EST, FAY, FUL, GAL, GRE, GRY, GUP, HLN, HRD, JEF, MCC, MEA, MRL, MUH, NEL, OLD, OWS, PER, POW, PUL, SHE, SPE, TRI, WHI, WOL, WOO. **DATES:** Apr. 6–Oct. 5. Very common.

Genus ***Cleora*** Curt.

6594 ***C. sublunaria*** (Gn.)
Double-lined Gray
PARKS AND RESERVES: BBL, BBM, BIP, BMF, CAC, CBF, GBT, KGC, LCP, LMP, OCP, PNM, RRG. **COUNTIES:** BEL, BOO, BUL, CAR, CRL, GRE, HLN, JEF, MEA, MEN, MUH, OLD, PIK, RUS, WHI. **DATES:** Mar. 14–24, Apr. 4–29, May 18, June 19. **REMARKS:** Common, but not as common as *Anavitrinella pampinaria*.

Genus ***Ectropis*** Hbn.

6597 ***E. crepuscularia*** (D. & S.)
Small Engrailed
PARKS AND RESERVES: BBM, BIP, BUC, CBB, CBF, HWS, JWS, KLK, LCP, OCP, PNM, RGR, RRG. **COUNTIES:** BEL, BRE, BRK, EST, FAY, FLO, GRE, GRY, HLN, JEF, MEA, MEN, MRL, OLD, PER, PIK, POW, RUS, WHI. **DATES:** Mar. 11–Oct. 16. Common.

Genus ***Protoboarmia*** McD.

6598 ***P. porcelaria*** (Gn.)
Porcelain Gray
PARKS AND RESERVES: CBF, HWS, LCP, NBP, OCP, RGR. **COUNTIES:** BRA, BRK, GRV, GRY, HLN, JEF, MCC, MEA, MEN, NEL,

OLD, POW, RUS, WHI, WOO. **DATES:** Apr. 19–30, May 15–28, June 8, July 3, 13, 21, 25, Aug. 5, 9, 18, 20, 30. Local and moderately common.

Genus *Epimecis* Hbn.

6599 ***E. hortaria*** (F.)
Tulip-tree Beauty
PARKS AND RESERVES: BAN, BBM, BIP, BUC, CAC, CBB, FKN, HWS, IRO, JCF, JWS, LBK, LBL, LCP, LWO, NBP, OCP, PMS, PNM, PYR, RRG. **COUNTIES:** BAR, BEL, BRE, BUL, CAD, CAR, CHR, FLO, FUL, HLN, HRD, JEF, LAR, LAU, LEE, LET, MCC, MEA, MEN, MOR, NEL, OLD, OWS, PER, PIK, POW, ROC, ROW, RUS, SPE, TRI, WHI, WOL. **DATES:** Mar. 24, Apr. 3, 13–Oct. 8. **REMARKS:** Common, including melanic form "*carbonaria*," which comprises nearly half of the individuals examined.

Genus *Melanolophia* Hulst

6620 ***M. canadaria*** (Gn.)
Canadian Melanolophia
PARKS AND RESERVES: BAN, BBM, BEC, BIP, BMF, BUC, CAC, CBB, CBF, FKN, HNK, HWS, JWS, KRC, LBK, LCP, LMP, LVJ, MCV, OCP, PMS, PNM, RGR, RRG. **COUNTIES:** BAR, BEL, BOD, BOL, BOY, BRE, BRK, BUL, CAL, CAR, EDM, EST, FLO, GAL, GRY, HLN, HRD, JEF, LAU, LEE, MAD, MEA, MEN, MUH, NEL, OLD, OWS, PER, PIK, POW, ROW, RUS, TRI, WHI, WOL. **DATES:** Apr. 11–Oct. 24. **REMARKS:** Very common. Our subspecies is *crama* Rindge.

6621 ***M. signataria*** (Wlk.)
Signate Melanophia
PARKS AND RESERVES: BEC, HWS, KGC, KLK, LCP, PNM, RGR. **COUNTIES:** BEL, BRK, GRE, GRY, HLN, JAC, JEF, MCC, MET, MRL, OLD, ROC, RUS. **DATES:** Mar. 30–Apr. 23, May 3, 6, 29, July 16, Aug. 18. Common.

Genus *Eufidonia* Pack.

6638 ***E. notataria*** (Wlk.)
Powder Moth
PARKS AND RESERVES: PNM, RRG. **COUNTIES:** BEL, JEF, LAU, MEN, OWS, POW. **DATES:** May 3–27. Very local and uncommon.

Genus *Biston* Leach

6640 ***B. betularia*** (L.)
Pepper-and-Salt Geometer
PARKS AND RESERVES: BIP, BMF, BUC, CMI, HWS, KGC, LVJ, NBP, OCP, PNM, WVP. **COUNTIES:** BEL, BOO, BRE, BUL, FAY, GAL, HLN, JEF, LAU, MEA, MOR, NEL, OLD, PER, PIK, POW, ROW. **DATES:** Apr. 12–Sept. 18. **REMARKS:** American subspecies is *cognataria* (Gn.). Common. Melanic specimens are frequent. Rindge (1975) has seven dots in Kentucky on his range map for *cognataria*.

Genus *Lycia* Hbn.

6652 ***L. ypsilon*** (S. A. Fbs.)
Woolly Gray
Boone Co., BBL, Mar. 25, Apr. 10, 25; Carroll Co., GBT, Apr. 2; Fayette Co., Mar. 29, Apr. 1–5, 27, May 5. Uncommon to rare.

Genus *Hypagyrtis* Hbn.

6654 ***H. unipunctata*** (Haw.)
One-spotted Variant
PARKS AND RESERVES: BAN, BBL, BBM, BEC, BOW, CBB, CBF, FKN, HWS, JCF, KGC, LBL, LCP, MCV, NBP, OCP, PMS, PNM, PYR, RBF, RGR, WVP. **COUNTIES:** BAR, BEL, BOO, BRA, BRE, BRK, BUL, CAD, CHR, EDM, GRY, GUP, HLN, HND, HRD, JEF, MCC, MEA, MEN, OLD, OWS, POW, ROW, RUS, TRI, WHI, WOL, WOO. **DATES:** Apr. 17–Sept. 10. **REMARKS:** Very common to abundant, and variable in size and pattern. Some individuals are melanic, but fewer than in Ohio.

6655 ***H. esther*** (Barnes)
Esther Moth
PARKS AND RESERVES: LBL, PYR, RRG. **COUNTIES:** CHR, EST, GUP, JEF, LAU, MCC, MEA, MEN, OWS, POW, ROW, TRI. **DATES:** June 5, 8, 19–21, 30, July 8, 13, 25, 27, Aug. 4–10, 22–27, Sept. 1, 18. Uncommon.

6657 ***H. brendae*** R. L. Heitzman
Brenda's Hypagyrtis
PARKS AND RESERVES: BMF, HWS, LCP, LMP, OCP. **COUNTIES:** BUL, JEF, MEA, MUH, NEL, OLD, PUL, RUS, WOO. **DATES:** Apr. 29, May 3–29, June 1–23, Aug. 27, Sept. 6. **REMARKS:** Locally common. Paratypes of this species, listed in Heitzman (1974), include the following from Kentucky: Nelson Co., Bardstown, 1 male, Apr. 29, 1970, G. F. Florence; Nelson Co., Leslie farm near Boston, 2 males, June 23, 1971, G. F. Florence; Oldham Co., HWS, 1 male, June 22, 1966, C. V. Covell, Jr.

Genus *Phigalia* Dup.

6658 ***P. titea*** (Cram.)
Half-wing
PARKS AND RESERVES: BBM, BMF, CAC, GBT, KGC, LMP, MCV, NBP, OCP, RRG. **COUNTIES:** BUL, CAR, CRL, EDM, HLN, JEF, MCC, MEA, MEN, MUH, POW, WHI. **DATES:**

Mar. 8–Apr. 2l (larvae: May 30, 31). **REMARKS:** Common. Ghent and Dull (1979) and Ghent et al. (1980) reported defoliation of hardwood trees by this species in Edmonson County (MCV area).

6659 ***P. denticulata*** Hulst
Toothed Phigalia
PARKS AND RESERVES: GBT, KGC, RRG. **COUNTIES:** BOO, CRL, FAY, GRV, HLN, JEF, MCC, MEN, PER, POW. **DATES:** Jan. 30, Feb. 15, 27, Mar. 9–19, Apr. 2, 20, Dec. 28. Common.

6660 ***P. strigataria*** (Minot)
Small Phigalia
PARKS AND RESERVES: BBL, CAC, GBT, HWS, KGC, KLK, LMP, NBP, PNM, RRG. **COUNTIES:** BEL, BOO, CAR, CRL, GRV, HLN, JEF, MCC, MEN, MRL, MUH, OLD, POW, TRM, WHI. **DATES:** Jan. 13, Feb. 24, 27, Mar. 5–29, Apr. 1–28. **REMARKS:** Common. Rindge (1975) has one dot in Kentucky on his range map for this species.

Genus ***Paleacrita*** Riley

6662 ***P. vernata*** (Peck)
Spring Cankerworm Moth
PARKS AND RESERVES: BBL, BMF, GBT, LBL, LMP, MCV, NBP. **COUNTIES:** BOO, BUL, CRL, EDM, FAY, JEF, LYO, MCC, MUH, POW, SHE, TRI, TRM. **DATES:** Jan. 10–Apr. 11 (larvae: May 7, 30). **REMARKS:** Common. Ghent et al. (1980) reported larvae of this species defoliating trees in Edmonson Co., MCV, on May 30, 1980. The same is reported for Lyon and Trigg counties, LBL, May 7, 1981 (Carothers 1981a).

6663 ***P. merriccata*** Dyar
White-spotted Cankerworm Moth
Boone Co., BBL, Mar. 21; Menifee Co., FS 9B East Fork Indian Creek, Kelley Branch, Mar. 15, 19; Powell Co., NBP, Mar. 18. Rare.

Genus ***Erannis*** Hbn.

6665 ***E. tiliaria*** (Harr.)
Winter Moth; Linden Looper Moth
PARKS AND RESERVES: BCN, BOW, LBL, LCP, MCV, NBP, OCP, PNM, RRG. **COUNTIES:** BEL, BOO, EDM, FAY, JEF, LYO, MEA, MEN, OLD, POW, RUS, TRI. **DATES:** Oct. 11, 23, Nov. 9–Dec. 1, 12 (larvae: May 7, 30, and 31). **REMARKS:** This species, which has wingless females, was unusually abundant in November to early December 1978 and 1979 in Louisville, and was more scarce in the same areas in 1980. Ghent and Dull (1979) and Ghent et al. (1980) reported basswood defoliation by larvae in Edmonson Co., MCV, May 30, 1979 and May 31, 1980. Carothers (1981a) reported similar activity in Lyon and Trigg counties, LBL, May 7, 1981.

Genus ***Lomographa*** Hbn.

6666 ***L. semiclarata*** (Wlk.)
Bluish Spring Moth
Bullitt County, BMF, Apr. 13; Jefferson Co., Valley Station, Apr. 19, 27, 29, May 5–18; Meade Co., OCP, Apr. 13; Nelson Co., Leslie farm, 9.7 km (6 mi) E of Boston, Apr. 18. Uncommon.

6667 ***L. vestaliata*** (Gn.)
White Spring Moth
PARKS AND RESERVES: AJJ, BBM, BEC, BMF, CBB, HWS, JCF, LBL, LCP, LMP, PMS, RBF, RRG. **COUNTIES:** BOO, BRE, BUL, CAM, EST, FAY, HLN, JEF, LAR, MCC, MEA, MEN, MUH, NEL, OLD, ROW, RUS, TRI. **DATES:** Apr. 8–Aug. 30. Common.

6668 ***L. glomeraria*** (Grt.)
Gray Spring Moth
PARKS AND RESERVES: BBM, BIP, BMF, CAC, CBB, LCP, LMP, MCV. **COUNTIES:** BRE, BUL, CAR, EDM, HLN, JEF, MCC, MUH, NEL, OLD, OWS, PIK, RUS. **DATES:** Mar. 19–Apr. 24. **REMARKS:** Common. Sometimes abundant in woodlands on sunny spring days.

Genus ***Cabera*** Tr.

6677 ***C. erythemaria*** Gn.
Yellow-dusted Cream Moth
Menifee Co., RRG, FS 9A, Indian Creek, 1 female, Aug. 29, 1984, J. S. Nordin. Uncommon to rare.

6678 ***C. variolaria*** Gn.
Pink-striped Willow Spanworm Moth
Nelson Co., Bardstown, May 22, G. F. Florence. Uncommon to rare.

6680 ***C. quadrifasciaria*** (Pack.)
Four-lined Cabera
Gallatin Co., Markland Dam, May 29, June 2, 15; Jefferson Co., Caperton Swamp, taken during day, May 29, and Greenwood Road, May 25. Very local and uncommon.

Genus ***Apodrepanulatrix*** Rindge

6693 ***A. liberaria*** (Wlk.)
Owsley Co., 4.8 km (3 mi) NE of Booneville, 1 male, Oct. 4, 1986, L. D. Gibson. Rare.

Genus ***Erastria*** Hbn.

6704 ***E. coloraria*** (F.)
Rowan Co., Clack Mtn., June 20, 1938, 2 males, A. F. Braun. Rare.

Genus ***Thysanopyga*** H.-S.

6711 ***T. intractata*** (Wlk.)
Black-dotted Ruddy
PARKS AND RESERVES: PNM, RRG. **COUNTIES:** BEL, FUL, JEF, MCC, POW, WHI. **DATES:** Apr. 11, 30, May 12, 21, July 5–12, 24, Sept. 8, Oct. 6, 27. Uncommon.

Genus ***Lytrosis*** Hulst

6720 ***L. unitaria*** (H.-S.)
Common Lytrosis
PARKS AND RESERVES: BMF, BUC, CBB, CBF, HWS, LCP, NBP, OCP, PNM, PYR, RRG, WVP. **COUNTIES:** BEL, BRE, BUL, CHR, JEF, LAU, MCC, MEA, MEN, OLD, OWS, PER, POW, RUS, WHI. **DATES:** May 18–June 26. Common.

6721 ***L. sinuosa*** Rindge
Sinuous Lytrosis
Bell Co., PNM, 1 male taken in light trap, June 9, 1981, C. Cornett. Rare and very local.

6723 ***L. permagnaria*** (Pack.)
Bell Co., PNM, in light trap, June 6, 9, C. Cornett; Edmonson Co., MCV, May 26, 1927, C. Krueger; Whitley Co., Corbin, 1 male, June 9, 1973, T. W. Carr. Rare. Listed as an endangered species in Kentucky (KSNPC 1996).

Genus ***Euchlaena*** Hbn.

6724 ***E. serrata*** (Drury)
Saw-wing
PARKS AND RESERVES: BMF. **COUNTIES:** BUL, MEA, NEL, ROW, TRM, WHI. **DATES:** Apr. 4, June 5, 12, 21–28, July 2, 5, Aug. 3. Common.

6726 ***E. obtusaria*** (Hbn.)
Obtuse Euchlaena
PARKS AND RESERVES: BBL, BMF, BUC, CBF, FKN, HWS, NBP, OCP, PNM, RGR. **COUNTIES:** BEL, BOO, BRE, BRK, BUL, FAY, GRY, GUP, JEF, MEA, NEL, OLD, OWS, PER, POW, ROW, WHI. **DATES:** Apr. 21–Sept. 8. Common.

6729 ***E. johnsonaria*** (Fitch)
Johnson's Euchlaena
PARKS AND RESERVES: CER, HWS, NBP, RRG. **COUNTIES:** BOO, FAY, FUL, GAL, MCC, NEL, OLD, POW, WHI. **DATES:** May 22–June 7, 19–28, Aug. 2, 6, 17–Sept. 2, 8. Common.

6731 ***E. madusaria*** (Wlk.)
Bullitt Co., SR 480, 11.4 km (6.9 mi) E of SR 61, May 28, Sept. 8; Grayson Co., RGR, Aug. 11; Jefferson Co., Valley Station, May 29; Nelson Co., Bardstown, May 24. Uncommon.

6733 ***E. amoenaria*** (Gn.)
Deep Yellow Euchlaena
PARKS AND RESERVES: BAN, BEC, BMF, BUC, HWS, LCP, MCV, NBP, OCP, PNM, RBF, RGR, RRG, WVP. **COUNTIES:** BAR, BEL, BRE, BRK, BUL, EDM, GAL, GRY, GUP, HND, JEF, MCC, MEA, NEL, OLD, OWS, PER, POW, ROW, RUS, WHI, WOO. **DATES:** Apr. 15–Sept. 26. Common.

6734 ***E. marginaria*** (Minot)
Kentucky is included in the range of this species Forbes (1948). Estill Co., Irvine, "in hills about 3000 feet," July 27, 1967, R. H. Leuschner. Uncommon.

6735 ***E. pectinaria*** (D. & S.)
Forked Euchlaena
PARKS AND RESERVES: CAC, CBF, HNK, HWS, JWS, LBL, OCP, PNM, RBF, RGR, RRG. **COUNTIES:** BEL, BRE, BRK, BUL, CAL, CAR, FLO, GRY, JEF, MCC, MEA, MEN, NEL, OLD, POW, ROW, TRI, WHI, WOL. **DATES:** Apr. 7–Aug. 30, Sept. 22, Oct. 8. Common.

6737 ***E. tigrinaria*** (Gn.)
Mottled Euchlaena
PARKS AND RESERVES: BBL, HWS, LCP, OCP. **COUNTIES:** BOO, BRA, BUL, EST, GAL, JEF, MCC, MEA, NEL, OLD, RUS. **DATES:** May 1, 14, June, July 21–28, Aug. 2–18. Common.

6739 ***E. irraria*** (B. & McD.)
Least-marked Euchlaena
PARKS AND RESERVES: BMF, JWS, OCP, PNM. **COUNTIES:** BEL, BRA, BUL, FLO, JEF, MEA, OWS, PIK, WHI. **DATES:** May 22–29, June 5–15. Locally common.

Genus ***Xanthotype*** Warr.

6740 ***X. urticaria*** Swett
False Crocus Geometer
PARKS AND RESERVES: BEC, BMF, CER, HWS, LBL, LCP, OCP, RGR. **COUNTIES:** BOO, BRK, BUL, FAY, FUL, GAL, GRY, HOP, JEF, MEA, MEN, OLD, OWS, POW, RUS, TRI, WOO. **DATES:** May 29, June 3, July 10–Sept. 13. Common.

6743 ***X. sospeta*** (Drury)
Crocus Geometer
PARKS AND RESERVES: BBM, BEC, BMF, CBB, FKN, HWS, OCP, RGR. **COUNTIES:** BOO, BRK, BUL, FAY, GRY, HLN, HRD, JEF, MCC, MEA, MEN, NEL, OLD, OWS, PIK, WHI,

WOO. **DATES:** Apr. 26–Oct. 6, Dec. 5. **REMARKS:** Common, but less so than *X. urticaria.*

6744 ***X. attenuaria*** Swett
Fulton Co., Willingham Bottoms, SR 94, 4.0 km (2.5 mi) E of Cayce, Aug. 3, Sept. 5, 8, 13; Graves Co., Bell farm near Kaler, Sept. 5. Locally common and flies during the daytime in swampy woods.

Genus *Pero* H.-S.

6752 ***P. zalissaria*** (Wlk.)
Fulton Co., in cypress swamp (Willingham Bottoms) 4.0 km (2.5 mi) E of Cayce, in light traps, Sept. 5–13; Henderson Co., SWR, Sept. 10, 11 (associated with bald cypress). Locally common.

6753 ***P. honestaria*** (Wlk.)
Honest Pero
PARKS AND RESERVES: CBB, LWO, RRG. **COUNTIES:** FAY, GAL, JEF, LET, MEN, OLD, ROW, WHI. **DATES:** Apr. 9–26, July 11, Sept. 3, 8. Common.

6754 ***P. hubneraria*** (Gn.)
Hübner's Pero
PARKS AND RESERVES: BBM, BIP, BMF, CAC, CCR, HWS, JWS, KGC, LWO, MCV, OCP, PMS, PNM, WVP. **COUNTIES:** BEL, BOO, BRA, BUL, CAR, EDM, FAY, FLO, GAL, GRE, GRV, HLN, JEF, LET, MEA, MEN, MOR, OLD, PIK, WHI, WOO. **DATES:** Apr. 3–15, May 5–18, July, Aug. 18. **REMARKS:** A very common species, probably found in wooded areas throughout the state, but easily confused with *P. honestaria* (Wlk.) (Covell 1984).

6755 ***P. morrisonaria*** (Hy. Edw.)
Morrison's Pero
Bell Co., PNM, June 6; Harlan Co., BBM, June 8, July 14; Jefferson Co., Valley Station, July 11, Aug. 9, 18; Meade Co., OCP, Aug. 3, 10; Menifee Co., May 26–27. Locally common.

Genus *Nacophora* Hulst

6763 ***N. quernaria*** (J. E. Smith)
Oak Beauty
PARKS AND RESERVES: AJJ, BBL, BBM, BIP, BMF, CAC, KGC, KLK, LCP, LMP, NBP, OCP, PNM, RBF, RRG. **COUNTIES:** BEL, BOO, BRE, BUL, CAM, CAR, HLN, JEF, MAD, MCC, MEA, MRL, MUH, NEL, PIK, POW, ROW, RUS, SPE. **DATES:** Mar. 20–Oct. 9. **REMARKS:** Common. Melanic individuals also fairly common.

Genus *Campaea* Lamarck

6796 ***C. perlata*** (Gn.)
Pale Beauty
PARKS AND RESERVES: ASC, BAN, BMF, BUC, CAC, FKN, GBT, HWS, JWS, LCP, LMP, NBP, OCP, PNM. **COUNTIES:** BAR, BEL, BRE, BUL, CAR, CRL, FLO, GAL, GRY, HRD, JEF, MEA, MUH, NEL, OLD, OWS, PER, POW, RUS, WHI. **DATES:** May 10–Sept. 19. Common.

Genus *Ennomos* Tr.

6797 ***E. magnaria*** Gn.
Maple Spanworm Moth
PARKS AND RESERVES: BIP, BMF, GBT, HWS, LCP, OCP, PNM, RGR. **COUNTIES:** BEL, BRK, BUL, CRL, FAY, GAL, GRY, JEF, MCC, MEA, NEL, OLD, PIK, RUS. **DATES:** July 30, Aug., Sept., Oct. 8–16, 26. Locally common.

6798 ***E. subsignaria*** (Hbn.)
Elm Spanworm Moth
Fayette Co., Lexington, in Malaise trap, June 16–18, July; Jefferson Co., Louisville, May 20, June 11–26; Muhlenberg Co., Bremen, reared from apple leaves, emerged June 14, 15; Shelby Co., Shelbyville, June 30. Very local and potentially a forest pest.

Genus *Homochlodes* Hulst

6812 ***H. fritillaria*** (Gn.)
Pale Homochlodes
Harlan Co., KGC, Apr. 20, and Trace Branch of Looney Creek, Apr. 21; Jefferson Co., Valley Station, Apr. 26, May 4; Oldham Co., HWS, May 17; Woodford Co., May 4. Uncommon.

6813 ***H. disconventa*** (Wlk.)
Dark Homochlodes
Bell Co., PNM, July 18; Harlan Co., BBM, June 8, July 9–15. Uncommon.

Genus *Gueneria* Pack.

6815 ***G. similaria*** (Wlk.)
PARKS AND RESERVES: KGC, RRG. **COUNTIES:** HLN, JEF, MEN, POW, ROW, WHI. **DATES:** May 11–12, 21, 24, June 2, 5, July 13, 21–22. Uncommon.

Genus *Selenia* Hbn.

6818 ***S. kentaria*** (G. & R.)
Kent's Geometer
PARKS AND RESERVES: BBL, BUC, CBB, CBF, KGC, OCP, RRG. **COUNTIES:** BOO, BRE, HLN, JEF, MCC, MEA, MEN, NEL, OLD,

PER, POW, WHI. **DATES:** Mar. 14, Apr. 11–26, May 7, 8, June 9, July 10, 19–23, Aug. 3. Common.

Genus *Metanema* Gn.

6819 ***M. inatomaria*** Gn.
Pale Metanema
Hickman Co., MUR, July 24; Jefferson Co., WVP, July 2; Meade Co., OCP, Apr. 11, Aug. 15, 17. Uncommon.

6820 ***M. determinata*** Wlk.
Dark Metanema
PARKS AND RESERVES: BBL, BMF, LBL, OCP. **COUNTIES:** BOO, BUL, MCC, MEA, NEL, TRI. **DATES:** May 1, 18, 19, June 8, July 16, Aug. 20. Moderately common.

Genus *Metarranthis* Warr.

6822 ***M. duaria*** (Gn.)
Ruddy Metarranthis
Bell Co., PNM, Apr. 28; Carter Co., Carter, June 30; Harlan Co., KGC, Apr. 20; Menifee Co., FS 9A Indian Creek, July 7. Uncommon.

6823 ***M. angularia*** B. & McD.
Angled Metarranthis
PARKS AND RESERVES: BMF, BUC, CBB, HWS, KGC, PNM, RRG, WVP. **COUNTIES:** BEL, BRE, BUL, HLN, JEF, LEE, MEN, OLD, OWS, PER, POW. **DATES:** Apr. 20, 26, May 4–8, 16–29, June 6–25, July 8–19, Aug. 6. Common.

6825 ***M. indeclinata*** (Wlk.)
Pale Metarranthis
Bullitt Co., BMF, in Malaise trap, May 23–26; Jefferson Co., Valley Station, Aug. 19; Russell Co., LCP, 1 female, May 30. Uncommon to rare.

6826 ***M. hypochraria*** (H.-S.)
Common Metarranthis
PARKS AND RESERVES: BAN, BMF, BUC, CBF, FKN, HWS, JWS, LMP, NBP, PNM, PYR. **COUNTIES:** BAR, BEL, BRE, BUL, CAD, CHR, FLO, HRD, JEF, MEA, MEN, MUH, NEL, OLD, PER, POW, ROC, WHI. **DATES:** Apr. 30–Aug. 6. Locally common.

6827 ***M. refractaria*** (Gn.)
Laurel Co., Hazel Patch, netted during daytime, June 5, 1981, C. V. Covell, Jr. Rare.

6828 ***M. homuraria*** (G. & R.)
Purplish Metarranthis
PARKS AND RESERVES: AJJ, BBL, CBB, CBF, FKN, HWS, LBL, NBP, OCP, PNM, WVP. **COUNTIES:** BEL, BOO, CAM, FAY, HRD, JEF, MCC, MEA, MEN, NEL, OLD, POW, TRI, WHI. **DATES:** Apr. 5–Sept. 19, Oct. 4. Common.

6832 ***M. obfirmaria*** (Hbn.)
Yellow-washed Metarranthis
PARKS AND RESERVES: BMF, BUC, CBF, NBP, PNM, RRG. **COUNTIES:** BEL, BRE, BUL, JEF, LAU, OWS, PER, POW, WHI, WOL. **DATES:** Apr. 15–May 25, Oct. 9. Locally common.

Genus *Cepphis* Hbn.

6835 ***C. armataria*** (H.-S.)
Scallop Moth
Bell Co., PNM, July 18, 19; Harlan Co., BBM, July 14, and PMS, July 9; Menifee Co., Indian Creek, June 21, July 8, 13, 22; Owsley Co., 4.8 km (3 mi) NE of Booneville, May 28, June 14; Rowan Co., Morehead, July 23. Locally common.

Genus *Anagoga* Hbn.

6836 ***A. occiduaria*** (Wlk.)
American Barred Umber
PARKS AND RESERVES: BBM, BMF, CAC, CBF, GBT, HWS, JWS, KGC, LCP, LWO, NBP, PNM, RRG. **COUNTIES:** BEL, BUL, CAR, CRL, EST, FLO, HLN, JEF, LET, MEN, NEL, OLD, POW, ROW, RUS, WHI. **DATES:** Apr. 6–May 4, 12–17, 29, July 8, 11, 27. Common.

Genus *Probole* H.-S.

6837.1 ***P. nyssaria*** (Gn.)
Dogwood Probole
PARKS AND RESERVES: BBM, BIP, BMF, CAC, CMI, HWS, KGC, LBL, LCP, MCV, NBP, OCP, PNM, RRG. **COUNTIES:** BEL, BOO, BUL, CAR, EDM, HLN, JEF, LAU, MCC, MEA, MEN, NEL, OLD, OWS, PIK, POW, RUS, TRI, WHI, WOL. **DATES:** Apr. 17–Aug. 6. **REMARKS:** Very common. This species was formerly known as *P. alienaria* H.-S.

6838 ***P. amicaria*** (H.-S.)
Friendly Probole
PARKS AND RESERVES: BBM, BEC, BIP, BMF, BUC, CAC, FKN, JCF, KGC, LBL, LCP, MCV, NBP, OCP, PMS, PNM, RRG. **COUNTIES:** BEL, BRE, BUL, CAR, EDM, EST, GRV, HLN, HRD, JEF, LEE, MEA, MEN, PER, PIK, POW, RUS, TRI, WHI, WOL. **DATES:** Apr. 11–May 5, 11–15, 26, 31, June 12, 28, July 3–18, 27, Aug. 3–10, 20, 23. Very common.

6839 ***P. nepiasaria*** (Wlk.)
Heath Probole
Bell Co., PNM, May 13; Menifee Co., RRG, Aug. 6, 8, 25–26; Oldham Co., HWS, Aug. 5; Powell

Co., RRG, Tunnel Ridge, Apr. 28; Russell Co., LCP, May 29. Uncommon to rare.

Genus *Plagodis* Hbn.

6840 ***P. serinaria*** H.-S.
Lemon Plagodis
PARKS AND RESERVES: BIP, CAC, NBP, OCP, PNM, RBF, RRG, WVP. **COUNTIES:** BEL, BRE, CAR, JEF, MEA, MEN, PIK, POW, WHI. **DATES:** Apr. 15–May 27, June 21. **REMARKS:** Very local, but moderately common when found.

6841 ***P. kuetzingi*** (Grt.)
Purple Plagodis
PARKS AND RESERVES: BBL, BUC, JCF, LCP, NBP, OCP, PNM. **COUNTIES:** BEL, BOO, BRA, BRE, BUL, EST, JEF, MEA, PER, POW, RUS. **DATES:** Apr. 22, May 5–18, June 3–9, 13–21, July 19–27, Aug. 7. Very common.

6842 ***P. phlogosaria*** (Gn.)
Straight-lined Plagodis
PARKS AND RESERVES: BBM, CAC, HWS, KRC, LBL, LCP, NBP, OCP, PNM, RBF, RRG. **COUNTIES:** BEL, BRE, CAR, EST, GAL, HLN, JEF, MEA, MEN, OLD, OWE, OWS, POW, ROW, RUS, TRI, WHI, WOL. **DATES:** Apr. 17–Aug. 28. Common.

6843 ***P. fervidaria*** (H.-S.)
Fervid Plagodis
PARKS AND RESERVES: BIP, CAC, HWS, JCF, KRC, LBL, LCP, LWO, MCV, NBP, OCP, PMS, PNM, RRG. **COUNTIES:** BEL, BUL, CAR, EDM, EST, HLN, JEF, LET, MCC, MEA, MEN, NEL, OLD, OWS, PIK, POW, ROW, RUS, TRI, WHI, WOL, WOO. **DATES:** Apr. 15–Sept. 27. Common.

6844 ***P. alcoolaria*** (Gn.)
Hollow-spotted Plagodis
PARKS AND RESERVES: BBL, BBM, BEC, BIP, CAC, HNK, HWS, KGC, KRC, LBL, LCP, NBP, OCP, PNM, RRG, WVP. **COUNTIES:** BEL, BOO, BUL, CAL, CAR, EST, HLN, JEF, MCC, MEA, OLD, PIK, POW, ROW, RUS, TRI, WOL. **DATES:** Apr. 13–Aug. 20. Common.

Genus *Lychnosea* Grt.

6858 ***L. intermicata*** (Wlk.)
PARKS AND RESERVES: BBL, HWS. **COUNTIES:** BOO, BRA, BUL, JEF, MAD, MEA, NEL, OLD, PEN. **DATES:** June 24–27, July 7, 9, 22, Aug. 1, 10, 24, 25, Sept. 8, 9, 18, Oct. 2–16. **REMARKS:** Conspicuous during the daytime. I found it locally common once in October at HWS.

Genus *Caripeta* Wlk.

6863 ***C. divisata*** Wlk.
Gray Spruce Looper Moth
Menifee Co., Indian Creek, Aug. 10, RRG, Aug. 6–26, Sept. 1, 2; Owsley Co., Aug. 26; Powell Co., NBP, Aug. 18; Rowan Co., DBNF, E side of SR 1274, 3.2 km (2 mi) W of SR 519, Aug. 18; Wolfe Co., 3.2 km (2 mi) SE of NBP, Aug. 21. Uncommon.

6869 ***C. aretaria*** (Wlk.)
Southern Pine Looper
Bath Co., CAV, FS 918, Aug. 28; Harlan Co., KGC, Apr. 25; Menifee Co., RRG, Aug. 22, 24; Owsley Co., 4.8 km (3 mi) NE of Booneville, Aug. 27; Rowan Co., DBNF, E side of SR 1274, 3.2 km (2 mi) W of SR 519, Aug. 18, 26, Morehead, Aug. 30, 31, Sept. 1–4. Uncommon.

Genus *Besma* Capps

6884 ***B. endropiaria*** (G. & R.)
Straw Besma
PARKS AND RESERVES: BAN, HWS, JWS, LCP, OCP, PNM. **COUNTIES:** BAR, BEL, FLO, JEF, MCC, MEA, MEN, NEL, OLD, RUS. **DATES:** May 20–June 20, July 19, 20, Aug. 4, 23. Common.

6885 ***B. quercivoraria*** (Gn.)
Oak Besma
PARKS AND RESERVES: BBL, BBM, BMF, BUC, CAC, HNK, HWS, JWS, KGC, LCP, LMP, NBP, OCP, PNM, RBF. **COUNTIES:** BEL, BOO, BRE, BUL, CAL, CAR, EST, FAY, FLO, GUP, HLN, JEF, MCC, MEA, MEN, MUH, NEL, OLD, OWS, PER, POW, ROW, RUS, WOL. **DATES:** Apr. 20–Oct. 4, 25. Common.

Genus *Lambdina* Capps

6888 ***L. fiscellaria*** (Gn.)
Hemlock Looper Moth
Bell Co., PNM, Oct. 10, 13, 17; Harlan Co., BBM, July 15. Locally common in southeastern Kentucky mountains.

6892 ***L. pellucidaria*** (G. & R.)
Yellow-headed Looper Moth
PARKS AND RESERVES: BUC, CAC, CBF, HWS, LCP, NBP, PNM, RRG. **COUNTIES:** BEL, BRE, CAR, HLN, JEF, MEN, OLD, PER, POW, RUS, WHI. **DATES:** Apr. 12, 20–May 1, 12–21. Common.

6894 ***L. fervidaria*** (Hbn.)
Curve-lined Looper Moth
PARKS AND RESERVES: AJJ, BBL, BBM, BIP, BMF, BUC, CAC, HWS, KGC, LBL, LCP, LWO, MCV, OCP, PNM, RRG. **COUNTIES:**

BAR, BEL, BOO, BRA, BRE, BUL, CAM, CAR, EDM, EST, GAL, HLN, JEF, LAU, LET, MCC, MEA, NEL, OLD, OWS, PER, PIK, POW, ROW, RUS, TRI, WHI, WOL. **DATES:** Mar. 20–Aug. 17. Very common.

Genus ***Nepytia*** Hulst

6906 ***N. canosaria*** (Wlk.)
False Hemlock Looper Moth
Bell Co., PNM, Oct. 11–23; Pike Co., BIP, Oct. 14–17; Powell Co., NBP, Oct. 15. Uncommon.

6908 ***N. semiclusaria*** (Wlk.)
Southern Nepytia
Whitley Co., SR 90 E of Cumberland Falls at Moonbow Cottages and Motel, 1 female, July 1, 1979, J. S. Nordin. Rare.

Genus ***Sicya*** Gn.

6912 ***S. macularia*** (Harr.)
Sharp-lined Yellow
Boone Co., BCN, June 15, 1984; Harlan Co., BBM, July 14, 1979. Rare.

Genus ***Eusarca*** Hbn.

6941 ***E. confusaria*** Hbn.
Confused Eusarca
PARKS AND RESERVES: ASC, BBM, BCN, BEC, BLA, BMF, CAC, CBF, FAO, FKN, HWS, IRO, JWS, LBK, LCP, MCV, OCP, PYR. **COUNTIES:** BOO, BUL, CAD, CAR, CHR, EDM, FAY, FLO, FUL, GRY, HLN, JEF, KNX, LAR, MAD, MAR, MEA, MEN, NEL, OLD, OWS, ROW, RUS, TRI, WHI. **DATES:** Apr. 8–Oct. 19. Very common.

Genus ***Tetracis*** Gn.

6963 ***T. crocallata*** Gn.
Yellow Slant-line
PARKS AND RESERVES: BAN, BEC, BMF, FKN, OCP, RGR. **COUNTIES:** BAR, BUL, FAY, GRY, HLN, JEF, MEA, NEL, OWS, ROW. **DATES:** Apr. 8–Aug. 10. Common.

6964 ***T. cachexiata*** Gn.
White Slant-line
PARKS AND RESERVES: BAN, BEC, BMF, CBF, OCP, PNM, RBF, RRG. **COUNTIES:** BAR, BEL, BRE, BUL, JEF, MEA, NEL, POW, ROW, WHI. **DATES:** Apr. 20–June 12, 20, 29, Aug. 14. Common.

Genus ***Eugonobapta*** Warr.

6965 ***E. nivosaria*** (Gn.)
Snowy Geometer
PARKS AND RESERVES: BBM, HWS, JWS, KGC, RBF. **COUNTIES:** BRE, FLO, HLN, LEE, LES, OLD, POW. **DATES:** June 3–July 14. Uncommon.

Genus ***Eutrapela*** Pack.

6966 ***E. clemataria*** (J. E. Smith)
Curve-toothed Geometer
PARKS AND RESERVES: BAN, BBL, BBM, BEC, BMF, BUC, CAC, CBF, HWS, JWS, KLK, LBK, LBL, LCP, LWO, NBP, OCP, PNM, PYR, RGR, RRG. **COUNTIES:** BAR, BEL, BOO, BRE, BRK, BUL, CAR, CHR, FLO, GRY, HLN, HOP, JEF, LAR, LET, MAR, MEA, MOR, MRL, NEL, OLD, PER, POW, ROW, RUS, TRI, WHI. **DATES:** Apr. 10–Sept. 2. Common.

Genus ***Patalene*** H.-S.

6974 ***P. olyzonaria*** (Wlk.)
Juniper Geometer
PARKS AND RESERVES: BEC, BMF, FKN, GBT, HWS, KLK, LBK, LBL, LCP, OCP, PYR, RGR. **COUNTIES:** BRK, BOO, BUL, CHR, CRL, EDM, EST, FAY, GRY, JEF, MCC, MEA, MRL, NEL, OLD, OWS, RUS, TRI. **DATES:** Apr. 10–Oct. 15. **REMARKS:** Our subspecies is *puber* (G. & R.). Common, especially late in the season.

Genus ***Prochoerodes*** Grt.

6982 ***P. transversata*** (Drury)
Large Maple Spanworm Moth
PARKS AND RESERVES: BIP, BLA, BMF, BUC, FKN, GBT, JWS, KRC, LCP, LWO, NBP, PNM, PYR, WVP. **COUNTIES:** BEL, BRE, BUL, CAD, CRL, FAY, FLO, FUL, GAL, HIC, HRD, JEF, LET, MAD, MCC, MEA, MEN, MOR, NEL, OLD, OWS, PER, PIK, POW, ROW, RUS, WOL. **DATES:** Apr. 11–Oct. 28. Very common.

Genus ***Antepione*** Pack.

6987 ***A. thisoaria*** (Gn.)
Variable Antepione
PARKS AND RESERVES: BBL, BBM, BMF, CAC, HNK, HWS, JCF, KGC, LCP, LMP, OCP, PNM, RRG. **COUNTIES:** BEL, BOO, BRA, BUL, CAL, CAR, FAY, GRV, HLN, JEF, MAD, MCC, MEA, MEN, MOR, MUH, OLD, OWS, ROW, RUS. **DATES:** Apr. 12–Oct. 15. Common.

Genus ***Nematocampa*** Gn.

7009 ***N. limbata*** (Haw.)
Horn Spanworm Moth
PARKS AND RESERVES: BEC, HWS, LCP, MUR, PNM. **COUNTIES:** BEL, FAY, HIC, HRD, HRT, JEF, OLD, OWS, RUS. **DATES:** May 25–July 2, 24–28, Aug. 8, 10. Locally common.

Subfamily **Geometrinae**

Genus ***Nemoria*** Hbn.

7029 ***N. elfa*** Fgn.
Hickman Co., MUR, Oct. 2; Jefferson Co., Caperton Swamp, May 29. Very local and uncommon in swamp habitats.

7030 ***N. tuscarora*** Fgn.
Jefferson Co., WVP, Aug., S. Scholz. Rare.

7033 ***N. lixaria*** (Gn.)
Red-bordered Emerald
Rowan Co., July 16, 21; Russell Co., LCP, Sept. 12. Uncommon.

7034 ***N. saturiba*** Fgn.
Trigg Co., LBL, 3 specimens, July 13, 16; Whitley Co., Goldbug, Aug. 2, 14. Rare.

7046 ***N. bistriaria*** Hbn.
PARKS AND RESERVES: BBL, BIP, BMF, CAC, CBB, CBF, HWS, KLK, LBL, LCP, LWO, NBP, OCP, PNM, PYR, RRG. **COUNTIES:** BAR, BEL, BOO, BUL, CAR, CHR, FAY, GAL, GRE, HLN, JEF, LET, MCC, MEA, MEN, MRL, OLD, OWS, PIK, POW, ROW, RUS, TRI, WOL, WHI. **DATES:** Mar. 14–Oct. 3. **REMARKS:** Common. Spring specimens are orangish brown and summer specimens are pale green.

7047 ***N. rubrifrontaria*** (Pack.)
Red-fronted Emerald
Floyd Co., JWS, June 5; Jefferson Co., Valley Station, July 7; Meade Co., OCP, July 29. Uncommon to rare.

7048 ***N. mimosaria*** (Gn.)
PARKS AND RESERVES: BBL, BIP, BMF, CBF, OCP, PNM, WVP. **COUNTIES:** BEL, BOO, BUL, JEF, MEA, OLD, PIK, ROW, WHI. **DATES:** Apr. 28–May 27. **REMARKS:** Common. Some specimens have white stripes on the red foretibia while others do not.

Genus ***Dichorda*** Warr.

7053 ***D. iridaria*** (Gn.)
Showy Emerald
PARKS AND RESERVES: BIP, BMF, CAC, FKN, HWS, LBL, LCP, NBP, OCP, PMS, PNM. **COUNTIES:** BEL, BUL, CAR, FUL, HLN, JEF, MCC, MEA, MEN, NEL, OLD, OWS, PER, PIK, POW, ROW, RUS, TRI. **DATES:** Apr. 21–Sept. 8. Common.

Genus ***Synchlora*** Gn.

7058 ***S. aerata*** (F.)
Wavy-lined Emerald
PARKS AND RESERVES: ASC, BMF, CBB, HWS, LBL, MUR, OCP, RGR. **COUNTIES:** BOO, BUL, EST, FAY, FUL, GRY, HIC, JEF, MCC, MEA, OLD, OWS, ROW, TRI, WOO. **DATES:** Apr. 27–Oct. 20. Common.

7059 ***S. frondaria*** Gn.
Southern Emerald
Trigg Co., LBL, July 12, 1970, C. V. Covell, Jr. Apparently rare in Kentucky.

Genus ***Chlorochlamys*** Hulst

7071 ***C. chloroleucaria*** (Gn.)
Blackberry Looper Moth
PARKS AND RESERVES: BBL, BEC, GBT, HWS, LBL, OCP, PMS, RGR. **COUNTIES:** BOO, BRK, BUL, CRL, FAY, GAL, GRY, HLN, HRT, JEF, MCC, MEA, NEL, OLD, ROW, TRI, WHI, WOL, WOO. **DATES:** Apr. 20–Nov. 18. Common.

Genus ***Chloropteryx*** Hulst

7075 ***C. tepperaria*** (Hulst)
Angle-winged Emerald
Jefferson Co., Valley Station, May 5, June 6, 16, Aug. 15, 16, 24, 28; Bullitt Co., BMF, May 23, Owsley Co., 4.8 km (3 mi) NE of Booneville, Aug. 27; Whitley Co., Goldbug, June 5. Uncommon.

Genus ***Hethemia*** Fgn.

7084 ***H. pistasciaria*** (Gn.)
Pistachio Emerald
PARKS AND RESERVES: BBL, BMF, HWS, LCP, OCP, RRG, WVP. **COUNTIES:** BOO, BUL, FAY, JEF, MEA, NEL, OLD, POW, RUS, WOO. **DATES:** Apr. 8, 17–30, May 1–26, June 5–17, July 17, 27. Common.

Subfamily **Sterrhinae**

Genus ***Euacidalia*** Pack.

7087 ***E. sericearia*** Pack.
Jefferson Co., Louisville, in light trap, Aug. 17, 1993, C. V. Covell, Jr. A rare migrant from farther south.

Genus ***Lobocleta*** Warr.

7094 ***L. ossularia*** (Gey.)
Drab Brown Wave
PARKS AND RESERVES: BBL, BMF, HWS, LBL, RGR. **COUNTIES:** BOO, BRK, BUL, FAY, FUL, GRY, HND, JEF, MCC, MRL, OLD, TRI. **DATES:** July 7–Sept. 10, 22. Very common, especially late in the season.

Genus ***Idaea*** Tr.

7102 ***I. bonifata*** (Hulst)
Caldwell Co., Aug. 14; Grayson Co., ASC, Sept. 7; McCracken Co., Ray Black farm, 9.7 km (6 mi)

SW of Paducah, June 21 and "Aug.–Oct." This species eats dried plant materials and may become an herbarium pest. Rare in Kentucky.

7105 ***I. scintillularia*** (Hulst)
Boone Co., BCN, 2 specimens taken at light, Aug. 15; Gallatin Co., Markland Dam, June 22; Jefferson Co., BEC, June 14, July 16; Oldham Co., Goshen, July 25. Rare. These records constitute a significant northwestern range extension for this species.

7108 ***I. furciferata*** (Pack.)
Notch-winged Wave
PARKS AND RESERVES: BMF, HWS, KGC, LCP, NBP, OCP, RBF, RRG. **COUNTIES:** BRE, BUL, CHR, EST, HLN, JEF, MCC, MEA, MEN, OLD, OWS, POW, RUS, WHI, WOO. **DATES:** Apr. 17, June 5–July 27. Locally common.

7109 ***I. celtima*** (Schaus)
Jefferson Co., Louisville, Sept. 3, 1965, C. V. Covell, Jr. This record constitutes a significant northern extension of its known range.

7114 ***I. demissaria*** (Hbn.)
Red-bordered Wave
PARKS AND RESERVES: HWS, KLK, LBK, PYR, RGR, SWR. **COUNTIES:** BRA, BRK, BUL, CHR, GRY, HND, JEF, MCC, MRL, OLD, OWS, TRI, WHI. **DATES:** May 9, 10, June 12–18, Aug. 2–Sept.14. Common.

7115 ***I. eremiata*** (Hulst)
Bell Co., PNM, June 8; Oldham Co., HWS, June 22; Powell Co., RRG, Tunnel Ridge, May 21; Whitley Co., Goldbug, at ultraviolet lights, June 5; all C. V. Covell, Jr. Local and uncommon.

7123 ***I. obfusaria*** (Wlk.)
Rippled Wave
PARKS AND RESERVES: BMF, CBF, HWS, OCP, PMS. **COUNTIES:** BUL, EST, HLN, JEF, MEA, OLD, OWS, WHI. **DATES:** June 29–July 22. **REMARKS:** Local and uncommon. May be found during daytime on low leaves in forests, but also comes to lights.

Genus *Pleuroprucha* Mösch.

7132 ***P. insulsaria*** (Gn.)
Common Tan Wave
PARKS AND RESERVES: BAN, BBL, BEC, BIP, BUC, CBB, GBT, HNK, HWS, LBK, LBL, LCP, LMP, KLK, NBP, OCP, PNM, PYR, RGR, SWR. **COUNTIES:** BAR, BEL, BOO, BRE, BRK, BUL, CAD, CAL, CAR, CHR, CRL, EST, FUL, GAL, GRY, HLN, HND, HOP, JEF, MCC, MEA, MER, MRL, MUH, NEL, OLD, PER, PIK, POW, ROW, RUS, SPE, TRI, WHI. **DATES:** May 4, 14, June 8–Oct. 27. Very common, especially in early fall.

Genus *Cyclophora* Hbn.

7136 ***C. packardi*** (Prout)
Packard's Wave
PARKS AND RESERVES: BBL, BBM, BEC, BUC, CBB, HWS, LCP, NBP, PMS, PNM, PYR. **COUNTIES:** BEL, BOO, BRE, CAD, HLN, JEF, LAU, MCC, MEN, OLD, OWS, PER, POW, ROW, RUS, WOL. **DATES:** Apr. 4, 15, 19, May 1, 2, 12–20, July 3–18, Aug. 6–15, Sept. 13. Common.

7139 ***C. pendulinaria*** (Gn.)
Sweetfern Geometer
PARKS AND RESERVES: BBM, NBP, PNM, RBF, RRG. **COUNTIES:** BEL, BRE, HLN, MEN, POW, ROW, WOL. **DATES:** Apr. 19, 26, July 9–15, Aug. 4, 18. Local and uncommon.

7140 ***C. nanaria*** (Wlk.)
Henderson Co., SWR, Sept. 10; Jefferson Co., onc specimen found dead in building at the University of Louisville, Feb.; Whitley Co., Goldbug, at lights, June 30. Rare.

Genus *Haematopis* Hbn.

7146 ***H. grataria*** (F.)
Chickweed Geometer
PARKS AND RESERVES: BAN, BEC, CBB, FKN, HWS, LBL, LCP, OCP, PYR, RBF, RGR. **COUNTIES:** BAR, BOO, BRA, BRE, BRK, BUL, CAD, CLK, FAY, FUL, GRY, HND, HRD, HRT, JEF, MAD, MCC, MEA, MOR, NEL, OLD, PIK, ROW, RUS, TRI, WHI. **DATES:** Apr. 13–Oct. 30. **REMARKS:** Very common, especially in old fields. Records for all-pink form "*annettearia*" Haimbach are: Bullitt Co., BMF, Aug. 7; Jefferson Co., July 5, Oct. 1; Fayette Co., Lexington, July 15, 19, 23, Aug. 5.

Genus *Timandra* Dup.

7147 ***T. amaturaria*** Wlk.
Cross-lined Wave
PARKS AND RESERVES: BBL, CBB. **COUNTIES:** BOO, FAY, FUL, JEF, NEL, OLD, PER. **DATES:** May 8–Oct. 9. **REMARKS:** May be common in old fields late in season.

Genus *Scopula* Schr.

7154 ***S. plantagenaria*** (Hulst)
Oldham Co., HWS, May 7, Aug. 6, C. V. Covell, Jr. A rare northern extension of range.

7157 **S. cacuminaria** (Morr.)
Frosted Tan Wave
Fulton Co., Willingham Bottoms, 4.0 km (2.5 mi) E of Cayce, in light trap, Sept. 10, 11; Henderson Co., SWR, in light trap, Sept. 9. Very local and uncommon in swamps.

7159 **S. limboundata** (Haw.)
Large Lace-border
PARKS AND RESERVES: BBL, BEC, BMF, BUC, CAV, CBB, CBF, HWS, JWS, KRC, LCP, LWO, NBP, OCP, PMS, PNM, PYR, RBF, RRG. **COUNTIES:** BAT, BEL, BOO, BRE, BUL, CHR, FAY, FLO, HLN, HRT, JAC, JEF, LET, MEA, MEN, MRL, NEL, OLD, OWS, PER, POW, ROW, RUS, WHI, WOL, WOO. **DATES:** Apr. 17–Sept. 21. **REMARKS:** Common in both its dark-shaded and unshaded forms.

7169 **S. inductata** (Gn.)
Soft-lined Wave
PARKS AND RESERVES: BBL, CAV, HWS, LBL, RRG. **COUNTIES:** BAT, BOO, BUL, FAY, GRV, HND, HRT, JEF, LAR, MCC, MEA, MEN, NEL, OLD, POW, TRI, WHI. **DATES:** May 9, 13, 25, June–Sept. 21. **REMARKS:** Common, but not as much so as *S. limboundata*.

Genus *Leptostales* Mösch.

7179 **L. rubromarginaria** (Pack.)
Dark-ribboned Wave
PARKS AND RESERVES: BBL, BEC, BMF, CBB, HWS, LBK, LBL, LMP, MCV, OCP. **COUNTIES:** BOO, BUL, CHR, CRI, EDM, FAY, GAL, JEF, MCC, MEA, MUH, NEL, OLD, ROW, TAY, TRI, WOO. **DATES:** Mar. 27–Aug. 18, Oct. 20. Common.

Genus *Lophosis* Hulst

7181 **L. labeculata** (Hulst)
PARKS AND RESERVES: BEC, PNM. **COUNTIES:** BEL, JEF, MCC, MEN, OWS, ROW, WHI. **DATES:** Apr. 27, May 15, 16, 26, 27, June 3–16, July 5, 16, Aug. 19, 29. Uncommon and very local.

Subfamily **Larentiinae**

Genus *Dysstroma* Hbn.

7187 **D. truncata** (Hufn.)
Marbled Carpet
Harlan Co., BBM in light trap, June 8, 1985, L. D. Gibson. Rare.

Genus *Eulithis* Hbn.

7196 **E. diversilineata** (Hbn.)
Lesser Grapevine Looper Moth
PARKS AND RESERVES: BBL, CER, HWS, LBK, LBL, LWO, NBP, OCP, PNM, RGR. **COUNTIES:** BEL, BOO, BRE, BRK, BUL, EST, FAY, GRY, JEF, LET, MCC, MEA, OLD, OWS, POW, PUL, ROW, TRI. **DATES:** May 18–Oct. 17. Common.

7197 **E. gracilineata** (Gn.)
Greater Grapevine Looper Moth
PARKS AND RESERVES: BBM, BEC, BIP, BLA, BMF, CAV, GBO, GBT, HWS, LCP, LWO, MEN, NBP, OCP, PMS, PNM, PYR, RGR, SGS. **COUNTIES:** BAT, BEL, BRK, BUL, CAD, CRL, FAY, GUP, GRY, HLN, JEF, LET, MCC, MEA, MEN, OLD, PIK, POW, RUS, WHI, WOO. **DATES:** Apr. 20, June 7, 11–Nov. 4. Common.

7206 **E. explanata** (Wlk.)
White Eulithis
Harlan Co., BBM, July 8–15; Letcher Co., head of Line Fork, June 18. Common on BBM summit.

Genus *Ecliptopera* Warr.

7214 **E. atricolorata** (G. & R.)
Dark-banded Geometer
PARKS AND RESERVES: BBM, BMF, BUC, CBB, HWS, JCF, LBL, LCP, LWO, NBP, OCP, PNM, RRG. **COUNTIES:** BEL, BRE, BUL, EST, HLN, JEF, LET, MCC, MEA, MEN, OLD, OWS, PER, POW, ROW, RUS, TRI, WOO. **DATES:** May 16–Aug. 2, 14. Common.

Genus *Hydriomena* Hbn.

7235 **H. divisaria** (Wlk.)
Black-dashed Hydriomena
PARKS AND RESERVES: BBM, BIP, KGC, OCP, PNM, RRG. **COUNTIES:** BEL, HLN, MEA, MEN, PIK, WHI. **DATES:** Apr. 11–25, May 1, 10–13. **REMARKS:** Our subspecies is *brunnescens* McD. Locally common.

7236 **H. renunciata** (Wlk.)
Renounced Hydriomena
Menifee Co., RRG, FS 9B, East Fork Indian Creek, Kelley Branch, Apr. 19; Nelson Co., Bardstown, Apr. 16. Our subspecies is *columbiata* Taylor. Uncommon.

7237 **H. transfigurata** Swett
Transfigured Hydriomena
PARKS AND RESERVES: BMF, HNK, JWS, KGC, KRC, OCP, RRG. **COUNTIES:** BUL, CAL, FLO, GRE, HLN, MEA, MEN, POW, WHI, WOL. **DATES:** Apr. 11–28, May 4, 12, July 15. **REMARKS:** Our subspecies is *manitoba* B. & McD. Common.

7239 **H. pluviata** (Gn.)
Sharp Green Hydriomena
PARKS AND RESERVES: BBL, KGC, MCV, OCP, RRG. **COUNTIES:** BOO, EDM, GRE,

HLN, JEF, MEA, MEN. **DATES:** Mar. 29–Apr. 26. **REMARKS:** Our subspecies is *meridianata* McD. Common. The holotype of *meridianata* McDunnough was collected by Carl Cook at Crailhope, Green Co., Mar. 29, 1949. This type is in the AMNH, New York (McDunnough 1954), and 7 paratypes with the same label data are in the NMNH.

7252 ***H. mississippiensis*** McD.
Mississippi Hydriomena
Bullitt Co., BMF, Apr. 17, 27; Meade Co., OCP, Apr. 25; Nelson Co., Bardstown, Apr. 16. Rare.

Genus ***Coryphista*** Hulst

7290 ***C. meadii*** (Pack.)
Barberry Geometer
Boone Co., Florence, Apr. 24; Fayette Co., June 11, July 13, 14, Sept. 8; Jefferson Co., Louisville, Apr. 29, May 3, June 5, 12, 16, July 1–9, 22–31, Aug. 4, 17, 31, Sept. 1, 10, 22, and Tom Wallace Lake, Apr. 11; Oldham Co., LaGrange, emerged from barberry indoors, Feb. 8. Common. It breeds on ornamental barberry bushes.

Genus ***Hydria*** Hbn.

7292 ***H. prunivorata*** (Fgn.)
Ferguson's Scallop Shell
PARKS AND RESERVES: AJJ, BBL, BBM, BMF, FKN, HWS, JWS, KGC, LCP, OCP, PNM, RBF, RRG. **COUNTIES:** BEL, BOO, BRE, BUL, CAM, FLO, HLN, HRD, JEF, MCC, MEA, MEN, OLD, POW, ROW, RUS, WOO. **DATES:** May 11–29, June 6–18, July 8–24. Common.

Genus ***Rheumaptera*** Hbn.

7293 ***R. hastata*** (L.)
Spear-marked Black
Harlan Co., BBM, July 1–2 1982, L.C. Koehn (several specimens of both forms); Oldham Co., Goshen, at lights, July 2, 1982, R.V. Gregg. Rare.

Genus ***Mesoleuca*** Hbn.

7307 ***M. ruficillata*** (Gn.)
White-ribboned Carpet
Bell Co., PNM, May 2; Harlan Co., BBM, July 12; Owsley Co., DBNF, E of Locust Gap, May 10. Uncommon in the mountains.

Genus ***Anticlea*** Steph.

7329 ***A. vasiliata*** Gn.
Bell Co., PNM, Apr. 11, 14; Harlan Co., BBM, Apr. 20, 21, and KGC, Apr. 20, 29; Jefferson Co., Louisville, Apr. 2, 8; Menifee Co., RRG, Apr. 4; Oldham Co., Goshen, May 1, and HWS, May 15. Uncommon.

7330 ***A. multiferata*** (Wlk.)
Many-lined Carpet
PARKS AND RESERVES: BBL, BMF, CAC, LCP, OCP, WVP. **COUNTIES:** BOO, BUL, CAR, HLN, JEF, MCC, MEA, RUS. **DATES:** Apr. 10–25, May 2. Common.

Genus ***Stamnodes*** Gn.

7333 ***S. gibbicostata*** (Wlk.)
Shiny Gray Carpet
Jefferson Co., WVP, Oct. 15, 17; Jefferson Co., no specific locality, "June" and Sept. 28; Menifee Co., RRG, FS 9B, East Fork Indian Creek, Oct. 12–13; Owsley Co., 4.8 km (3 mi) NE of Booneville, Oct. 8. Uncommon.

Genus ***Xanthorhoe*** Hbn.

7368 ***X. labradorensis*** (Pack.)
Menifee Co., RRG, along Indian Creek, in light trap, Sept. 7–8, 14, Oct. 3, L. D. Gibson. Rare.

7369 ***X. packardata*** McD.
Whitley Co., CBF, Oct. 4, 1975, P. H. Freytag. Rare.

7371 ***X. iduata*** (Gn.)
Menifee Co., RRG, FS 9A, Indian Creek, Sept. 15, 21–22; Woodford Co., along Kentucky River, 1.6 km (1 mi) W of Highway 68, 198 m (650 ft) elevation, May 22, June 6, Sept. 8–9; all collected by J. S. Nordin. Rare.

7388 ***X. ferrugata*** (Cl.)
Red Twin-spot
Boone Co., BBL, Apr. 25; Grayson Co., RGR, Aug. 11; Jefferson Co., BEC, Apr. 13, July 16. Uncommon.

7390 ***X. lacustrata*** (Gn.)
Toothed Brown Carpet
PARKS AND RESERVES: BBL, BBM, BMF, CAC, CAH, CBB, HWS, LCP, NBP, OCP, PNM, RGR, RRG. **COUNTIES:** BEL, BOO, BUL, CAR, EST, FAY, GRY, HLN, JEF, MEA, MEN, OLD, POW, ROW, RUS. **DATES:** Mar. 6–Oct. 7.

Genus ***Epirrhoe*** Hbn.

7394 ***E. alternata*** (Müller)
White-banded Toothed Carpet
Bell Co., PNM, May 3; Harlan Co., BBM, Apr. 21, July 9; Jefferson Co., Valley Station, July 16. Uncommon.

Genus ***Euphyia*** Hbn.

7399 ***E. unangulata*** (Haw.)
Sharp-angled Carpet
PARKS AND RESERVES: BBL, BBM, HWS, OCP, PNM. **COUNTIES:** BEL, BOO, FAY,

GUP, HLN, JEF, MEA, MEN, OLD. **DATES:** Apr. 25, May 2–19, June 14, July 14, 16, Aug. 8, 27. **REMARKS:** Our subspecies is *intermediata* (Gn.). Common.

Genus ***Orthonama*** Hbn.

7414 ***O. obstipata*** (F.)
The Gem
PARKS AND RESERVES: ASC, BEC, BMF, CBB, FKN, GRY, HWS, KLK, KRC, LBK, LCP, NBP, OCP, PNM, RGR, RRG. **COUNTIES:** BEL, BOO, BRE, BUL, CAM, FAY, GAL, GRY, HRD, HRT, JEF, MCC, MEA, MEN, MRL, NEL, OLD, OWS, PIK, POW, ROW, RUS, TRI, WOL. **DATES:** Apr. 3–Oct. 20, Nov. 22. Very common.

7416 ***O. centrostrigaria*** (Woll.)
Bent-line Carpet
PARKS AND RESERVES: BMF, FKN, HWS, LBK, LBL, LCP, LMP, OCP, PNM, RGR. **COUNTIES:** BAR, BEL, BOO, BRK, BUL, EST, FAY, GRY, HLN, HRD, JEF, MCC, MEA, MUH, NEL, OLD, ROC, RUS, TRI. **DATES:** Apr. 21–Oct. 24. Common.

Genus ***Disclisioprocta*** Wallgr.

7417 ***D. stellata*** (Gn.)
Somber Carpet
PARKS AND RESERVES: BIP, PNM, RGR, RRG. **COUNTIES:** BEL, BRK, FAY, FUL, GRY, JEF, MCC, MEN, OLD, OWS, PIK, ROW. **DATES:** July 2, 25, Aug. 14, 26–Nov. 4. **REMARKS:** Sometimes common late in the season.

Genus ***Hydrelia*** Hbn.

7422 ***H. inornata*** (Hulst)
Unadorned Carpet
PARKS AND RESERVES: BBM, BEC, BMF, BUC, CAC, JCF, JWS, LCP, OCP, PMS, PNM, RRG, WVP. **COUNTIES:** BEL, BRE, BUL, CAR, FAY, FLO, HLN, JEF, MEA, MEN, OWS, PER, POW, RUS. **DATES:** Apr. 20–27, May 6–June 7, 17, July 3–29, Aug. 9–8, Sept. 12. Common.

7423 ***H. albifera*** (Wlk.)
Fragile White Carpet
Oldham Co., Goshen, June 11, Aug. 7; Powell Co., NBP, May 27. Rare.

Genus ***Venusia*** Curt.

7428 ***V. comptaria*** (Wlk.)
Brown-shaded Carpet
PARKS AND RESERVES: BBL, BBM, CAC, IRO, KGC, LCP, MCV, NBP, OCP, PNM. **COUNTIES:** BEL, BOO, CAR, EDM, HLN, HRD, JEF, MEA, NEL, POW, RUS, WHI. **DATES:** Apr. 4–June 24. Common.

Genus ***Trichodezia*** Warr.

7430 ***T. albovittata*** (Gn.)
White-striped Black
PARKS AND RESERVES: BBM, BEC, CBB, PMS. **COUNTIES:** BOO, BUL, CAM, HLN, JAC, JEF, KNX, MEN, OLD, PIK, WOO. **DATES:** Apr. 11–Aug. 21. **REMARKS:** Common and often seen during the daytime in woodlands during spring.

Genus ***Eubaphe*** Hbn.

7440 ***E. mendica*** (Wlk.)
The Beggar
PARKS AND RESERVES: BMF, FKN, HWS, JWS, LCP, NBP, OCP, RGR, RRG. **COUNTIES:** BOO, BRE, BRK, BUL, FAY, FLO, FUL, GRY, HRD, JEF, LEW, MEA, MEN, NEL, OLD, PIK, POW, ROW, RUS, WHI, WOO. **DATES:** Apr. 18, May 3, 13–Sept. 6, 16–19, Oct. 15. Common.

7441 ***E. meridiana*** (Slosson)
Little Beggar
Rowan Co., Morehead, 5 specimens at light, July 25; Whitley Co., Goldbug, June 5; both C. V. Covell, Jr. Local and rare.

Genus ***Horisme*** Hbn.

7445 ***H. intestinata*** (Gn.)
Brown Bark Carpet
PARKS AND RESERVES: BBL, BMF, CBF, PMS, RGR. **COUNTIES:** BOO, BRK, BUL, FAY, GRY, HLN, JEF, NEL, OLD, OWS, WHI. **DATES:** May 10–July 16, 25, 26, Aug. 18, 20. Common.

Genus ***Eupithecia*** Curt.

Pugs

7449 ***E. palpata*** Pack.
Owsley Co, 4.8 km (3 mi) NE of Booneville, May 27; Rowan Co., Morehead, May 21; Russell Co., LCP, May 23. Uncommon.

7451 ***E. slossonata*** McD.
Russell Co., LCP, May 23, 1977, C. Cornett. Rare.

7453 ***E. peckorum*** Heitzman & Enns
Laurel Co., FS 615A, Apr. 6, and Holly Bay Campground, Apr. 19; L. D. Gibson and D. J. Wright. Uncommon.

7459 ***E. columbiata*** (Dyar)
PARKS AND RESERVES: CAC, GBT, KGC, MCV, NBP, RRG. **COUNTIES:** CAR, CRL, EDM, HLN, MEN, POW. **DATES:** Apr. 2–May 2,

10. **REMARKS:** Our subspecies is *erpata* Pears. Uncommon.

7474 ***E. miserulata*** Grt.
Common Eupithecia; Common Pug
PARKS AND RESERVES: BBL, BBM, BEC, BIP, BMF, CAC, GBT, HWS, LCP, NBP, PNM. **COUNTIES:** BEL, BOO, BUL, CAR, CRL, EST, FAY, FUL, HLN, HRD, JEF, JES, MCC, MEN, OLD, OWS, PIK, POW, RUS, WOO. **DATES:** Mar. 13–Oct. 25. Very common.

7475 ***E. chlorofasciata*** Dietze
Menifee Co., RRG, N of Slade along Indian Creek, at black-light trap, May 2, 1979, J. S. Nordin. Identified provisionally as this species by D. C. Ferguson.

7481 ***E. coloradensis*** (Hulst)
Owsley Co., 4.8 km (3 mi) NE of Booneville, July 18, and Caney Fork Creek near Stay, Aug. 16; both L. D. Gibson. Rare.

7486 ***E. jejunata*** McD.
PARKS AND RESERVES: BMF, CAC, LCP, NBP, OCP, PNM, RRG. **COUNTIES:** BEL, BUL, CAR, MEA, MEN, POW, RUS, WOO. **DATES:** Mar. 20–25, Apr. 3–27, May 2, 11. Common.

7489 ***E. luteata*** Pack.
Bell Co., PNM, May 2; Carter Co., CAC, Apr. 24; Menifee Co., RRG, FS 9A, Leatherwood Fork of Indian Creek, May 12. Uncommon.

7491 ***E. fletcherata*** Tayl.
Harlan Co., Apr. 20, 21; Menifee Co., RRG, Mar. 29, Apr. 2, 4, 13, 26, May 10. Uncommon.

7495 ***E. affinata*** Pears.
Jefferson Co., Valley Station, Sept. 3; Meade Co., OCP, Aug. 31; Menifee Co., RRG, FS 9A, Indian Creek, Sept. 21–22; Woodford Co., Sept. 8, 12. Uncommon.

7509.1 ***E. matheri*** Rindge
Mather's Eupithecia; Mather's Pug
PARKS AND RESERVES: BMF, CAC, GBT, LMP, NBP, OCP, PNM. **COUNTIES:** BEL, BUL, CAR, CRL, JEF, MEA, MEN, MUH, POW. **DATES:** Mar. 21–May 13. **REMARKS:** Common. This moth was formerly known as *E. herefordaria* (C. & S.).

7525 ***E. grata*** Tayl.
Woodford Co., 1.6 km (1 mi) W of SR 68, along Kentucky River, 198 m (650 ft) elevation, at black-light trap, May 4, July 24, J. S. Nordin. Rare.

7526 ***E. russeliata*** Swett
Harlan Co., BBM, at light, July 11, 1987, C. V. Covell, Jr. Rare.

7529 ***E. coagulata*** Gn.
Fayette Co., Sept. 7–18; Jefferson Co., Valley Station, Aug. 14; Menifee Co., Indian Creek, NE of Slade, Sept. 19, 24. Uncommon.

7530 ***E. swettii*** Grossb.
PARKS AND RESERVES: CAC, CBF, LCP, LMP, MCV, NBP, RRG. **COUNTIES:** CAR, EDM, HLN, JEF, MEN, MUH, POW, ROW, RUS, WHI. **DATES:** Mar. 19–Apr. 30, July 25. Common.

7543 ***E. annulata*** (Hulst)
Powell Co., RRG, Tunnel Ridge Rd., Mar. 11, 1989, D. J. Wright. Uncommon.

Genus *Acasis* Dup.

7635 ***A. viridata*** (Pack.)
Olive-and-Black Carpet
Carter Co., CAC, Apr. 18; Harlan Co., BBM, July 12; Menifee Co., RRG, FS 9A, Leatherwood Fork of Indian Creek, Apr. 14; Powell Co., NBP, July 13. Uncommon.

Genus *Cladara* Hulst

7637 ***C. limitaria*** (Wlk.)
Mottled Gray Carpet
PARKS AND RESERVES: BIP, CAC, CBF, GBT, KGC, NBP, PNM, RRG. **COUNTIES:** BEL, CAR, CRL, HLN, MEN, PIK, POW, WHI. **DATES:** Mar. 29, Apr. 11–May 3. Common.

7638 ***C. anguilineata*** (G. & R.)
Angle-lined Carpet
PARKS AND RESERVES: BIP, CAC, CBF, FKN, GBT, KGC, NBP, PNM, RRG. **COUNTIES:** BEL, CAR, CRL, HLN, HRD, MEA, MEN, PIK, POW, WHI. **DATES:** Apr. 10–29, May 16. Common.

7639 ***C. atroliturata*** (Wlk.)
The Scribbler
PARKS AND RESERVES: BBM, BIP, BMF, CAC, KGC, LMP, MCV, NBP, OCP, PNM, RRG. **COUNTIES:** BEL, BUL, CAR, EDM, HLN, JEF, LAU, MEA, MEN, MUH, NEL, OWS, PIK, POW. **DATES:** Mar. 15, 28–May 3. Common.

Genus *Heterophleps* H.-S.

7645 ***H. refusaria*** (Wlk.)
Three-patched Bigwing
Bell Co., PNM, July 18; Harlan Co., BBM, June 8, July 14; Menifee Co., RRG, FS 9A, Indian Creek, May 19, July 8; Powell Co., RRG, July 7. Uncommon.

7647 ***H. triguttaria*** H.-S.
Three-spotted Fillip
PARKS AND RESERVES: BBL, BCN, CBB, FKN, HWS, LBL. **COUNTIES:** BOO, FLN, GAL, HRT, JEF, MEA, NEL, OLD, SHE, SPE, TRI. **DATES:** May 7–June 21, July 12, 22, Aug. 5–15, Sept. 7. In spring it is seen often during the day in forests. Common.

Genus ***Dyspteris*** Hbn.

7648 ***D. abortivaria*** (H.-S.)
The Bad-wing
PARKS AND RESERVES: BBL, BEC, BMF, BUC, CAC, CBB, CBF, FKN, HWS, LBK, LCP, NBP, OCP, PMS, RGR, RRG. **COUNTIES:** BOO, BRE, BRK, BUL, CAR, FAY, GAL, GRY, HLN, HRD, JEF, LAU, MCC, MEA, MEN, MER, NEL, OLD, PER, POW, RUS, TRI, WHI, WOO. **DATES:** Apr. 13–Sept. 23, Nov. 17. Common.

Family **EPIPLEMIDAE**

Epiplemid Moths

Genus ***Callizzia*** Pack.

7650 ***C. amorata*** Pack.
Gray Scoopwing
PARKS AND RESERVES: ASC, BEC, CBB, HWS, LCP, OCP. **COUNTIES:** BRA, GRY, JEF, MCC, MEA, MER, NEL, OLD, ROC, RUS. **DATES:** May 9–20, June 6, 14–22, July 14–23, Aug. 14, 30, Sept. 7, 18. Common.

Genus ***Calledapteryx*** Grt.

7653 ***C. dryopterata*** Grt.
Brown Scoopwing
Boone Co., BBL, May 29, June 25; Jefferson Co., WVP, June 18, July 1, Aug. 27; "Kentucky near Cincinnati, Ohio," emerged July 12, 17, 24; McCracken Co., "Aug.–Oct."; Oldham Co., HWS, June 14. Uncommon.

Superfamily **MIMALLONOIDEA**

Family **MIMALLONIDAE**

Sack-bearer Moths

Genus ***Lacosoma*** Grt.

7659 ***L. chiridota*** Grt.
Scalloped Sack-bearer
PARKS AND RESERVES: BAN, BBL, BMF, BUC, CBB, CBF, LBL, LCP, OCP, WVP. **COUNTIES:** BAR, BOO, BRA, BRE, BUL, JEF, MCC, MEA, NEL, OLD, OWS, PER, RUS, TRI, WHI. **DATES:** Apr. 5, 18, May 6, 18–June 27, July 18, 26, Aug. 4, Sept. 20–30. Common.

Genus ***Cicinnus*** Blanchard

7662 ***C. melsheimeri*** (Harr.)
Melsheimer's Sack-bearer
PARKS AND RESERVES: BBM, BMF, BUC, LWO. **COUNTIES:** BUL, HLN, LET, JEF, OWS, PER. **DATES:** May 15–22, June 3, 8, 16, July 11, 14. Moderately common.

Superfamily **BOMBYCOIDEA**

Family **APATELODIDAE**

Apatelodid Moths

Genus ***Apatelodes*** Pack.

7663 ***A. torrefacta*** (J. E. Smith)
Spotted Apatelodes
PARKS AND RESERVES: BAN, BBM, BMF, BUC, CAC, CBB, CBF, JWS, LBL, LVJ, LWO, NBP, OCP, PMS, PNM, RRG, WVP. **COUNTIES:** BAR, BEL, BRE, BUL, CAR, CRL, FLO, GRV, HLN, JEF, LAR, LAU, LET, MEA, MEN, NEL, OLD, OWS, PER, PIK, POW, ROC, ROW, TAY, TRI, WAR, WHI. **DATES:** May 22–Aug. 18. Common.

Genus ***Olceclostera*** Butler

7665 ***O. angelica*** (Grt.)
The Angel
PARKS AND RESERVES: BUC, HWS, KGC, KRC, LCP, NBP, OCP, PNM, WVP. **COUNTIES:** ADA, BEL, BRE, GAL, JEF, JES, LAR, HLN, MEA, MEN, NEL, OLD, PER, POW, ROW, RUS, WOL. **DATES:** May 5–Sept. 3, Nov. 23. **REMARKS:** Common, but less so than than *A. torrefacta*.

Family **LASIOCAMPIDAE**

Tent Caterpillar and Lappet Moths

Genus ***Tolype*** Hbn.

7670 ***T. velleda*** (Stoll)
Large Tolype
PARKS AND RESERVES: BIP, BOW, GBT, LCP, MUR, PNM, RRG. **COUNTIES:** BEL, BOO, CAM, CRL, FAY, GRY, HIC, HLN, JEF, MAD, MCC, PIK, RUS. **DATES:** July 30, Sept. 21, 24–Oct. 22. Common.

7674 ***T. notialis*** Franc.
Small Tolype
PARKS AND RESERVES: BBM, BMF, BUC, LCP, LWO, RRG. **COUNTIES:** BRE, BUL, HLN, LET, NEL, PER, POW, RUS, WHI. **DATES:** June 5, 15–23, July 7, 11, Sept. 2, 13. Common.

7675 ***T. minta*** Dyar
McCracken Co., Mayfield Creek bottoms, off Clinton Road 14.5 km (9 mi) SW of Paducah, at UV light, 3 males, Oct. 6, 1989, C. V. Covell, Jr. This record represents a substantial northward extension of the previously known range (Franclemont 1973).

Genus ***Artace*** Wlk.

7683 ***A. cribraria*** (Ljungh)
Dot-lined White
PARKS AND RESERVES: BIP, BMF, BUC, FKN, GBT, LBL, LCP, PNM, PYR, RGR. **COUNTIES:** BEL, BRE, BUL, CAD, CRL, EST, FUL, GRY, HRD, JEF, MCC, MEA, NEL, OWS, PER, PIK, ROC, ROW, RUS, TRI, WHI. **DATES:** June 3–Nov. 4. Locally common.

Genus ***Heteropacha*** Harv.

7685 ***H. rileyana*** Harv.
Riley's Lappet Moth
PARKS AND RESERVES: BAN, HWS, LBL. **COUNTIES:** BAR, BRA, JEF, MAD, MCC, NEL, OLD, TRI. **DATES:** Apr. 14, 18, May 3, 14–June 16, July 12, 15, 29, Aug. 1, 3, Sept. 19, 21, "Oct." Locally common.

Genus ***Phyllodesma*** Hbn.

7686 ***P. occidentis*** (Wlk.)
Bullitt Co., BMF, May 4, 1978, A. J. Brownell. Rare.

7687 ***P. americana*** (Harr.)
Lappet Moth
PARKS AND RESERVES: BBL, BIP, BMF, CAC, KGC, NBP, OCP, PNM, RRG. **COUNTIES:** BEL, BOO, BUL, CAR, HLN, JEF, MAD, MCC, MEA, NEL, OWS, PIK, POW. **DATES:** Mar. 4, 16–May 4, 24, Sept. 22. Common.

Genus ***Malacosoma*** Hbn.

7698 ***M. disstria*** Hbn.
Forest Tent Caterpillar Moth
PARKS AND RESERVES: BMF, HWS, KLK, LCP, PNM, PYR. **COUNTIES:** BEL, BUL, CAD, CHR, JEF, MRL, NEL, OLD, RUS. **DATES:** Apr. 7, June 6–20, July 5, 11, Aug. 9, Sept. 1, 25. Locally common.

7701 ***M. americanum*** (F.)
Eastern Tent Caterpillar Moth
PARKS AND RESERVES: BAN, BBM, BEC, BMF, BUC, FKN, HWS, JWS, LCP, LMP, MCV, NBP, OCP, PNM, PYR. **COUNTIES:** BAR, BEL, BRE, BUL, CAD, CHR, EDM, FLO, HLN, JEF, MCC, MEA, MUH, NEL, OLD, OWS, PER, POW, RUS, WHI. **DATES:** May 12–June 27, July 16, 30, Aug. 6. **REMARKS:** Common. Often a serious defoliator of *Prunus* and other trees and shrubs throughout the state. The larvae form their tents in crotches of trees in early April. Abundance seems to follow cycles. Rauschenberger et al. (1966) reported egg masses at MCV on Aug. 16, 1966 on black cherry *(Prunus serotina)*.

Family **SATURNIIDAE**

Giant Silkworm and Royal Moths

Subfamily **Citheroniinae**

Genus ***Eacles*** Hbn.

7704 ***E. imperialis*** (Drury)
Imperial Moth
PARKS AND RESERVES: BMF, CBF, HWS, JCF, LBL, LWO, NBP, OCP, PMS, PNM, RGR, RRG. **COUNTIES:** BEL, BUL, CLI, CRL, EST, FAY, GRV, GRY, HLN, HND, HRD, JEF, KNX, LAR, LET, MAD, MAR, MAS, MCC, MCY, MEA, MEN, NEL, OLD, OWS, PIK, POW, ROW, SIM, SPE, TRI, UNI, WHI. **DATES:** July 4–Oct. 8. **REMARKS:** Common. Adults of this species and other Citheroniinae come readily to lights, often remaining perched near them during the following day.

Genus ***Citheronia*** Hbn.

7706 ***C. regalis*** (F.)
Royal Walnut Moth; Regal Moth
PARKS AND RESERVES: BMF, BUC, CBF, HWS, KGC, LBL, LCP, MUR, NBP, OCP, PMS, RBF, RGR, RRG. **COUNTIES:** BOU, BRE, BUL, FAY, GRY, HIC, HLN, JEF, JES, KNX, LYO, MAD, MCC, MEA, MEN, MOR, NEL, OLD, OWS, PER, PIK, POW, ROC, ROW, RUS, SCO, SIM, TRI, UNI, WHI. **DATES:** June 3–Sept. 27 (mostly July–Aug.). **REMARKS:** Common, especially in the mountains. The larva (hickory horned devil) is often found leaving trees in late summer to make a pupation cell underground. Almost every year several citizens in the Louisville area call me to describe these caterpillars and ask what they are.

7708 ***C. sepulcralis*** G. & R.
Pine Devil Moth
PARKS AND RESERVES: KGC, PMS, PNM, RRG. **COUNTIES:** BEL, CAR, FLO, HLN, KNO, KNX, MEN, OWS, PIK, POW, ROW. **DATES:** June 11, 24–Aug. 2. **REMARKS:** More local and less common than *C. regalis*. Also less common in western two-thirds of Kentucky.

Genus ***Sphingicampa*** Walsh

7709 ***S. bicolor*** (Harr.)
Honey Locust Moth
PARKS AND RESERVES: BAN, BMF, CBB, HWS, JCF, OCP, RGR. **COUNTIES:** BAR, BUL, FAY, GAL, GRV, GRY, JEF, JES, MCC, MEA, NEL, OLD, SPE, UNI, WAR. **DATES:** Apr. 20, May 7–Sept. 21. Common.

7712 ***S. bisecta*** (Lint.)
Bisected Honey Locust Moth
PARKS AND RESERVES: BMF, BOW, HWS, PYR, RGR. **COUNTIES:** BOO, BUL, CAD, CHR, FAY, GAL, GRY, JEF, JES, MEA, NEL, OLD. **DATES:** Apr. 20, June 6–15, July 5, 26, Aug. 1–20, Sept. 1. Common.

Genus ***Dryocampa*** Harr.

7715 ***D. rubicunda*** (F.)
Rosy Maple Moth
PARKS AND RESERVES: BAN, BBM, BIP, BMF, BUC, CBF, HWS, JWS, KGC, LBK, LCP, LWO, NBP, OCP, PNM, PYR, RBF, RGR. **COUNTIES:** BAR, BEL, BRE, BUL, CAD, CHR, FAY, FLO, GAR, GRY, HLN, JEF, JES, KNX, LAR, LAU, LET, MAR, MEA, NEL, OLD, OWS, PER, PIK, POW, ROW, RUS, TRI, WAR, WAY, WHI. **DATES:** Apr. 19–Sept. 2. **REMARKS:** Common. I took a rare individual with entirely pink forewings at OCP, Meade Co., Aug. 9, 1989.

Genus ***Anisota*** Hbn.

7716 ***A. stigma*** (F.)
Spiny Oakworm Moth
PARKS AND RESERVES: BMF, HWS, JCF, KEC, KGC, LBL, LWO, NBP, PNM, RRG. **COUNTIES:** BEL, BUL, GRY, HLN, HOP, HRD, JEF, LET, MCC, MOR, NEL, OLD, PIK, POW, ROW, TRI, UNI, WAY, WHI. **DATES:** June 11–Sept. 10. Common.

7719 ***A. senatoria*** (J. E. Smith)
Orange-striped Oakworm Moth
Breathitt Co., larvae reported damaging oak trees, July, 1975, R. Scheibner; Jefferson Co., Valley Station, June 28; Knox Co., larvae photographed by John MacGregor; Letcher Co., Pine Mountain, July 1; Meade Co., OCP, June 21; Owsley Co., 4.8 km (3 mi) NE of Booneville, May 6. Adults very rarely seen; more easily found as larvae.

7720 ***A. peigleri*** Riotte
Peigler's Oakworm Moth
Whitley Co., Corbin, last instar larvae abundant on pin oak on Aug. 18, 1986, James P. Tuttle. Adults not yet recorded from Kentucky.

7723 ***A. virginiensis*** (Drury)
Pink-striped Oakworm Moth
COUNTIES: GRE, JEF, MCC, MON, OWS, UNI, WHI. **DATES:** May 6, 17–20, June 23–27, July 2, 13–27, Aug. 10, Sept. 1. Common.

Subfamily **Hemileucinae**

Genus ***Hemileuca*** Wlk.

7730 ***H. maia*** (Drury)
Buck Moth
PARKS AND RESERVES: FKN, LBL, OCP, PCP. **COUNTIES:** BUL, EDM, FLE, HLN, HOP, HRD, HRT, JEF, KNO, MAD, MEA, OWS, ROC, TRI. **DATES:** Oct. 3–Nov. 14. **REMARKS:** Fleming Co., larvae reared on oak, emerged Oct. 28, Nov. 2, 4, and 5, A. F. Braun; Jefferson Co., larva kept in lab, emerged Nov. 11, 1974. Adults are common, but must be caught with nets since they do not usually come to light. They fly during a short period from about 1200 to 1400 hours. The late S. Scholz took one female at light in Jefferson Co., Valley Station. Occurrence varies from year to year, but it is sometimes locally common. Larvae have stinging bristles.

Genus ***Automeris*** Hbn.

7746 ***A. io*** (F.)
Io Moth
PARKS AND RESERVES: BMF, BUC, CBB, FAO, HWS, JCF, KRC, LBL, LVJ, NBP, OCP, PMS, PNM, SGS. **COUNTIES:** BAR, BEL, BOO, BRE, BUL, EST, FAY, GAL, HLN, JEF, JES, LAU, LET, LYO, MAD, MCC, MEA, NEL, OLD, OWS, PEN, PER, POW, ROW, RUS, SHE, TRI, UNI, WAY, WOL. **DATES:** May 3–Oct. 3. **REMARKS:** Common. The caterpillars have stinging bristles.

Subfamily **Saturniinae**

Genus ***Antheraea*** Hbn.

7757 ***A. polyphemus*** (Cram.)
Polyphemus Moth
PARKS AND RESERVES: BAN, BBM, BMF, BUC, JWS, KRC, LBL, LCP, NBP, OCP, PNM, RGR. **COUNTIES:** ADA, BAR, BEL, BOO, BRE, BUL, FAY, FLO, GAL, GRY, HLN, HND, JEF, JES, KNX, MAD, MCC, MEA, NEL, OLD, OWS, PIK, PER, POW, RUS, TRI, UNI, WAR, WOL. **DATES:** Apr. 11–Sept. 26. Common.

Genus ***Actias*** Leach

7758 ***A. luna*** (L.)
Luna Moth
PARKS AND RESERVES: BAN, BBM, BMF, BUC, CBB, CBF, JWS, KRC, LBL, LCP, MCV,

NBP, OCP, PNM, PYR, RRG, RGR. **COUNTIES:** ADA, BAR, BEL, BRE, BRK, BUL, CAD, CAS, CHR, CLI, EDM, FLO, GAL, GRE, GRY, HLN, HRD, JEF, LAU, MAD, MAR, MCC, MEA, MEN, MON, NEL, OWS, OLD, PER, POW, RUS, SIM, TRI, UNI, WAR, WHI, WOL. **DATES:** Mar. 24–Aug. 15, Sept. 2. Locally common.

Genus ***Samia*** Hbn.

7759 ***S. cynthia*** (Drury)
Ailanthus Silkmoth
One unconfirmed record from Louisville. Alan Barron (*pers. comm.*) found a caterpillar of what he believes was this species on an *Ailanthus* tree at 5th and York Streets in Aug., 1971. The species was introduced into this country in the 19th century, and has spread to many urban areas, feeding on *Ailanthus* and other trees in waste areas like vacant lots and railroad yards. It is seldom encountered as an adult because it is diurnal and does not come to light.

Genus ***Callosamia*** Pack.

7764 ***C. promethea*** (Drury)
Promethea Moth
PARKS AND RESERVES: BMF, KGC, NBP. **COUNTIES:** BOO, BUL, CAS, HLN, JEF, LAR, PIK, POW, ROW, RUS, WOL. **DATES:** Apr. 21–30, May 1–11, 20–27, June 9, July 16, 18, 30, Aug. 8–15. **REMARKS:** More common than adult records indicate. I found cocoons abundant in Jefferson Co., along Crittenden Drive and Outer Loop on small sassafras saplings (winter 1971). Adults emerged indoors on Apr. 15, 1972.

7765 ***C. angulifera*** (Wlk.)
Tulip Tree Silkworm Moth
PARKS AND RESERVES: BAN, BMF, CAC, CBF, LBL, LMP, NBP, OCP, PNM, RRG. **COUNTIES:** BAR, BEL, BUL, CAD, CAR, FAY, GAL, JEF, LAU, MAS, MCC, MEA, MEN, MUH, NEL, OWS, POW, RUS, TRI, WAR, WHI. **DATES:** Apr. 2–Sept. 11. Common.

Genus ***Hyalophora*** Duncan

7767 ***H. cecropia*** (L.)
Cecropia Moth
PARKS AND RESERVES: BMF, LCP. **COUNTIES:** BUL, FAY, JEF, MAD, MAS, MCC, NEL, OWS, RUS, TRM. **DATES:** Apr. 7, May 25, June 6, 10, 19, July, Aug. 19, Sept. 13. **REMARKS:** I have received many phone calls and specimens (usually females) during June–July in the Louisville area, where it seems well established, feeding on tree leaves. More common and generally distributed than our records show. Often represented in 4-H collections from around the state exhibited in the Kentucky State Fair.

Superfamily **SPHINGOIDEA**

Family **SPHINGIDAE**

Sphinx or Hawk Moths

Subfamily **Sphinginae**

Genus ***Agrius*** Hbn.

7771 ***A. cingulata*** (F.)
Pink-spotted Hawkmoth
PARKS AND RESERVES: BOW. **COUNTIES:** BOO, CAL, CAM, FAY, JEF, MCC, TRI, WAR. **DATES:** Sept. 1–Oct. 3, 13–29, Nov. 3. **REMARKS:** Moderately common, and found only in autumn.

Genus ***Manduca*** Hbn.

7775 ***M. sexta*** (L.)
Carolina Sphinx; Tobacco Hornworm Moth
PARKS AND RESERVES: BAN, BMF, BOW, HWS, OCP, WVP. **COUNTIES:** BAR, BOO, BUL, CAM, CRL, FAY, JEF, JES, MCC, MEA, NEL, OLD, ROW, WAR. **DATES:** May 24–Sept. 6. **REMARKS:** Common. The tobacco hornworm larva is a major pest of tobacco. This species has been studied extensively at the University of Kentucky by Dr. Douglas Dahlman and his associates.

7776 ***M. quinquemaculata*** (Haw.)
Five-spotted Hawkmoth
PARKS AND RESERVES: BOW. **COUNTIES:** BOO, FAY, HND, HRD, JEF, MCC, NEL, ROW, WHI. **DATES:** June 16, July 8, 15–Sept. 1, Nov. 3. **REMARKS:** Less common than the Carolina sphinx. The tomato hornworm larva is a pest of solonaceous plants, including tomato and tobacco.

7778 ***M. rustica*** (F.)
Rustic Sphinx
PARKS AND RESERVES: BOW, OCP. **COUNTIES:** BOO, BRA, CHR, MEA, MET, MRL. **DATES:** May 8, June 25, July 3, 26, 27, Sept. 6. Uncommon.

7783 ***M. jasminearum*** (Guer.)
Ash Sphinx
PARKS AND RESERVES: BAN, BBL, BBM, BMF, RGR. **COUNTIES:** BAR, BOO, BRK, BUL, GAL, GRA, GRY, HLN, JEF, JES, KNO, MCC, MEA, NEL, OWS, ROW, UNI, WAR. **DATES:** Apr. 22–Aug. 29. Common.

Genus ***Dolba*** Wlk.

7784 ***D. hyloeus*** (Drury)
Papaw Sphinx
Boone Co., BBL, June 26; Jefferson Co., Valley Station, June 30; Knott Co., Hindman, Aug., Vest, May 31; Menifee Co., RRG, FS 9A, Aug. 12; Trigg Co., LBL, July 15. Uncommon.

Genus ***Ceratomia*** Harr.

7786 ***C. amyntor*** (Geyer)
Four-horned Sphinx
PARKS AND RESERVES: BBM, BMF, HWS, JWS, WVP. **COUNTIES:** BRA, BUL, FAY, FLO, GRA, HLN, JEF, KNO, MCC, NEL, OLD, ROW. **DATES:** Apr. 6, 10, May 3, 4, 15, June 7–July 20, Aug. 10, Sept. 9. Uncommon.

7787 ***C. undulosa*** (Wlk.)
Waved Sphinx
PARKS AND RESERVES: BAN, BBL, BBM, BMF, BUC, HWS, LBL, LCP, LMP, NBP, OCP, PNM, PYR, RGR, WVP. **COUNTIES:** BAR, BEL, BOO, BRE, BUL, CAD, FAY, GRV, GRY, HLN, JEF, KNO, MCC, MEA, MUH, NEL, OLD, PER, POW, RUS, SPE, TRI, WAR, WAY. **DATES:** Apr. 2–Oct. 18. **REMARKS:** One of our most common sphinx moths. A striking semi-melanic form with wide postmedian bands on the forewings was taken several times at Valley Station, June 29–July 21, by S. Scholz.

7789 ***C. catalpae*** (Bdv.)
Catalpa Sphinx
PARKS AND RESERVES: HWS, JWS, LMP, OCP, PYR, RGR. **COUNTIES:** CAD, CHR, FAY, FLO, FRA, FUL, GRY, JEF, JES, KNX, MEA, MUH, OLD, ROW. **DATES:** May 21–Sept. 17. **REMARKS:** May be locally common, as it was in Louisville, 1994. Caterpillars ("Catalpa" or "Catawba" worms) are used for fish bait.

7790 ***C. hageni*** Grt.
Hagen's Sphinx
PARKS AND RESERVES: AJJ, BAN, BMF, FKN, HWS, SWR. **COUNTIES:** BAR, BUL, CAM, CHR, GAL, GRV, HND, HRD, JEF, JES, MCC, NEL, OLD, UNI, WHI. **DATES:** Apr. 20–Oct. 1. Common.

Genus ***Isoparce*** R. & J.

7791 ***I. cupressi*** (Bdv.)
Cypress Sphinx
Fulton Co., Willingham Bottoms, 4.0 km (2.5 mi) E of Cayce on SR 94, in light trap, Sept. 6, 1985, C. V. Covell, Jr. Found only in cypress swamps.

Genus ***Paratrea*** Grt.

7793 ***P. plebeja*** (F.)
Plebeian Sphinx
PARKS AND RESERVES: BAN, BMF, BUC, HWS, LBL, OCP, PNM, RGR. **COUNTIES:** BAR, BEL, BOO, BRE, BRK, BUL, FAY, GRV, GRY, JEF, KNO, LAW, MCC, MEA, NEL, OLD, OWS, PER, ROW, TRI, WAR, WAY. **DATES:** Apr. 30–Sept. 28. Common.

Genus ***Sphinx*** L.

7796 ***S. eremitus*** (Hbn.)
Hermit Sphinx
Harlan Co., PMS near Bledsoe, at light, July 8, 1977, C. V. Covell, Jr., and BBM, larva in early stage collected feeding on leaves of *Monarda clinopodia*, Sept. 22, 1984, M. Medley. Rare.

7802 ***S. chersis*** (Hbn.)
Great Ash Sphinx
PARKS AND RESERVES: BMF, BOW, RGR. **COUNTIES:** BOO, BUL, GAL, GRY, JEF, NEL, ROW. **DATES:** May 27, 30, June 6, 12, 29, July 13, 25, Aug. 5–11, 20. Uncommon.

7807 ***S. canadensis*** Bdv.
Canadian Sphinx
PARKS AND RESERVES: BBL, BOW, HWS, OCP. **COUNTIES:** BOO, FRA, GAL, JEF, JES, MEA, NEL, OLD. **DATES:** May 17, 20, June 14, July 1–30, Aug. 8–13. Uncommon.

7808 ***S. franckii*** Neum.
Franck's Sphinx
PARKS AND RESERVES: BBM, BMF, LCP, OCP, PYR, RRG, WVP. **COUNTIES:** BUL, CAD, HLN, JEF, KNO, MEA, MEN, NEL, OWS, POW, ROW, RUS. **DATES:** June 17–July 23. May be locally common.

7809 ***S. kalmiae*** J. E. Smith
Laurel Sphinx
PARKS AND RESERVES: BBM, BMF, BOW, BUC, CBF, CER, LCP, LMP, OCP. **COUNTIES:** BEL, BOO, BRE, BUL, HLN, JEF, MEA, MUH, NEL, PER, RUS, WHI. **DATES:** May 2–Sept. 1. Common.

7810 ***S. gordius*** Cram.
Apple Sphinx
Harlan Co., KGC, near Cumberland, July 6, 1977, L. D. Gibson. Rare.

7812 ***S. drupiferarum*** J. E. Smith
Wild-cherry Sphinx
Jefferson Co., (no date), D. A. Young; Fayette Co., Lexington (UK), May 26, 1932, June 15, 1902; Rowan Co., Morehead, June 21. No recent captures.

Genus ***Lapara*** Wlk.

7816 ***L. coniferarum*** (J. E. Smith)
Pine Sphinx
PARKS AND RESERVES: BMF, BUC, JCF, LCP, LWO, NBP, PNM, RRG, WVP. **COUNTIES:** BEL, BRE, BUL, JEF, LET, MOR, NEL, OWS, PER, POW, ROW, RUS, WAY. **DATES:** June 16–Aug. 20. May be locally common.

Genus ***Smerinthus*** Latr.

7821 ***S. jamaicensis*** (Drury)
Twin-spotted Sphinx
PARKS AND RESERVES: BAN, HWS, JWS, LBK, PYR, SWR. **COUNTIES:** BAR, CAD, CHR, FAY, FLO, GAL, HND, JEF, JES, MCC, NEL, OLD, OWS, PIK, POW, ROW, TRI, WAR. **DATES:** May 6–Sept. 10. Common.

Genus ***Paonias*** Hbn.

7824 ***P. excaecatus*** (J. E. Smith)
Blinded Sphinx
PARKS AND RESERVES: BBM, BMF, BUC, JCF, LBK, LCP, NBP, OCP, PMS, PNM, RGR. **COUNTIES:** BEL, BRE, BRK, BUL, CAR, FAY, GAL, GRY, HLN, JEF, MCC, MEA, OWS, PER, POW, RUS, TRI, UNI, WAY. **DATES:** Apr. 15–Sept. 2. Common.

7825 ***P. myops*** (J. E. Smith)
Small-eyed Sphinx
PARKS AND RESERVES: BAN, BBM, BEC, BMF, BUC, CBB, HWS, KLK, LCP, LMP, OCP, PNM, RBF, RGR. **COUNTIES:** BAR, BEL, BRE, BRK, BUL, FAY, FUL, GAL, GRY, HLN, JEF, JES, LAR, MCC, MEA, MOR, MRL, MUH, NEL, OLD, PER, ROW, RUS, UNI, WAR. **DATES:** Apr. 15–Sept. 1. Common.

7826 ***P. astylus*** (Drury)
Huckleberry Sphinx
PARKS AND RESERVES: KGC, PNM, RRG. **COUNTIES:** BEL, BOD, GUP, HLN, KNO, MEN, MET, ROW. **DATES:** June 8, 13, 28, July 12–20, Aug. 6. **REMARKS:** Uncommon and apparently restricted to the mountains.

Genus ***Laothoe*** F.

7827 ***L. juglandis*** (J. E. Smith)
Walnut Sphinx
PARKS AND RESERVES: BAN, BMF, BUC, CBF, JCF, LBL, LCP, OCP. **COUNTIES:** BAR, BRA, BRE, BUL, FAY, GAL, HLN, HRD, JEF, JES, LAR, MAD, MCC, MEA, NEL, PER, ROB, RUS, TRI, WHI. **DATES:** Apr. 17–Sept. 20. Common.

Genus ***Pachysphinx*** R. & J.

7828 ***P. modesta*** (Harr.)
Big Poplar Sphinx
PARKS AND RESERVES: COB, OCP. **COUNTIES:** BRA, HIC, HOP, JEF, MEA, NEL. **DATES:** Apr. 20, 24, May 4–14, June 11–20, 29–July 9, 27. **REMARKS:** Local and uncommon. L. D. Gibson reared it from eggs laid on poplar at OCP.

Subfamily **Macroglossinae**

Genus ***Erinnyis*** Hbn.

7834 ***E. ello*** (L.)
Ello Sphinx
Gallatin Co., Glencoe exit of I-71, Sept. 1, 1975, L. D. Gibson; Meade Co., OCP, June 16, 1979, R. A. Henderson. A rare stray from farther south.

7837 ***E. obscura*** (F.)
Obscure Sphinx
Boone Co., BOW, Oct. 1, 1977, found on a tennis court by L. D. Gibson. A rare stray from southward.

Genus ***Aellopos*** Hbn.

7849 ***A. titan*** (Cram.)
Titan Sphinx
McCracken Co., Ray Black farm, 9.7 km (6 mi) SW of Paducah, taken at light July 1, 1973, by W. R. Black, Jr. A rare stray from the deep south.

Genus ***Enyo*** Hbn.

7851 ***E. lugubris*** (L.)
Mournful Sphinx
Fulton Co., 12 miles (19.3) W of Hickman, Sept. 9, 1995, W. R. Black, Jr; Harlan Co., PMS, Aug. 8, 1992, L. E. Smith. A rare migrant from farther south.

Genus ***Hemaris*** Dalm.

7853 ***H. thysbe*** (F.)
Hummingbird Clearwing
PARKS AND RESERVES: AJJ, BMF. **COUNTIES:** BRA, BUL, CAM, FAY, HLN, JEF, LET, LEW, MAS, NEL, ROW, SPE, TRM. **DATES:** Apr. 11, 12, 26, June 2, 18, 27–Aug. 17. **REMARKS:** Common at flowers during the day; hovers while feeding.

7855 ***H. diffinis*** (Bdv.)
Snowberry Clearwing
PARKS AND RESERVES: BLA, BMF, CBB, HWS, OCP, SGS, WKM. **COUNTIES:** AND, BUL, FAY, GRY, HLN, HRD, JEF, JES, KEN, LAR, LET, MAD, MAR, MAS, MCC, MEA, NEL, OLD, OWS, PEN, SCO, SIM, SPE, TRI,

TRM, UNI, WAR, WOO. **DATES:** Mar. 23, Apr. 10–Sept. 29. **REMARKS:** More common than hummingbird clearwing, but has similar habits.

Genus ***Eumorpha*** Hbn.

7859 ***E. pandorus*** (Hbn.)
Pandorus Sphinx
PARKS AND RESERVES: BBM, BUC, CBF, FAO, KGC, LWO, OCP, SWR. **COUNTIES:** BOO, BRA, BRE, FAY, GAL, HLN, HND, JEF, KNX, LET, MAD, MCC, MEA, MOR, NEL, OLD, PER, PIK, ROC, WHI. **DATES:** June 9–Oct. 1. Common.

7861 ***E. achemon*** (Drury)
Achemon Sphinx
Christian Co., July 19, 1974, J. Askew (in state fair 4-H collection, 1977); Fayette Co., Lexington, on a stone wall, June 23, 1930; Grant Co., June 24; Marshall Co., Lebanon, early Sept., 1992; Nicholas Co., no date. Uncommon.

7865 ***E. fasciata*** (Sulz.)
Banded Sphinx
PARKS AND RESERVES: SWR. **COUNTIES:** FUL, HEN, HIC, HND, JEF, MCC, WAR. **DATES:** July 30, Sept. 9, 11, 27, Oct. 12. Uncommon.

Genus ***Sphecodina*** Blanchard

7870 ***S. abbottii*** (Swainson)
Abbot's Sphinx
PARKS AND RESERVES: BBM, BMF, BUC, JWS, KGC, LCP, NBP, OCP, RGR, RRG, WVP. **COUNTIES:** BRA, BRE, BRK, BUL, FLO, GRY, HLN, JEF, LET, MCC, MEA, NEL, PER, POW, RUS. **DATES:** Apr. 13, 20–Aug. 24, Sept. 10. Locally common.

Genus ***Deidamia*** Clem.

7871 ***D. inscripta*** (Harr.)
Lettered Sphinx
PARKS AND RESERVES: BBM, BIP, BMF, CAC, CBF, GBT, KGC, LCP, OCP, PNM, RRG, WVP. **COUNTIES:** BEL, BUL, CAR, CRI, CRL, GAL, HLN, JEF, LET, MAR, MEA, MEN, NEL, PIK, ROW, RUS, WHI. **DATES:** Apr. 10–July 2. **REMARKS:** Very common in late spring.

Genus ***Amphion*** Hbn.

7873 ***A. floridensis*** B. P. Clark
Nessus Sphinx
PARKS AND RESERVES: HWS, LCP, OCP, PNM, RRG, SGS. **COUNTIES:** BEL, BUL, JEF, KEN, MEA, MEN, OLD, RUS, UNI. **DATES:** Apr. 14–26, May 27, June 14–27, July 30, Aug. 8. **REMARKS:** Locally common; a day flier. Known until recently as *A. nessus* (Cram.), a preoccupied name.

Genus ***Darapsa*** Wlk.

7884 ***D. versicolor*** (Harr.)
Hydrangea Sphinx
PARKS AND RESERVES: BBM, KGC, RRG. **COUNTIES:** BOY, HLN, LET, MEN, OWS, ROW. **DATES:** July 5–14, 24, 25, Aug. 29. **REMARKS:** Very local and uncommon; only in the mountains.

7885 ***D. myron*** (Cram.)
Hog Sphinx
PARKS AND RESERVES: BAN, BBM, BEC, BMF, BOW, BUC, CAC, HWS, JWS, LBK, LCP, LMP, NBP, OCP, PNM, RGR. **COUNTIES:** BAR, BEL, BOL, BOO, BRE, BUL, CAR, FAY, FLO, GAL, GRV, GRY, HLN, JEF, KEN, MAD, MAR, MCC, MEA, MUH, NEL, OLD, PER, POW, ROW, RUS, TRI, WHI. **DATES:** Apr. 4–Sept. 28. **REMARKS:** One of our most common sphingids.

7886 ***D. pholus*** (Cram.)
Azalea Sphinx
PARKS AND RESERVES: BMF, BUC, CBF, HWS, KGC, LWO, NBP, OCP, PNM, PYR, RRG, WVP. **COUNTIES:** BEL, BUL, CHR, GAL, HLN, JEF, LET, MAR, MEA, NEL, OLD, OWS, PER, POW, WAY, WHI. **DATES:** Apr. 16–Aug. 28. Common.

Genus ***Xylophanes*** Hbn.

7890 ***X. tersa*** (L.)
Tersa Sphinx
PARKS AND RESERVES: BAN, BMF, BOW, KLK, LCP, RGR WVP. **COUNTIES:** BAR, BOO, BRK, BUL, CAL, FAY, FUL, GRY, JEF, KNO, MCC, MRL, MUH, NEL, OWS, ROW, RUS, WAR, WAY. **DATES:** June 14–Oct. 3, 18. Uncommon; a southern species.

Genus ***Hyles*** Hbn.

7894 ***H. lineata*** (F.)
White-lined Sphinx
PARKS AND RESERVES: BBM. **COUNTIES:** BOO, BUL, CAM, CRL, FAY, GRV, HLN, HRD, JEF, ROW, UNI. **DATES:** Apr. 4–Oct. 2. Moderately common.

Superfamily **NOCTUOIDEA**

Family **NOTODONTIDAE**

Prominents

Genus ***Clostera*** Samouelle

7895 ***C. albosigma*** Fitch
Sigmoid Prominent
PARKS AND RESERVES: BAN, CAC, LBK, LCP, OCP, PNM, RGR. **COUNTIES:** BAR, BEL, BRK, CAR, GRY, JEF, MEA, OLD, RUS, TRI. **DATES:** Apr. 11–30, May 8, 9, 28–June 16, July 18, 28, Aug. 11–20, Sept. 14. Common.

7896 ***C. inclusa*** (Hbn.)
Angle-lined Prominent
PARKS AND RESERVES: BBM, CAC, KGC, LBL, LMP, OCP. **COUNTIES:** CAR, FAY, FUL, GAL, HLN, JEF, MCC, MEA, MUH, OWS, ROW, TRI. **DATES:** Apr. 12–30, May 6, 7, 17–June 20, July 12–Aug. 8, 20–30, Sept. 18. Common.

Genus ***Datana*** Wlk.

7902 ***D. ministra*** (Drury)
Yellow-necked Caterpillar Moth
PARKS AND RESERVES: CMI, RRG. **COUNTIES:** BOO, FAY, GAL, KNX, MCC, MEN, MET, MOR, OWS. **DATES:** June 8, 25–July 3, 13, 22, 24, Aug. 1–6, 15, 27, Sept. 17. Locally common.

7903 ***D. angusii*** G. & R.
Angus's Datana
PARKS AND RESERVES: BBL, BMF, HWS, LBL, LWO, NBP, PMS, PNM, RRG. **COUNTIES:** BEL, BOO, BUL, HLN, JEF, LET, MCC, MEN, MOR, NEL, OLD, POW, TRI. **DATES:** Apr. 15–Sept. 19. Common.

7904 ***D. drexelii*** Hy. Edw.
Drexel's Datana
PARKS AND RESERVES: BMF, KRC, LWO, PMS, PNM. **COUNTIES:** BEL, BUL, HLN, JEF, LET, ROW, WOL. **DATES:** May 17, July 8–22, Sept. 8. Common.

7906 ***D. contracta*** Wlk.
Contracted Datana
PARKS AND RESERVES: BBL, BMF, KRC, LBL, LVJ, LWO, PMS, PNM. **COUNTIES:** BEL, BOO, BUL, HLN, JEF, LAU, LET, ROW, TRI, WOL. **DATES:** Apr. 15, June 10, 19–27, July 9–29, Aug. 17. Common.

7907 ***D. integerrima*** G. & R.
Walnut Caterpillar Moth
PARKS AND RESERVES: BAN, CBB, CBF, HWS, KRC, LBL, LWO, OCP, PYR, RGR, RRG, SGS. **COUNTIES:** BAR, BRK, BUL, CAD, FAY, GAL, GRY, JEF, LAU, LET, MCC, MEA, OLD, POW, ROW, SHE, TRI, WHI, WOL. **DATES:** May 8–Sept. 19. Common.

7908 ***D. perspicua*** G. & R.
Spotted Datana
PARKS AND RESERVES: CAV, RGR. **COUNTIES:** BAT, BOO, BRK, CAD, FAY, GRY, JEF, LAU, MOR, ROW. **DATES:** July 16–25, Aug. 3–18, 26, Sept. 17. Moderately common.

Genus ***Nadata*** Wlk.

7915 ***N. gibbosa*** (J. E. Smith)
White-dotted Prominent
PARKS AND RESERVES: AJJ, BAN, BIP, BMF, BUC, CAC, CBB, CBF, FKN, HNK, HWS, KGC, KRC, LBL, LCP, LMP, NBP, OCP, PNM, PYR, RGR, WVP. **COUNTIES:** BAR, BEL, BRE, BRK, BUL, CAD, CAL, CAM, CAR, FAY, GRY, HLN, HRT, JEF, LAU, MCC, MEA, MUH, NEL, OLD, PER, PIK, POW, ROW, RUS, TRI, WHI, WOL. **DATES:** Apr. 24–Oct. 16. Common.

Genus ***Hyperaeschra*** Butler

7917 ***H. georgica*** (H.-S.)
Georgian Prominent
PARKS AND RESERVES: BIP, BMF, CAC, CBF, HNK, HWS, LBL, LCP, NBP, OCP, PNM, PYR. **COUNTIES:** BEL, BUL, CAD, CAL, CAR, JEF, MEA, NEL, OLD, OWS, PIK, POW, ROW, RUS, TRI, WHI. **DATES:** Apr. 8–Aug. 14. Common.

Genus ***Peridea*** Steph.

7919 ***P. basitriens*** (Wlk.)
Oval-based Prominent
PARKS AND RESERVES: BAN, BMF, BUC, CAC, CBF, HWS, JWS, KLK, LCP, LWO, NBP, OCP, PMS, PNM, PYR. **COUNTIES:** BAR, BEL, BOO, BRE, BUL, CAD, CAR, FLO, HLN, JEF, LET, MCC, MEA, MRL, OLD, OWS, PER, POW, ROW, RUS, WHI. **DATES:** Apr. 2–Oct. 28. Common.

7920 ***P. angulosa*** (J. E. Smith)
Angulose Prominent
PARKS AND RESERVES: BAN, BBM, BMF, BUC, CAV, HWS, KLK, KRC, LBK, LBL, LCP, LWO, NBP, PMS, PNM, PYR, RRG. **COUNTIES:** BAR, BAT, BEL, BRE, BRK, BUL, CAD, GRV, HLN, JEF, LET, MCC, MER, MRL, OLD, PER, POW, ROW, RUS, TRI, WOL. **DATES:** May 7, 17–Oct. 6. Common.

7921 ***P. ferruginea*** (Pack.)
Chocolate Prominent
PARKS AND RESERVES: BBM, KGC, KRC, LCP, NBP, PMS, PNM, RGR, RRG. **COUNTIES:** BEL, BRK, GRV, GRY, HLN, OWS, POW, RUS, WOL. **DATES:** May 12, June 14, July, Aug. 12–20. Moderately common.

Genus *Pheosia* Hbn.

7922 ***P. rimosa*** Pack.
Black-rimmed Prominent
Barren Co., BAN, May 31; Fulton Co., Williams Road, 4.5 miles E of Cayce, Sept. 7, 11; Jefferson Co., Valley Station, May 20–July 18, Aug. 5–31, Sept. 7, Oct. 9; Meade Co., OCP, June 21, July 6; Russell Co., LCP, Apr. 21. Local and uncommon.

Genus *Nerice* Wlk.

7929 ***N. bidentata*** Wlk.
Double-toothed Prominent
PARKS AND RESERVES: BAN, BBL, BMF, CAC, HNK, JWS, LMP, PYR, RGR, RRG. **COUNTIES:** BAR, BOO, BRK, BUL, CAD, CAL, CAR, FAY, FLO, GAL, GRY, HND, JEF, MCC, MEA, MEN, MUH, OWS. **DATES:** Apr. 12–Sept. 1. Common.

Genus *Ellida* Grt.

7930 ***E. caniplaga*** (Wlk.)
Linden Prominent
PARKS AND RESERVES: BAN, BBM, BUC, CAC, HWS, KGC, LBK, LCP, NBP, OCP, PNM, RRG. **COUNTIES:** BAR, BEL, BOO, BRA, BRE, CAR, HLN, JEF, MEA, OLD, PER, POW, RUS, TRI, WOL. **DATES:** Mar. 22, Apr. 20–Aug. 19, Oct. 12. Common.

Genus *Gluphisia* Bdv.

7931 ***G. septentrionis*** Wlk.
Common Gluphisia
PARKS AND RESERVES: BBL, BUC, CBF, HWS, LBK, LCP, OCP. **COUNTIES:** BOO, BRA, BRE, FAY, JEF, MCC, MEA, OLD, PER, RUS, TOD, TRI, WHI. **DATES:** Apr. 11–Sept. 29. Common.

Genus *Furcula* Lamarck

7936 ***F. borealis*** (Guér.-Méneville)
White Furcula
PARKS AND RESERVES: BBL, BBM, BMF, BOW, CBF, HWS, OCP, RGR. **COUNTIES:** BOO, BRK, BUL, FAY, GRY, HLN, JEF, MCC, MEA, NEL, OLD, WHI. **DATES:** Apr. 5–Aug. 24. Locally common.

7937 ***F. cinerea*** (Wlk.)
Gray Furcula
PARKS AND RESERVES: CBB, LBK, OCP, RGR, WVP. **COUNTIES:** BRK, FAY, GRY, JEF, MCC, MEA, OLD, OWS, ROW, TRI. **DATES:** Apr. 5–July 6, 24, 26, Aug. 9, 17, Sept. 15. Locally common.

Genus *Cerura* Schr.

7942 ***C. scitiscripta*** Wlk.
Black-etched Prominent
Boone Co., BBL, May 25; Bracken Co., SR 8 W of Foster, June 16, July 3; Grayson Co., RGR, Aug. 18; Jefferson Co., WVP, May 2, June 3–31, July, Aug. 8, 13; Owsley Co., 4.8 km (3 mi) NE of Booneville, July 30. Uncommon.

Genus *Symmerista* Hbn.

7951 ***S. albifrons*** (J. E. Smith)
White-headed Prominent
PARKS AND RESERVES: BAN, BBM, BMF, BUC, CAC, CBB, CBF, HWS, KGC, LCP, LMP, NBP, OCP, PNM. **COUNTIES:** BAR, BEL, BRE, BUL, CAR, HLN, JEF, KGC, MCC, MEA, MUH, NEL, OLD, OWS, PER, POW, ROW, RUS, WHI. **DATES:** Mar. 25–Aug. 8, Oct. 6. Common.

7952 ***S. canicosta*** Franc.
Oldham Co., HWS, Aug. 8; Russell Co., LCP, June 8. Rare.

7953 ***S. leucitys*** Franc.
PARKS AND RESERVES: BAN, BUC, CBB, PNM. **COUNTIES:** BAR, BEL, BRE, JEF, OLD, PER, RUS. **DATES:** Apr. 20, 26, May 17, June 3–18. Uncommon.

Genus *Dasylophia* Pack.

7957 ***D. anguina*** (J. E. Smith)
Black-spotted Prominent
PARKS AND RESERVES: BMF, CAV, LCP, NBP, OCP, PNM. **COUNTIES:** BAT, BEL, BOO, BUL, FAY, JEF, MCC, MEA, POW, RUS. **DATES:** Apr. 22–Sept. 24. Common.

7958 ***D. thyatiroides*** (Wlk.)
Gray-patched Prominent
PARKS AND RESERVES: BAN, BBM, BMF, BUC, CAC, CBF, HWS, JWS, LCP, OCP, PNM, RRG. **COUNTIES:** BAR, BEL, BRE, BUL, CAR, FLO, HLN, JEF, MEA, OLD, OWS, PER, POW, RUS, WHI. **DATES:** Apr. 20–Aug. 22. Common.

Genus *Misogada* Wlk.

7974 ***M. unicolor*** (Pack.)
Drab Prominent
PARKS AND RESERVES: BCN, BEC, BMF, BUC, HWS, OCP, PYR. **COUNTIES:** BOO, BRE, BUL, CAD, FAY, GAL, JEF, KNX, MCC, MEA, OLD, PER. **DATES:** Apr. 30–Sept. 24. Moderately common.

Genus *Macrurocampa* Dyar

7975 ***M. marthesia*** (Cram.)
Mottled Prominent
PARKS AND RESERVES: BAN, BMF, BUC, HNK, HWS, KLK, KRC, LBK, LCP, LMP, NBP, OCP, PMS, PNM, PYR. **COUNTIES:** BAR, BEL, BRE, BRK, BUL, CAD, CAL, HLN, JEF, MEA, MRL, MUH, NEL, OLD, PER, POW, ROW, RUS, TRI, WOL. **DATES:** Apr. 24–Nov. 18. Common, including melanic individuals.

Genus *Heterocampa* Doubleday

7983 ***H. obliqua*** Pack.
Oblique Heterocampa
PARKS AND RESERVES: BMF, BUC, HWS, KGC, KRC, LBK, LBL, LCP, LVJ, OCP, PNM, PYR, RRG. **COUNTIES:** BEL, BOO, BRE, BUL, CAD, EST, HLN, JEF, LAU, MEA, OLD, PER, POW, ROW, RUS, TRI, WAY, WOL. **DATES:** May 3–Nov. 1. Common.

7985 ***H. subrotata*** Harv.
Small Heterocampa
PARKS AND RESERVES: BMF, CBB, HWS, LBK, LBL, LCP, OCP, RGR, WVP. **COUNTIES:** BRA, BRK, BUL, GRY, JEF, MCC, MEA, OLD, RUS, TRI. **DATES:** Apr. 21–Aug. 24. Moderately common.

7990 ***H. umbrata*** Wlk.
White-blotched Heterocampa
PARKS AND RESERVES: AJJ, BAN, BBM, BIP, BMF, BUC, CAC, CBF, KGC, LBK, LCP, LMP, OCP, PNM, PYR, RGR, RRG. **COUNTIES:** BAR, BEL, BRE, BRK, BUL, CAD, CAM, CAR, CLI, EST, GAL, GRY, HLN, JEF, MEA, MOR, MUH, PER, PIK, POW, ROW, RUS, TRI, WHI. **DATES:** Apr. 18–Oct. 30. Common, including melanics.

7994 ***H. guttivitta*** (Wlk.)
Saddled Prominent; Maple Prominent
PARKS AND RESERVES: BAN, BMF, BUC, CAC, CBB, CBF, HNK, JWS, KRC, LBK, LBL, LCP, LMP, NBP, OCP, PMS, PNM, PYR, RGR. **COUNTIES:** BAR, BEL, BRE, BRK, BUL, CAD, CAL, CAR, EST, FLO, GRY, HLN, JEF, MEA, MUH, NEL, OLD, PER, POW, ROW, RUS, TRI, WHI, WOL. **DATES:** Apr. 19–Sept. 25, Oct. 17. Common.

7995 ***H. biundata*** Wlk.
Wavy-lined Heterocampa
PARKS AND RESERVES: BAN, BBM, BMF, BUC, HWS, PNM, PYR, RGR, RRG. **COUNTIES:** BAR, BEL, BRE, BRK, BUL, CAD, GRY, HLN, JEF, MCC, NEL, OLD, PER, POW, WHI. **DATES:** May 17–Aug. 30. Moderately common.

Genus *Lochmaeus* Doubleday

7998 ***L. manteo*** Doubleday
Variable Oakleaf Caterpillar Moth
PARKS AND RESERVES: BMF, NBP, RGR. **COUNTIES:** BRK, BUL, FAY, GRY, JEF, MCC, POW, ROW. **DATES:** May 12–Oct. 17. Common.

7999 ***L. bilineata*** (Pack.)
Double-lined Prominent
PARKS AND RESERVES: BAN, BBL, BEC, BMF, BUC, CBB, HWS, JWS, LBK, LBL, LCP, OCP, PYR, RGR. **COUNTIES:** BAR, BOO, BRA, BRE, BRK, BUL, CAD, FAY, FLO, FUL, GRY, JEF, MEA, MOR, NEL, OLD, PER, RUS, TRI. **DATES:** May 4–Sept. 8. Common.

Genus *Schizura* Doubleday

8005 ***S. ipomoeae*** Doubleday
Morning-glory Prominent
PARKS AND RESERVES: BAN, BBL, BMF, HWS, KRC, LBL, LCP, NBP, PNM, PYR, RGR, WVP. **COUNTIES:** BAR, BEL, BOO, BRA, BRK, BUL, CAD, FAY, GRY, JEF, MCC, MEN, OLD, POW, ROW, RUS, TRI, WAR, WOL. **DATES:** Apr. 12–Oct. 24. Common.

8006 ***S. badia*** (Pack.)
Chestnut Schizura
Boone Co., BBL, May 30, July 18, 31; Bracken Co., 0.8 km (0.5 mi) W of Foster, May 14; Bullitt Co., BMF, May 11; Jefferson Co., WVP, July 14; Meade Co., Aug. 3. Uncommon.

8007 ***S. unicornis*** (J. E. Smith)
Unicorn Caterpillar Moth
PARKS AND RESERVES: BBL, BMF, CAV, HWS, LBK, LCP, PMS, PNM. **COUNTIES:** BAT, BEL, BOO, BUL, CLI, FAY, FLE, FUL, HLN, HND, JEF, MCC, OLD, OWS, ROW, RUS, TRI, WHI. **DATES:** Apr. 20–Sept. 8. Common.

8009 ***S. apicalis*** (G. & R.)
Plain Schizura
Breckinridge Co., Irvington, Aug.; Jefferson Co., Valley Station, May 7, 15, June 10, 15, July 9, 28, Aug. 15–21, Sept. 23; McCracken Co., Ray Black farm, 9.7 km (6 mi) SW of Paducah, June 1; Owsley Co., 4.8 km (3 mi) NE of Booneville, June 10, 14, July 18. Locally common.

8010 ***S. concinna*** (J. E. Smith)
Red-humped Caterpillar Moth
Daviess Co., Owensboro, emerged July 31; Fayette Co., Lexington, emerged July 20; Grayson Co., RGR, Aug. 11; Jefferson Co., Valley Station, July 1, 20, Aug. 4, 18 and BEC, July 10; Rowan Co., Morehead, July 17. Uncommon.

8011 ***S. leptinoides*** (Grt.)
Black-blotched Schizura
PARKS AND RESERVES: BAN, BMF, BUC, LBL, LCP, NBP, PNM, PYR, RGR. **COUNTIES:** BAR, BEL, BRE, BRK, BUL, CAD, GRY, JEF, MEN, NEL, PER, POW, ROW, RUS, TRI. **DATES:** May 1–Aug. 25. Common.

Genus ***Oligocentria*** H.-S.

8012 ***O. semirufescens*** (Wlk.)
Red-washed Prominent
PARKS AND RESERVES: BAN, BBM, BMF, CBB, HWS, KGC, LBK, LCP, NBP, OCP. **COUNTIES:** BAR, BOO, BUL, GAL, HLN, JEF, MCC, MEA, OLD, POW, RUS, TRI, WHI. **DATES:** May 10–Sept. 6. Common.

8017 ***O. lignicolor*** (Wlk.)
White-streaked Prominent
PARKS AND RESERVES: BBM, BMF, CBF, HWS, KGC, KRC, LBL, LCP, LWO, NBP, OCP, PNM, WVP. **COUNTIES:** BEL, BUL, GAL, HLN, JEF, LET, MCC, MEA, OLD, OWS, POW, RUS, TRI, WHI, WOL. **DATES:** May 6–Oct. 18. Common.

Genus ***Hyparpax*** Hbn.

8022 ***H. aurora*** (J. E. Smith)
Pink Prominent
Caldwell Co., PYR, Aug. 11; Jefferson Co., May 8, June 14; Livingston Co., Carrsville, July; McCracken Co., Ray Black farm, 9.7 km (6 mi) SW Paducah, Aug. 8, 16. Rare.

Family **ARCTIIDAE**

Tiger, Lichen and Wasp Moths

Subfamily **Lithosiinae**

Genus ***Crambidia*** Pack.

8045 ***C. lithosioides*** Dyar
Dark Gray Lichen Moth
Powell Co., NBP, Sept. 19; Whitley Co., Goldbug, June 5. Rare.

8045.1 ***C. pallida*** Pack.
Pale Lichen Moth
PARKS AND RESERVES: ASC, BAN, CAV, KLK, LCP, NBP, OCP, PNM. **COUNTIES:** BAR, BAT, BEL, FAY, GRY, HLN, JEF, MCC, MEA, MEN, MRL, PIK, POW, ROC, ROW, RUS, WHI. **DATES:** June 5–Sept. 17, Oct. 2. **REMARKS:** Common.

8046 ***C. uniformis*** Dyar
Uniform Lichen Moth
PARKS AND RESERVES: NBP, PNM. **COUNTIES:** BEL, BUL, MEN, POW, ROC, WHI. **DATES:** June 5, 25, July 8, 13, Sept. 1, 2, 19. Local and uncommon.

8051 ***C. casta*** (Pack.)
Pearly-winged Lichen Moth
Barren Co., Diamond Caverns, Sept. 1; Bullitt Co., N side of SR 480, 11.0 km (6.9 mi) E of SR 61, Sept. 6; Laurel Co., Sept. 24; Meade Co., OCP, July 25; Owsley Co., 4.8 km (3 mi) NE of Booneville, July 24. Uncommon. While these specimens key to *casta* according to Forbes (1960), they may be another species since Forbes stated that he did not see this species south of northern New York and Manitoba.

8052 ***C. pura*** B. & McD.
Pure Lichen Moth
Bullitt Co., Sept. 6; Meade Co., Doe Run area, July 21; Menifee Co., Indian Creek, Aug. 25–26. Rare.

8053 ***C. cephalica*** (G. & R.)
Yellow-headed Lichen Moth
Harlan Co., KGC, Aug. 9; Meade Co., OCP, Aug. 8; Owsley Co., 4.8 km (3 mi) NE of Booneville, July 24, and Caney Fork Creek near Stay, Aug. 16. Rare. Forbes (1960) indicates this name might not apply to eastern white *Crambidia* specimens with front of head yellow; but this problem has not yet been resolved.

Genus ***Cisthene*** Wlk.

8061 ***C. kentuckiensis*** (Dyar)
Kentucky Lichen Moth
PARKS AND RESERVES: ASC, LCP, RRG. **COUNTIES:** FUL, GRY, MCC, OWS, POW, RUS, WHI. **DATES:** June 28, 30, Aug. 27, Sept. 1–14. **REMARKS:** The only moth named for Kentucky. The type citation in the original description was "One specimen, Kentucky, (Geo. Franck)" (Dyar 1904). Uncommon.

8067 ***C. plumbea*** Stretch
Lead-colored Lichen Moth
PARKS AND RESERVES: BMF, HWS, KRC, LCP, NBP, OCP, PNM, RRG, SWR. **COUNTIES:** BEL, BUL, FAY, FUL, GUP, HND, JEF, MCC, MEA, MEN, OLD, POW, RUS, WOL, WOO. **DATES:** June 11, July 4–Sept. 15. Common.

8071 ***C. subjecta*** Wlk.
Subject Lichen Moth
Letcher Co., Pine Mountain, June 10, A. F. Braun. Rare this far north.

8072 ***C. packardii*** (Grt).
Packard's Lichen Moth
PARKS AND RESERVES: BBL, BMF, CER, HWS, LCP, OCP, PNM. **COUNTIES:** BEL,

BOO, BUL, GUP, JEF, MCC, MEA, MEN, NEL, OLD, OWS, ROW, RUS, WHI. **DATES:** June 2–Sept. 8. Common.

Genus ***Lycomorpha*** Harr.

8087 ***L. pholus*** (Drury)
Black-and-yellow Lichen Moth
PARKS AND RESERVES: AJJ, BBL, HWS, OCP, RRG. **COUNTIES:** BOO, BOU, CAM, FAY, JEF, KEN, LIN, MEA, MEN, NEL, OLD, WOO. **DATES:** June 6, 30–July 9, Aug. 11, 26–Sept. 2, 10–30, Oct. 2, Nov. 2. Common.

Genus ***Hypoprepia*** Hbn.

8089 ***H. miniata*** (Kby.)
Scarlet-winged Lichen Moth
PARKS AND RESERVES: BBL, BMF, LCP, PNM. **COUNTIES:** BEL, BOO, BRA, BUL, EST, FAY, JEF, LAR, NEL, ROW, RUS, WAY, WHI. **DATES:** Apr. 29–Aug. 29. Common.

8090 ***H. fucosa*** Hbn.
Painted Lichen Moth
PARKS AND RESERVES: AJJ, BAN, BEC, BMF, CBB, HWS, KGC, LBK, LCP, LWO, OCP, PMS, PYR, SWR. **COUNTIES:** BAR, BUL, CAD, CAM, CHR, FAY, FUL, GAL, HLN, HND, JEF, LAR, LET, MAR, MCC, MEA, NEL, OLD, OWS, PER, ROW, RUS, TRI, WOO. **DATES:** Apr. 30–Oct. 6. Very common.

Genus ***Clemensia*** Pack.

8098 ***C. albata*** Pack.
Little White Lichen Moth
PARKS AND RESERVES: BBL, BIP, BMF, HWS, LCP, NBP, PNM, RBF. **COUNTIES:** BEL, BOO, BRE, BUL, FAY, JEF, MCC, MEA, MEN, OLD, OWS, PIK, POW, RUS. **DATES:** Apr. 30, May 22–June 17, July 25, Aug. 6, 7, 15–Sept. 19, Nov. 14. Common.

Genus ***Pagara*** Wlk.

8099 ***P. simplex*** Wlk.
Mouse-colored Lichen Moth
PARKS AND RESERVES: KLK, LBL, PCP. **COUNTIES:** BUL, HRD, JEF, LAU, LIV, MRL, OWS, ROC, ROW, TRI, WHI. **DATES:** Apr. 20, 28, May 7, 28, June 12–19, July 1, 2, 10, 24, 31, Aug. 25, Sept. 2, 6, 8. Common.

Genus ***Comachara*** Franc.

8104 ***C. cadburyi*** Franc.
Cadbury's Lichen Moth
Carter Co., CAC, Apr. 21; Jefferson Co., Valley Station, May 2, 5, June 16; Meade Co., OCP, May 3; Perry Co., BUC, May 1. Uncommon.

Subfamily **Arctiinae**

Genus ***Utetheisa*** Hbn.

8106 ***U. bella*** (L.)
Bella Moth; Beautiful Utetheisa
COUNTIES: FAY, GRV, JEF, MAS, MCC, MRL, NEL, OLD, PIK. **DATES:** July 20–Oct. 18. **REMARKS:** Uncommon. A day-flier that is common farther south.

Genus ***Haploa*** Hbn.

8107 ***H. clymene*** (Brown)
Clymene Moth
PARKS AND RESERVES: AJJ, BBL, BBM, JCF, KGC, LBL, LCP, LMP, LWO, MCV, NBP, OCP, PMS, PNM, RGR, RRG. **COUNTIES:** AND, BAT, BEL, BOO, BRK, BUL, CAM, CLI, EDM, FUL, FRA, GRY, HLN, JEF, LAR, LAU, LET, LYO, MCC, MEA, MOR, MUH, NEL, OLD, OWS, POW, ROW, RUS, SHE, TRI, UNI. **DATES:** Apr. 29–Aug. 27. **REMARKS:** Females in this genus are entirely white and can best be identified by association with males in the field. Common.

8108 ***H. colona*** (Hbn.)
Colona Moth
PARKS AND RESERVES: BMF, FKN, HWS, LBK, PYR, SGS. **COUNTIES:** BUL, CAD, CHR, HRD, JEF, JES, LOG, MEA, NEL, OHI, OLD, TRI. **DATES:** Apr. 16–July 30. Common.

8110 ***H. contigua*** (Wlk.)
The Neighbor
PARKS AND RESERVES: BBM, BMF, CAC, CBB, HWS, JWS, LWO, PNM. **COUNTIES:** BEL, BOO, BUL, CAR, FLO, HLN, JAC, JEF, LET, MEA, NEL, OLD, OWS, ROW. **DATES:** May 17–July 20. Common.

8111 ***H. lecontei*** (Guér.-Méneville)
Leconte's Haploa
PARKS AND RESERVES: BAN, BBL, BBM, BMF, BUC, CBB, FKN, HWS, JWS, NBP, OCP, PNM, SGS. **COUNTIES:** BAR, BEL, BOO, BRE, BUL, FLO, GAL, HLN, JAC, JEF, KNX, MCY, MEA, MEN, OLD, OWS, PER, POW, WHI, WOO. **DATES:** May 8–July 30, Aug. 11. Common.

Genus ***Holomelina*** H.-S.

8114 ***H. laeta*** (Guér.-Méneville)
Joyful Holomelina
PARKS AND RESERVES: BBL, BCN, CCR, HWS, OCP. **COUNTIES:** BOO, GAL, JEF, MEA, MCC, OLD. **DATES:** May 13–June 18, Aug. 6–Sept. 6. Uncommon.

8118 ***H. opella*** (Grt.)
Tawny Holomelina
PARKS AND RESERVES: ASC, BMF, BUC, GRY, JCF, KGC, KLK, LWO, OCP, PMS, RBF. **COUNTIES:** BRE, BUL, GRY, HLN, JEF, LET, MEA, MEN, MRL, NEL, OWS, PER, WHI. **DATES:** Apr. 12–Sept. 22. Locally common.

8121 ***H. aurantiaca*** (Hbn.)
Orange Holomelina
PARKS AND RESERVES: BBL, BMF, CAV, GBO, HWS. **COUNTIES:** BAT, BOO, BUL, FUL, FAY, GUP, JEF, LAR, MCY, MEN, NEL, OLD, WHI. **DATES:** May 26–June 7, 22, 27, July 24, Aug. 6–18, 27–Sept. 15, 26. Common.

8123 ***H. ferruginosa*** (Wlk.)
Rusty Holomelina
Boone Co., BBL, June 18, 1979, L. D. Gibson. Rare.

8124 ***H. immaculata*** (Reak.)
Immaculate Holomelina
PARKS AND RESERVES: BMF, CAH, HWS, LCP, OCP. **COUNTIES:** BUL, CAM, FAY, JEF, MCC, MEA, MEN, OLD, ROW, RUS. **DATES:** May 18–Sept. 23. **REMARKS:** Common and often found during daytime. This name is being used tentatively at the suggestion of D. C. Ferguson (*pers. comm.*).

Genus *Pyrrharctia* Pack.

8129 ***P. isabella*** (J. E. Smith)
Isabella Tiger Moth
PARKS AND RESERVES: BAN, BEC, BMF, BUC, CBF, GBT, HWS, IRO, JWS, LCP, LWO, NBP, PNM, RBF, RGR, SGS. **COUNTIES:** ADA, BAR, BEL, BRE, BRK, BUL, CRL, FAY, FLO, FUL, GAL, GRY, JAC, JEF, JES, LAR, LET, MAD, MEA, MOR, NEL, OLD, OWS, PER, POW, ROW, RUS, WAS, WHI. **DATES:** Apr. 11–Oct. 11. **REMARKS:** Common. Two broods. The larva is the well-known black and brown "woolly bear" caterpillar which migrates in autumn to find hibernation sites. It may be locally abundant and conspicuous as it crosses roads in its wanderlust. Pupation occurs in spring.

Genus *Estigmene* Hbn.

8131 ***E. acrea*** (Drury)
Salt Marsh Moth
PARKS AND RESERVES: BAN, BBL, BMF, HWS, LBL, OCP, RGR, SWR. **COUNTIES:** BAR, BOO, BRK, BUL, CAS, FAY, GRY, HLN, HND, JEF, MAD, MCC, MEA, MRL, NEL, OLD, ROW, SHE, TRI, UNI. **DATES:** Apr. 10, 25–Sept. 11. **REMARKS:** Common. The larva (salt marsh caterpillar) is a pest of many crops and other plants.

Genus *Spilosoma* Curt.

8133 ***S. latipennis*** Stretch
Pink-legged Tiger Moth
Bullitt Co., BMF, June 17; Grayson Co., RGR, Aug. 17; Jefferson Co., Louisville, June 10, 17, 29, July 18; Menifee Co., Indian Creek, June 8–22, July 8; Oldham Co., CBB, June 16. Uncommon.

8134 ***S. congrua*** Wlk.
Agreeable Tiger Moth
PARKS AND RESERVES: BEC, BMF, BUC, CBF, FKN, HWS, JWS, LBL, LCP, LWO, PNM, RGR, WVP. **COUNTIES:** BEL, BRE, BRK, BUL, CAD, FLO, GRY, HRD, JEF, LET, MCC, MEA, NEL, OLD, PER, ROW, RUS, TRI, WHI, WOL, WOO. **DATES:** Apr. 1–Sept. 17. Common.

8137 ***S. virginica*** (F.)
Virginian Tiger Moth; Yellow Bear Moth
PARKS AND RESERVES: BAN, BMF, BUC, FKN, GBT, JWS, KLK, LBK, LBL, LCP, LMP, LWO, OCP, RBF, RGR, SWR, WVP. **COUNTIES:** BAR, BOO, BRE, BRK, BUL, CRL, FAY, FLO, FUL, GAL, GRV, GRY, HND, HRT, JEF, LAU, LET, MCC, MEA, MEN, MOR, MRL, MUH, OLD, PER, ROW, RUS, SHE, TRI, WAS, WHI, WOO. **DATES:** Apr. 4–Oct. 15. Common.

Genus *Hyphantria* Harr.

8140 ***H. cunea*** (Drury)
Fall Webworm Moth
PARKS AND RESERVES: BBL, BLA, BMF, BUC, CAC, CBF, COB, LBK, LBL, LCP, MCV, NBP, OCP, PNM, RGR, RRG. **COUNTIES:** BEL, BOO, BRE, BRK, BUL, CAR, EDM, FAY, FUL, GRY, HIC, JEF, JES, MEA, MEN, NEL, OLD, OWS, PER, POW, ROW, RUS, TAY, TRI, WHI. **DATES:** Apr. 11–Oct. 8 (larvae: July–Sept.) **REMARKS:** Common. Rauschenberger et al. (1966) reported larvae defoliating trees in MCV, Edmonson Co., Aug. 16, 1966. I have noted fluctuations of this species from widespread heavy infestations in some years to widely dispersed and very light, as in 1982–83. This is an important pest of shade trees throughout the Commonwealth.

Genus *Euerythra* Harv.

8141 ***E. phasma*** Harv.
Red-tailed Specter
Grayson Co., RGR, Aug. 4; Warren Co., Western Kentucky University, July 3. A rare species this far north.

Genus ***Ecpantheria*** Hbn.

8146 ***E. scribonia*** (Stoll)
Giant Leopard Moth
PARKS AND RESERVES: BMF, BUC, CBB, HWS, JWS, KGC, LBL, LCP, LWO, NBP, OCP, PNM, RBF, SGS. **COUNTIES:** AND, BEL, BOO, BRE, BUL, EST, FAY, FLO, HLN, LAR, LAU, LET, MEA, MEN, OLD, OWS, PER, PIK, POW, ROW, ROC, RUS, SPE, TRI, WHI, WOO. **DATES:** Apr. 29–Aug. 6, Sept. 1. **REMARKS:** Common. Larvae of this species have been found by L. D. Gibson in October.

Genus ***Apantesis*** Harr.

8169 ***A. phalerata*** (Harr.)
Harnessed Moth
PARKS AND RESERVES: AJJ, BAN, BMF, BUC, CBF, FKN, GBT, HWS, JCF, KLK, LBL, LCP, LMP, OCP, PNM, PYR, RGR, SWR, WVP. **COUNTIES:** BAR, BEL, BOO, BRE, BUL, CAD, CAM, CRL, FAY, FUL, GRV, GRY, HND, HRD, JEF, LAR, MCC, MEA, MOR, MRL, MUH, NEL, OLD, PER, ROW, RUS, SHE, TRI, UNI, WHI. **DATES:** Apr. 21–Oct. 22. **REMARKS:** Our most common tiger moth.

8170 ***A. vittata*** (F.)
Banded Tiger Moth
Fayette Co., Lexington, May 23, 1929, June 10, 1892, Aug. 3, 1928, Sept. 6–12, 1889 (all UK); Hopkins Co., Nortonville, Mar. 8, 1891 (UK). Rowan Co., Morehead, Aug. 26–Sept. 2. Uncommon.

8170.1 ***A. carlotta*** Fgn.
Charlotte's Tiger Moth
PARKS AND RESERVES: HWS, RGR. **COUNTIES:** BOO, BRK, BUL, FAY, GRE, GRY, JEF, MCC, OLD. **DATES:** May 14, 26, June 12, July 7, 8, Aug. 1–18, Sept. 27. **REMARKS:** Specimens with yellow hindwing and no pink tinting are definitely this species. Some with pink are also, but they are difficult to separate from other *Apantesis* species.

8171 ***A. nais*** (Drury)
Nais Tiger Moth
PARKS AND RESERVES: BAN, BMF, HWS, LBL, RGR. **COUNTIES:** BAR, BRK, BUL, FAY, GAL, GRY, HND, JEF, MCC, OLD, TRI. **DATES:** Apr. 8–Oct. 21. Uncommon.

Genus ***Notarctia*** M. E. Smith

8181 ***N. proxima*** (Guér.-Méneville)
Mexican Tiger Moth
Jefferson Co., Louisville, May 2, 1984. Live male found on lettuce in Paul's Market by Ms. Teri Toole. The lettuce – and presumably the moth – was from California.

NOTE: The number for the preceding species is out of normal sequence because of recent taxonomic changes (Ferguson 1985b), which also introduced the following generic changes.

Genus ***Grammia*** Rambur

8175 ***G. virguncula*** (W. Kby.)
Little Virgin Moth
Listed from "Ky." (Smith 1938). It is not clear if she meant typical form, or form "*otiosa*" (Neumoegen & Dyar). Rare.

8176 ***G. anna*** (Grt.)
Anna Tiger Moth
PARKS AND RESERVES: AJJ, BBL, BMF, BUC, FKN, OCP, PMS, WVP. **COUNTIES:** BOO, BUL, BRE, CAM, GAL, HLN, HRD, JEF, MEA, MEN, MOR, OWS, PER. **DATES:** May 3–Aug. 6. Moderately common.

8188 ***G. figurata*** (Drury)
Figured Tiger Moth
PARKS AND RESERVES: BMF, LCP, LMP, OCP, PYR. **COUNTIES:** BUL, CHR, EDM, HRD, JEF, MCC, MEA, MOR, MUH, OWS, RUS, WHI. **DATES:** Apr. 28–Sept. 5. Locally common.

8191 ***G. placentia*** (J. E. Smith)
Placentia Tiger Moth
Barren Co., BAN, May 21, 1973, C. Cornett. Rare.

8194 ***G. phyllira*** (Drury)
Phyllira Tiger Moth
Jefferson Co., Louisville, May 31, 1955, collected by McCord. Identification is tentative.

8195 ***G. oithona*** (Stkr.)
Oithona Tiger Moth
PARKS AND RESERVES: BAN, KLK, LBL, LMP, MCV, OCP. **COUNTIES:** BAR, BRA, BRE, BUL, CAD, CHR, EDM, FAY, GRE, JEF, MCC, MRL, MEA, MUH, TRI. **DATES:** May 2–Sept. 21. Uncommon.

8196 ***G. parthenice*** (W. Kby.)
Parthenice Tiger Moth
PARKS AND RESERVES: ASC, BBL, BUC, CAC, CAV, GBO, OCP, PYR, SWR, WVP. **COUNTIES:** BAT, BOO, BRE, BUL, CAD, CAR, EDM, GRE, GRY, GUP, HND, JEF, LAU, MEA, MEN, OWS, PER. **DATES:** Feb. 22, June 8, Aug. 3, Sept. 3–24, Oct. 14. Uncommon.

8197 ***G. virgo*** (L.)
Virgin Tiger Moth
PARKS AND RESERVES: ASC, BBL, BBM, RRG. **COUNTIES:** BOO, GRY, HLN, JEF,

MEN, MOR, NEL, OWS, RRG, ROW, SHE, WHI. **DATES:** July 1–Aug. 7, Sept. 4, 24. Locally common in some years.

8199 ***G. arge*** (Drury)
Arge Tiger Moth
PARKS AND RESERVES: GBT, HWS, OCP, RRG. **COUNTIES:** BOO, BRA, BUL, CRL, GAL, HND, JEF, MEA, MOR, NEL, OLD, OWS, POW, ROW, UNI, WAY. **DATES:** Apr. 22–Oct. 12. Locally common.

Genus ***Halysidota*** Hbn.

8203 ***H. tessellaris*** (J. E. Smith)
Banded Tussock Moth
PARKS AND RESERVES: BAN, BBL, BBM, BEC, BMF, BUC, CAC, HWS, JCF, JWS, KRC, LCP, LMP, LVJ, LWO, NBP, OCP, PNM, PYR, RBF, SGS. **COUNTIES:** BAR, BEL, BOO, BRE, BRK, BUL, CAR, CHR, FAY, FLO, FUL, GAL, HLN, HOP, JEF, LAU, LET, MAD, MEA, MUH, NEL, OLD, OWS, PER, POW, ROW, RUS, SPE, WOL. **DATES:** May 6–Sept. 12. **REMARKS:** Very common. Females are difficult to distinguish from *H. harrisii* Walsh.

8204 ***H. harrisii*** Walsh
Harris's Tussock Moth
PARKS AND RESERVES: NBP, PYR, RGR. **COUNTIES:** BRE, BRK, CHR, FAY, FLO, GAL, GRY, JEF, MEA, MER, POW. **DATES:** May 12–23, June 3, July 1, 7, 26, Aug. 2–20. **REMARKS:** Common. Females are difficult to distinguish from *H. tessellaris* (J. E. Smith).

Genus ***Lophocampa*** Harr.

8211 ***L. caryae*** Harr.
Hickory Tussock Moth
PARKS AND RESERVES: BBL, BBM, BIP, BUC, JWS, NBP, PNM. **COUNTIES:** BEL, BOO, BRE, FLO, HLN, KNX, MEN, OWS, PER, PIK, POW, ROW. **DATES:** Apr. 24–June 8. **REMARKS:** Locally common where found in eastern Kentucky.

8214 ***L. maculata*** Harr.
Spotted Tussock Moth
Harlan Co., BBM, at light, July 8–14. Moderately common on BBM.

Genus ***Cycnia*** Hbn.

8228 ***C. inopinatus*** (Hy. Edw.)
Unexpected Cycnia
Boone Co., BBL, 1 male, May 19; Bullitt Co., PCP, 2 males, Apr. 16, and SR 480, 11.0 km (6.9 mi) E of SR 61, May 28; Campbell Co., AJJ, 1 female, July 15. Uncommon to rare.

8230 ***C. tenera*** Hbn.
Delicate Cycnia
PARKS AND RESERVES: BEC, BMF, BUC, CBF, FKN, HWS, LCP, RGR, WVP. **COUNTIES:** BRE, BRK, BUL, FAY, GRY, HND, HRD, JEF, JES, MEA, NEL, OLD, PER, ROW, RUS, WHI. **DATES:** Apr. 12–Sept. 31. Common.

8231 ***C. oregonensis*** (Stretch)
Oregon Cycnia
PARKS AND RESERVES: AJJ, LMP, OCP, RGR. **COUNTIES:** BOO, CAM, FAY, GRY, HRD, JEF, MEA, MUH. **DATES:** Apr. 12, 25, June 15, July 1–8, 20, 25, Aug. 9–12. Uncommon.

Genus ***Euchaetes*** Harr.

8238 ***E. egle*** (Drury)
Milkweed Tussock Moth
PARKS AND RESERVES: BBL, BMF, BOW, BUC, HWS, KGC, LBL, NBP, OCP. **COUNTIES:** BOO, BRE, BUL, FAY, FUL, HLN, JEF, JES, LET, MCC, MEA, NEL, OLD, PER, POW, ROW, TRI. **DATES:** Apr. 6–Aug. 25. Local.

Subfamily **Ctenuchinae**

Genus ***Cisseps*** Franc.

8267 ***C. fulvicollis*** (Hbn.)
Yellow-collared Scape Moth
PARKS AND RESERVES: ASC, BBM, BLA, CBB, CMI, FKN, GBT, HWS, LWO, NBP, PYR, RGR. **COUNTIES:** BEL, BOL, BOO, BRK, BUL, CAD, CLI, CLK, CRL, FAY, FUL, GRY, HEN, HLN, HOP, HRD, JEF, LAR, LET, MAD, MCY, OLD, PER, POW, ROC, ROW, SHE, SPE, WHI. **DATES:** Apr. 9–Dec. 5. **REMARKS:** Common to abundant on flowers during daytime, especially on goldenrod in early fall.

Genus ***Antichloris*** Hbn.

8274 ***A. viridis*** Druce
One specimen in the UK Collection is labelled: "collected in Louisville from bananas from the Dominican Republic" (no date or collector's name). Another was brought to me by Richard Roberts after hatching from a pupa on a Dole brand banana in Louisville, February 4, 1992. This tropical species is frequently brought in by accident to the U.S.

Family **LYMANTRIIDAE**

Tussock Moths

Genus ***Dasychira*** Hbn.

8292 ***D. tephra*** Hbn.
Tephra Tussock Moth
PARKS AND RESERVES: ASC, BMF, CAV, NBP, PMS, RRG. **COUNTIES:** BAT, BUL,

GRY, HLN, JEF, MEA, MEN, OLD, POW, WHI. **DATES:** May 22, June 5, July 9, Aug. 10, 12, Sept. 7–11, 21, Oct. 3–15. Locally common.

8293 ***D. dorsipennata*** (B. & McD.)
Sharp-lined Tussock Moth
Bullitt Co., BMF, May 22; Powell Co., NBP, July 14; Russell Co., LCP, May 29. Uncommon.

8294 ***D. vagans*** (B. & McD.)
Variable Tussock Moth
Barren Co., BAN, May 16; Owsley Co., 4.8 km (3 mi) NE of Booneville, June 16; Whitley Co., CBF, July 12. Uncommon.

8296 ***D. basiflava*** (Pack.)
Yellow-based Tussock Moth
PARKS AND RESERVES: CBF, KGC, PNM. **COUNTIES:** BEL, HLN, JEF, LEE, OWS, ROW, WHI, WOL. **DATES:** June 28–July 25, Aug. 29, Oct. 26. Common in mountain counties.

8302 ***D. obliquata*** (G. & R.)
Streaked Tussock Moth
PARKS AND RESERVES: BBL, BCN, LBL, LCP, NBP, OCP, PMS, PNM, PYR, RGR, RRG, WVP. **COUNTIES:** BEL, BOO, BRK, CAD, GRY, HLN, JEF, MEA, MEN, MET, MOR, NEL, OLD, OWS, POW, ROW, RUS, TRI, WHI. **DATES:** May 22, 24, June 16, 24–Sept. 1, Oct. 15. Common.

8304 ***D. plagiata*** (Wlk.)
Rowan Co., Morehead, Aug. 24, 1962, T.N. Freeman. Rare.

8305 ***D. pinicola*** (Dyar)
Bell Co., PNM, June 7; Bullitt Co., BMF, May 15–28, June 5–19, Aug. 7, 17, and S side of SR 480, 11.0 km (6.9 mi) E of SR 61, June 23, Aug. 30, and PCP, May; Jefferson Co., Aug. 24; Nelson Co., June 13, 14, 23. Locally common.

8307 ***D. manto*** (Stkr.)
Manto Tussock Moth
PARKS AND RESERVES: BMF, LCP, NBP, PNM, RRG. **COUNTIES:** BEL, BUL, HLN, NEL, OWS, POW, ROW, RUS, WHI. **DATES:** Apr. 21, May 7, 8, 18–June 6, 23–27, July 4, 8, Aug. 21–27. Common.

Genus *Orgyia* Ochs.

8314 ***O. definita*** Pack.
Definite Tussock Moth
PARKS AND RESERVES: BIP, BMF, GBT, HWS, KRC, LCP, NBP, PNM. **COUNTIES:** BEL, BOO, BUL, CRL, FAY, GAL, JEF, MEN, OLD, PIK, POW, RUS, WOL. **DATES:** Apr. 17, 25, May 6, 13, 29, June 8, 14, 27, July 13, 15, Aug. 21–Oct. 20, 30. Common.

8316 ***O. leucostigma*** (J. E. Smith)
White-marked Tussock Moth
PARKS AND RESERVES: BAN, BIP, BMF, GBT, HWS, KRC, LCP, LWO, NBP, PNM, PYR. **COUNTIES:** BAR, BEL, BRK, BUL, CAD, CRL, FAY, FUL, GRV, JEF, KEN, LET, MCC, OLD, PIK, POW, ROW, RUS, WOL. **DATES:** Feb. 29, June 10, 29–Nov. 17. Common.

Genus *Lymantria* Hbn.

8318 ***L. dispar*** (L.)
Gypsy Moth
The first state record of this important forest pest was a male from Doe Valley, Meade Co., collected by a Kentucky Division of Forestry worker on Sept. 5, 1973. Surveillance has continued to date by means of thousands of bait traps set around the state. A small colony was discovered in Anchorage, Jefferson Co., in 1985; but as of 1998 no major outbreak has occurred in Kentucky. Males have been taken in bait traps in the following counties each of the following years by USDA personnel: 1983, Fayette, Jefferson, Oldham, and Perry; 1984, Fayette, Grant, Jefferson, Kenton, Madison, and Rowan; 1985, Jefferson (29 males in Anchorage); 1986, Boyd, Bracken, Greenup, and Jefferson (11 larvae in Anchorage); 1987, Boone, Bullitt, Fayette, Kenton, Oldham, Rowan, and Scott; 1988, Greenup, Jefferson and Pulaski; 1989, Boyd, Elliott, Fayette, Hardin, Jefferson, Kenton, Lawrence, Owen, Rockcastle, Rowan, Shelby, and Spencer; 1991, Boone, Carroll, Jefferson, Kenton, Laurel (LVJ), and Oldham. Specific dates of capture are not available.

Family **NOCTUIDAE**

Owlet or Noctuid Moths

Subfamily **Herminiinae**

Genus *Idia* Hbn.

8322 ***I. americalis*** (Gn.)
American Idia
PARKS AND RESERVES: BAN, BBL, BBM, BEC, BIP, CBF, GBT, HWS, KLK, LBL, LCP, LMP, LWO, NBP, PNM, PYR, WVP. **COUNTIES:** BAR, BEL, BOO, CAD, CRL, FAY, HLN, JEF, LET, MCC, MEA, MRL, MUH, OLD, PIK, POW, RUS, TRI, WHI. **DATES:** May 5–Nov. 23. Common.

8323 ***I. aemula*** Hbn.
Common Idia
PARKS AND RESERVES: BAN, BBL, BEC, BIP, BMF, BUC, CAC, CBF, GBO, GBT, HWS, KGC, KLK, LBL, LCP, LMP, OCP, PNM, PYR, RGR. **COUNTIES:** BAR, BEL, BOO, BRE,

BRK, BUL, CAD, CAR, CRL, FAY, GRY, GUP, HLN, JEF, MCC, MEA, MRL, MUH, OLD, OWS, PER, PIK, RUS, TRI, WHI, WOO. **DATES:** Apr. 27–Oct. 18. Common.

8326 ***I. rotundalis*** (Wlk.)
Rotund Idia
PARKS AND RESERVES: ASC, BBL, CAV, HWS, OCP, PCP, PNM. **COUNTIES:** BAT, BEL, BOO, BUL, GRY, GUP, JEF, MEA, MEN, OLD, ROW, WHI. **DATES:** June 5–14, 27–July 9, 18, 25, Aug. 10–27, Sept. 5–11. Common.

8327 ***I. forbesi*** (French)
PARKS AND RESERVES: ASC, BMF, BUC, CAV, HWS, LCP, NBP, OCP, PMS, PNM. **COUNTIES:** BAT, BEL, BRE, BUL, GRY, GUP, HLN, JEF, MEA, OLD, PER, POW, RUS, WHI. **DATES:** May 25, June 8–15, 25, 30, July 7–9, 18, 19, Aug. 2–14, 25, 27, Sept. 5, 7, 21. Common.

8328 ***I. julia*** (B. & McD.)
PARKS AND RESERVES: KLK, LBK, PMS, PNM, PYR. **COUNTIES:** BEL, CAD, HLN, JEF, MRL, TRI, WHI. **DATES:** May 11, June 9–17, July 7, Aug. 18, 21, Sept. 24. Moderately common.

8329 ***I. diminuendis*** (B. & McD.)
Orange-spotted Idia
Caldwell Co., PYR, June 12; Harlan Co., PMS, July 8; Jefferson Co., Valley Station, June 11–20; Marshall Co., KLK, June 14; Oldham Co., Prospect, July 15. Probably common.

8330 ***I. scobialis*** (Grt.)
Smoky Idia
PARKS AND RESERVES: BBM, HWS, LCP, OCP, PNM. **COUNTIES:** BEL, HLN, JEF, MEA, OLD, RUS. **DATES:** May 3, 27, June 7, 21, 27, July 9, 19, Aug. 8, Sept. 10, 13. Moderately common.

8333 ***I. denticulalis*** (Harv.)
Fayette Co., July 8, 15; Harlan Co., PMS, July 8; Menifee Co., RRG, June 21, 22; Oldham Co., HWS, June 29; Powell Co., RRG, June 7. Probably common.

8334 ***I. lubricalis*** (Gey.)
Glossy Black Idia
PARKS AND RESERVES: BBM, BEC, BUC, CAV, HWS, LBL, PMS, PNM, WVP. **COUNTIES:** BAT, BEL, BRE, FAY, HLN, JEF, OLD, PER, ROW, TRI. **DATES:** May 13, June 7–15, 25–29, July, Aug. 18, Sept. 1–16. **REMARKS:** Common. Attracted to sugar bait as well as lights.

Genus ***Phalaenophana*** Grt.

8338 ***P. pyramusalis*** (Wlk.)
Dark Banded Owlet
PARKS AND RESERVES: BBL, BMF, CBB, HWS, LCP, NBP, PNM. **COUNTIES:** BEL, BOO, BUL, FAY, JEF, OLD, POW, ROW, RUS. **DATES:** Apr. 21–Aug. 22. Common.

Genus ***Zanclognatha*** Led.

8340 ***Z. lituralis*** (Hbn.)
Lettered Zanclognatha
PARKS AND RESERVES: BBL, BMF, BUC, HWS, LCP, OCP, RRG. **COUNTIES:** BOO, BRA, BRE, BUL, FAY, HLN, HRT, JEF, LEE, MEA, MEN, OLD, PER, RUS, WHI. **DATES:** Apr. 26–31, May 11–June 13, July 8, 18–Aug. 8, 18. Common.

8341 ***Z. theralis*** (Wlk.)
Bell Co., PNM, July 19; Bullitt Co., SR 480 11.3 km (7 mi) E of Shepherdsville, July 8; Oldham Co., HWS, June 14. Identification tentative; they may be either *Z. minoralis* Smith or an unnamed species.

8343 ***Z. minoralis*** Smith
Whitley Co., Goldbug, June 5, C. V. Covell, Jr. Identification is tentative.

8345 ***Z. laevigata*** (Grt.)
Variable Zanclognatha
PARKS AND RESERVES: BBM, NBP, OCP, PMS, PNM, RBF. **COUNTIES:** BEL, BRE, EST, HLN, LEE, MEA, MEN, POW, ROW, WHI. **DATES:** June 5, July 8–25, Aug. 9–18. Common.

8347 ***Z. obscuripennis*** (Grt.)
Dark Zanclognatha
Bullitt Co., Mt. Washington, Sept. 7; Caldwell Co., PYR, June 5; Jefferson Co., Valley Station, May 24; McCracken Co., Ray Black farm, 9.7 km (6 mi) SW of Paducah, July 24; Oldham Co., HWS, Sept. 6. Common.

8348 ***Z. pedipilalis*** (Gn.)
Grayish Zanclognatha
PARKS AND RESERVES: BMF, HWS, LBL, RRG. **COUNTIES:** BUL, FUL, GAL, JEF, LEE, MEN, OLD, TRI. **DATES:** Mar. 10, May 7, 21, 27, June, July 15, 25, 31, Aug. 5, 18, 25, Sept. 7. Common.

8349 ***Z. protumnusalis*** (Wlk.)
PARKS AND RESERVES: CAC, HWS, LCP, RGR. **COUNTIES:** BRK, CAR, GRY, HIC, JEF, MEN, OLD, RUS. **DATES:** May 16, 28, 29, June

14, 15, July 13, 31, Aug. 9–18, Sept. 11, 13. **REMARKS:** Common. Records here include those of *Z. inconspicualis* (Grt.), which I consider to be identical to *protumnusalis* (Wlk.).

8350 ***Z. martha*** Barnes
Bell Co., PNM, July 18, 1975, C. V. Covell, Jr.

8351 ***Z. cruralis*** (Gn.)
Early Zanclognatha
PARKS AND RESERVES: BMF, CAV, CBB, CBF, COB, HWS, NBP, OCP, PNM, RRG. **COUNTIES:** BAT, BEL, BUL, HIC, JEF, MCC, MEA, OLD, POW, WHI. **DATES:** Apr. 19–Sept. 7. Common.

8352 ***Z. jacchusalis*** (Wlk.)
Yellowish Zanclognatha
PARKS AND RESERVES: CAV, HWS, NBP. **COUNTIES:** BAT, BUL, HND, MEA, OLD, POW, WHI. **DATES:** June 5–27, July 14, Aug. 17, 18, Sept. 5, 6. Common.

8353 ***Z. ochreipennis*** (Grt.)
Wavy-lined Zanclognatha
PARKS AND RESERVES: KRC, LCP, PMS, PYR, RRG. **COUNTIES:** CHR, HLN, JEF, MEA, MEN, POW, RUS, WHI, WOL, WOO. **DATES:** May 29, June 8, 21, 27, July 3–15, Aug. 9, 17, 25, Sept. 1, 8, 26. Common.

Genus *Chytolita* Grt.

8355 ***C. morbidalis*** (Gn.)
Morbid Owlet
PARKS AND RESERVES: BMF, HWS, LCP, LMP, PNM, PYR, RBF. **COUNTIES:** BEL, BRE, BUL, CAD, JEF, MEN, MUH, OLD, OWS, RUS, WOO. **DATES:** May 5–June16, Aug. 18, Sept. 21. Common.

8356 ***C. petrealis*** Grt.
PARKS AND RESERVES: BBL, COB, OCP. **COUNTIES:** BOO, FUL, HIC, JEF, MEA, OLD. **DATES:** May 4–27, June 4, 7, July 1–Sept. 8. Common.

Genus *Macrochilo* Hbn.

8357 ***M. absorptalis*** Wlk.
Slant-lined Owlet
Boone Co., BBL, June 1; Fayette Co., Lexington, June 9, July 23–28.

8357.1 ***M. hypocritalis*** Fgn.
PARKS AND RESERVES: BBL, BMF, LBL. **COUNTIES:** BOO, BUL, FUL, GAL, JEF, TRI. **DATES:** Apr. 29–June 2, July 7–22. **REMARKS:** Paratype male (NMNH): Bullitt Co., BMF, July 7, 1977, A. J. Brownell. Another specimen in the ULK Collection has these data but is not a paratype.

8358 ***M. litophora*** (Grt.)
Jefferson Co., Valley Station, Aug. 7, 15, Sept. 3, 24; Meade Co., OCP, July 25; Menifee Co., Indian Creek, July 9; Oldham Co., HWS, June 14 and Goshen, May 22, 31, June 6–26, July 17–29, Aug. 1–21; Owsley Co., Caney Fork Creek, near Stay, Aug. 16. Locally common.

8360 ***M. orciferalis*** Wlk.
Jefferson Co., Valley Station, July 8, 1975, S. Scholz.

8361 ***M. louisiana*** (Fbs.)
Livingston Co., Carrsville, Aug.; Oldham Co., Goshen, June 5, 21, July 31, Aug. 7, 10. Uncommon.

Genus *Phalaenostola* Grt.

8362 ***P. metonalis*** (Wlk.)
Oldham Co., HWS, June 8, 1968, C. V. Covell, Jr. Listed from Kentucky by Crumb (1956).

8363 ***P. eumelusalis*** (Wlk.)
Fayette Co., Lexington, July 28; McCracken Co., near Paducah, July 24; Spencer Co., Taylorsville, June 4.

8364 ***P. larentioides*** Grt.
Black-banded Owlet
PARKS AND RESERVES: BBL, BMF, CAV, HWS, LBL, PYR. **COUNTIES:** BAT, BOO, BUL, CAD, FAY, HND, JEF, MCC, MEN, OLD, OWS, TRI. **DATES:** May 26–Sept. 12. Common.

Genus *Tetanolita* Grt.

8366 ***T. mynesalis*** (Wlk.)
Smoky Tetanolita
PARKS AND RESERVES: BAN, HWS, LBK, LCP, RGR. **COUNTIES:** BAR, BRK, FAY, FUL, GRY, HOP, HRT, JEF, MCC, MEA, MEN, OLD, RUS, TRI, WHI. **DATES:** May 13–Nov. 6. **REMARKS:** Very common. Yeargan (1988) documented this species as prey of the bolas spider, *Mastophora hutchinsoni*, which apparently attracts the male moths by releasing chemicals similar to sex pheromones of several moth species, including this one. K. F. Haynes has reared this species on grass clippings in Fayette Co.

8368 ***T. floridana*** (Sm.)
Florida Tetanolita
PARKS AND RESERVES: ASC, BMF, GBT, HWS, KRC, LCP, RGR. **COUNTIES:** BRK, BUL, CRL, FAY, GRY, HND, JEF, OLD, RUS, WOL. **DATES:** Apr. 26, July 11–17, Aug. 4–Sept. 18, Oct. 4–14. **REMARKS:** Our subspecies is *fulata* Sm. Common.

Genus ***Bleptina*** Gn.

8370 ***B. caradrinalis*** Gn.
Bent-winged Owlet
PARKS AND RESERVES: BAN, BBM, BUC, CBB, COB, LCP, PMS. **COUNTIES:** BAR, BRE, BUL, FAY, GAL, HIC, HLN, JEF, OLD, PER, RUS, WOO. **DATES:** May 5–June 8, July 14, 28–Aug. 3, 14–17. Common.

8371 ***B. inferior*** Grt.
Trigg Co., LBL, July 10, 1965, C. V. Covell, Jr.

8372 ***B. sangamonia*** B. & McD.
Laurel Co., powerline cut E side of S end of FS 775, May 30, 1997, L.D. Gibson.

Genus ***Renia*** Gn.

8378 ***R. salusalis*** (Wlk.)
Jefferson Co., Valley Station, May 24, June 6; Owsley Co., 4.8 km (3 mi) NE of Booneville, May 29; Russell Co., LCP, May 23. Uncommon.

8379 ***R. factiosalis*** (Wlk.)
Listed from Kentucky by Crumb (1956).

8380 ***R. nemoralis*** B. & McD.
Chocolate Renia
Meade Co., Aug. 3; Owsley Co., 4.8 km (3 mi) E of Booneville, Sept. 1, 7. Uncommon.

8381 ***R. discoloralis*** Gn.
Discolored Renia
PARKS AND RESERVES: BMF, CAV, HWS, LCP, NBP, OCP, RGR. **COUNTIES:** BAT, BRK, BUL, EST, FAY, GRY, JEF, LAU, MEA, MEN, OLD, OWS, POW, RUS. **DATES:** May 29–Sept. 12. Common.

8384.1 ***R. flavipunctalis*** (Gey.)
Bullitt Co., PCP, July 8; Fayette Co., Lexington, June 22; Green Co., Crailhope, May 23; Jefferson Co., WVP, June 29; Trigg Co., LBL, July 14. Uncommon.

8386 ***R. adspergillus*** (Bosc)
Gallatin Co., Markland Dam, June 2; Jefferson Co., Valley Station, May 22; McCracken Co., Paducah, May 11; Meade Co., OCP, Aug. 12.

8387 ***R. sobrialis*** (Wlk.)
Sober Renia
Grayson Co., RGR, Aug. 12; Jefferson Co., Valley Station, May 3. Uncommon.

Genus ***Lascoria*** Wlk.

8393 ***L. ambigualis*** Wlk.
Ambiguous Moth
PARKS AND RESERVES: BMF, CAC, CBB, CBF, HWS, LCP, NBP, OCP, RGR. **COUNTIES:** BRK, BUL, CAR, FAY, GAL, GRY, HRT, JEF, MCC, MEA, OLD, OWS, POW, PUL, ROC, RUS, WHI. **DATES:** Mar. 19, Apr. 15–May 24, June 23–July 9, 18–24, Aug. 10–14, 24–31. Common.

Genus ***Palthis*** Hbn.

8397 ***P. angulalis*** (Hbn.)
Dark-spotted Palthis
PARKS AND RESERVES: BMF, BOW, CAC, CBF, HWS, JWS, LCP, PMS, PNM. **COUNTIES:** BEL, BOO, BUL, CAR, FAY, FLO, HLN, JEF, OLD, OWS, RUS, WHI. **DATES:** Apr. 20–July 14, 28, Sept. 12. Very common.

8398 ***P. asopialis*** (Gn.)
Faint-spotted Palthis
PARKS AND RESERVES: BAN, CBB, COB, GBT. **COUNTIES:** BAR, CRL, FAY, HIC, HRD, HRT, JEF, MCC, MEN, OLD, SPE. **DATES:** Jan. 12, May 5–Nov. 4. Common.

Genus ***Redectis*** Nye

8401 ***R. vitrea*** (Grt.)
Fulton Co., SR 94, 4.0 km (2.5 mi) E of Cayce, Sept. 7; Jefferson Co., Louisville, Nov. 14; Menifee Co., RRG, Aug. 25; Oldham Co., Goshen, June 30, July 7, Aug. 6, 28, Sept. 6, and HWS, Aug. 18. Uncommon.

Subfamily **Rivulinae**

Genus ***Rivula*** Gn.

8404 ***R. propinqualis*** Gn.
Spotted Grass Moth
PARKS AND RESERVES: ASC, BBL, KRC, LCP, OCP, SWR. **COUNTIES:** BOO, BUL, GRY, HND, JEF, MEA, OWS, RUS, WOL. **DATES:** Apr. 27, 29, June 20–Sept. 11. Common.

Genus ***Oxycilla*** Grt.

8407 ***O. malaca*** (Grt.)
Bent-lined Tan
Bell Co., PNM, June 9; Jefferson Co., Valley Station, Apr. 20, May 16, July 16; Perry Co., BUC, June 8; Powell Co., 1.6 km (1 mi) E of Clay City, May 13. Uncommon.

8408 ***O. mitographa*** (Grt.)
Owsley Co., 4.8 km (3 mi) NE of Booneville, May 26–28, June 14, 16, July 18; all L. D. Gibson.

Genus ***Colobochyla*** Hbn.

8411 ***C. interpuncta*** (Grt.)
Yellow-lined Owlet
PARKS AND RESERVES: CBB, RGR, RNR. **COUNTIES:** BRK, FUL, GRY, HRT, JEF, MEN, OLD, OWS. **DATES:** May 6–14, June 12, 26, July 8, Aug. 3–16, 27, Sept. 5, 8. Common.

Genus ***Melanomma*** Grt.

8412 ***M. auricinctaria*** Grt.
Fulton Co., SR 94, 4.0 km (2.5 mi) E of Cayce, in black-light trap, Sept. 7; Gallatin Co., Markland Dam, June 7; Oldham Co., Goshen, Aug. 10. Uncommon.

Subfamily **Hypenodinae**

Genus ***Hypenodes*** Doubleday

8421 ***H. fractilinea*** (Sm.)
Broken-line Hypenodes
Fulton Co., 4.0 km (2.5 mi) E of Cayce, Sept. 7; Menifee Co., RRG, FS 9B, East Fork Indian Creek, Kelley Branch, 220 m (720 ft) elevation, Aug. 23–24; Oldham Co, CBB, May 14, and HWS, June 29. Common.

Genus ***Dyspyralis*** Warr.

8427 ***D. puncticosta*** (Sm.)
Bullitt Co., BMF, July 24, 1977, A. J. Brownell.

8428 ***D. nigella*** (Stkr.)
Powell Co., RRG, Tunnel Ridge Rd., at light, July 8, 1988, D. J. Wright.

Genus ***Quandara*** Nye

8432 ***Q. brauneata*** (Swett)
Oldham Co., HWS, Aug. 25; Powell Co., NBP, May 2. This species was, until recently, listed in the Geometridae. Uncommon.

Genus ***Nigetia*** Wlk.

8440 ***N. formosalis*** Wlk.
Thin-winged Owlet
PARKS AND RESERVES: BBL, BMF, CER, HWS, SWR, WVP. **COUNTIES:** BOO, BUL, FAY, GAL, HND, JEF, OLD, WOO. **DATES:** May 27, 28, June 2–4, 14, 17, Aug. 12, 17, 25, Sept. 8, 12. Common.

Subfamily **Hypeninae**

Genus ***Bomolocha*** Hbn.

8441 ***B. manalis*** (Wlk.)
Flowing-line Bomolocha
PARKS AND RESERVES: AJJ, BMF, BUC, HWS, LBL. **COUNTIES:** BRE, BUL, CAM, GAL, JEF, OLD, PER, TRI. **DATES:** Apr. 29–Oct. 10. Common.

8442 ***B. baltimoralis*** (Gn.)
Baltimore Bomolocha
PARKS AND RESERVES: BMF, CAC, KRC, LCP, PNM, RBF, RRG. **COUNTIES:** BEL, BOO, BRE, BUL, CAR, EST, FAY, GAL, JEF, LAU, MEN, OWS, POW, RUS, WOL. **DATES:** Jan. 28, Apr. 20–Sept. 6, 17, Oct. 5. Common.

8443 ***B. bijugalis*** (Wlk.)
Dimorphic Bomolocha
PARKS AND RESERVES: AJJ, BAN, BMF, CAC, CBF, GBT, JWS, KGC, LCP, NBP, OCP, PNM, RBF. **COUNTIES:** BAR, BEL, BRE, BUL, CAM, CAR, CRL, FAY, FLO, GAL, HLN, JEF, LAU, LEE, MEA, MEN, MCC, MRL, OWS, POW, RUS, WHI. **DATES:** Apr. 20–Sept. 10, Nov. 22. Common.

8444 ***B. palparia*** (Wlk.)
Mottled Bomolocha
PARKS AND RESERVES: BBM, BMF, BUC, CAC, CBF, LCP, NBP, OCP, PMS, PNM, RGR. **COUNTIES:** ADA, BEL, BOO, BRA, BRE, BRK, BUL, CAR, EST, GAL, GRY, HLN, JEF, MEA, PER, POW, RUS, WHI. **DATES:** Apr. 15–Sept. 6. Common.

8445 ***B. abalienalis*** (Wlk.)
White-lined Bomolocha
PARKS AND RESERVES: BBL, BMF, JWS, LCP, LMP, RGR. **COUNTIES:** BOO, BRA, BRK, BUL, EST, FLO, GAL, GRY, JEF, MEA, MEN, MUH, NEL, OWS, ROW, RUS. **DATES:** Apr. 17–Aug. 27, Oct. 17, Dec. 6. Common.

8446 ***B. deceptalis*** (Wlk.)
Deceptive Bomolocha
Bell Co., PNM, June 9; Boone County, Florence, Apr. 16; Jefferson Co., May 14, June 2, 10–25, July 20, 28, Aug. 5, 21, 22, Oct. 17; Whitley Co., Goldbug, Apr. 16; Woodford Co., May 25. Locally common.

8447 ***B. madefactalis*** (Gn.)
Gray-edged Bomolocha
PARKS AND RESERVES: BBL, BMF, BUC, CAC, CBB, LBL, LCP, LMP, NBP, OCP, PMS, PNM, PYR. **COUNTIES:** BEL, BOO, BRE, BUL, CAD, CAR, EST, GAL, HLN, JEF, JES, MCC, MEA, MUH, NEL, OLD, OWS, PER, POW, RUS, TRI, WHI. **DATES:** Apr. 26–Oct. 8, Dec. 8. Common.

8448 ***B. sordidula*** (Grt.)
Sordid Bomolocha
Bell Co., PNM, July 19; Carter Co., CAC, Apr. 29; Gallatin Co., Markland Dam, June 29; Jefferson Co., Valley Station, May 15, July 9, 14, 29, Nov. 18; McCracken Co., Paducah, July 24. Separation of this species from *madefactalis* (Gn.) is very difficult, and some of these may in fact be that species. The Pine Mountain record, a male, seems to be the most nearly typical *B. sordidula*.

8448.1 ***B. appalachiensis*** L. Butler
Appalachian Bomolocha
A paratype male was designated by Butler (1987)

from Harlan Co., KGC, collected July 13, 1984, by L. D. Gibson. Rare.

8452 ***B. edictalis*** (Wlk.)
Large Bomolocha
Harlan Co., BBM, July 6, 9, 14, and KGC (no date); Letcher Co., LWO, July 11; Powell Co., NBP, July 14. Apparently restricted to the eastern Kentucky mountains.

Genus *Hypena* Schr.

8461 ***H. humuli*** Harr.
Hop Hypena
Harlan Co., KGC, Apr. 20, 1979, C. V. Covell, Jr; Livingston Co., Carrsville, Aug. 1998, J.A. Bess. Rare.

Genus *Plathypena* Grt.

8465 ***P. scabra*** (F.)
Green Cloverworm Moth
PARKS AND RESERVES: BAN, BBM, BEC, BIP, BMF, BUC, CBF, FKN, GBO, GBT, HWS, JWS, LBK, LBL, LCP, LMP, NBP, OCP, PNM, PYR, RBF, RGR, WVP. **COUNTIES:** BAR, BEL, BRE, BRK, BUL, CAD, CRL, DAV, FAY, FLO, FUL, GRY, GUP, HLN, HND, HOP, HRD, JEF, LAU, MAD, MCC, MEA, MRL, MUH, NEL, OLD, PER, PIK, POW, PUL, ROC, ROW, RUS, SHE, TRI, WHI. **DATES:** Jan. 14, 18, Feb. 21, Mar. 9, 21–Dec. 15. **REMARKS:** One of the most common Kentucky moths. Often abundant in early autumn, especially in wet woods.

Genus *Tathorhynchus* Hamp.

8466 ***T. exsiccatus*** (Led.)
Jefferson Co., Valley Station, Oct. 22; Meade Co., OCP, Sept. 13. Rare, and east of previously known range.

Genus *Hemeroplanis* Hbn.

8467 ***H. scopulepes*** (Haw.)
Variable Tropic
PARKS AND RESERVES: LCP, LWO. **COUNTIES:** BAR, HRD, JEF, LET, OWS, ROW, RUS, WHI. **DATES:** Apr. 14, 16, May 10, 29, June 21, July 2, 11–16, 30, Aug. 15, Sept. 21. Locally common.

Genus *Spargaloma* Grt.

8479 ***S. sexpunctata*** Grt.
Six-spotted Gray
PARKS AND RESERVES: BMF, HWS, OCP, PNM, PYR, WVP. **COUNTIES:** BEL, BUL, CAD, GAL, JEF, MCC, MEA, OLD, OWS. **DATES:** Apr. 27, 28, May 6–26, June 3, 5, 20–23, July 18, 29, Aug. 24. Common.

Genus *Phytometra* Haw.

8481 ***P. rhodarialis*** (Wlk.)
Pink-bordered Yellow
PARKS AND RESERVES: BMF, FKN, HWS. **COUNTIES:** BUL, GUP, JEF, MEA, OLD, OWS, WHI. **DATES:** Mar. 17, Apr. 5, May 11, 28–Sept. 22. Common in some years.

Genus *Hormoschista* Mösch.

8488 ***H. latipalpis*** (Wlk.)
Double-lined Brown
Bullitt Co., BMF, in Malaise trap, June 6–10, 1976, A. J. Brownell.

Genus *Ommatochila* Butler

8489 ***O. mundula*** (Zell.)
Fulton Co., Sept. 19; Jefferson Co., Valley Station, Sept. 3. Apparently rare.

Subfamily **Catocalinae**

Genus *Pangrapta* Hbn.

8490 ***P. decoralis*** Hbn.
Decorated Owlet
PARKS AND RESERVES: BBM, BIP, BMF, CBF, KGC, LCP, LMP, NBP, PNM, PYR. **COUNTIES:** BEL, BUL, CAD, HLN, HRT, JEF, MCY, MEN, MUH, NEL, OWS, PIK, POW, ROW, RUS, WHI, WOL. **DATES:** Apr. 17–Sept. 17. **REMARKS:** Common. Often mistaken for a geometrid moth.

Genus *Ledaea* Druce

8491 ***L. perditalis*** (Wlk.)
Lost Owlet
Fulton Co., SR 94, 4.0 km (2.5 mi) E of Cayce, Sept. 5; Grayson Co., RGR, Aug. 13; Jefferson Co., Valley Station, Apr. 17; Owsley Co., Caney Fork Creek near Stay, Aug. 16; Trigg Co., LBL, July 12–16. Common.

Genus *Isogona* Gn.

8493 ***I. tenuis*** (Grt.)
Thin-lined Owlet
PARKS AND RESERVES: BBL, BMF, CBB, HWS, LCP, OCP. **COUNTIES:** BOO, BUL, FAY, GAL, JEF, MCC, MEA, OLD, RUS, WOO. **DATES:** Apr. 8–24, May 2, 14–30, June 2, 14, 29, Aug. 9–24. Common.

Genus *Metalectra* Hbn.

8499 ***M. discalis*** (Grt.)
Common Fungus Moth
PARKS AND RESERVES: BMF, CBB, HWS, LBK, LBL, LCP, OCP. **COUNTIES:** BOO, BUL, FAY, FUL, HRD, JEF, MEA, MEN, OLD, OWS,

RUS, TAY, TRI. **DATES:** May 11–Sept. 22. Common.

8500 ***M. quadrisignata*** (Wlk.)
Four-spotted Fungus Moth
PARKS AND RESERVES: BBM, OCP. **COUNTIES:** FAY, FUL, HLN, HRT, JEF, MEA, TAY. **DATES:** May 13, 21, June 18, July 12–30, Aug. 2–8, 24, Sept. 3. **REMARKS:** Common. Listed from Kentucky by Crumb (1956).

8502 ***M. tantillus*** (Grt.)
Black Fungus Moth
Fayette Co., Aug. 3; Jefferson Co., Valley Station, May 21, June 3. Uncommon.

8505 ***M. richardsi*** Brower
Richard's Fungus Moth
PARKS AND RESERVES: BMF, BUC, CAV, LCP, OCP. **COUNTIES:** BAT, BRE, BUL, JEF, MEA, OWS, PER, RUS, WHI. **DATES:** Apr. 29, May 18–27, June 4–25, July 27, Aug. 1–19, Sept. 19, 22. Locally common.

Genus *Arugisa* Wlk.

8509 ***A. latiorella*** (Wlk.)
Common Arugisa
PARKS AND RESERVES: KLK, NBP, OCP, PYR, RGR. **COUNTIES:** BRK, CAD, GAL, GRY, JEF, LAU, LIV, MEA, MRL, OWS, POW. **DATES:** May 29, June 2–16, 26–July 1, Aug. 19. Common.

Genus *Scolecocampa* Gn.

8514 ***S. liburna*** (Gey.)
Dead-wood Borer Moth
PARKS AND RESERVES: BMF, CAV, LBL, LCP, LWO, NBP, OCP, PMS, PNM. **COUNTIES:** BAT, BEL, BUL, HLN, JEF, LET, MCC, MEA, MEN, OWS, POW, RUS, TRI. **DATES:** June 2–Oct. 5. Common.

Genus *Gabara* Wlk.

8522 ***G. subnivosella*** Wlk.
Hardin Co., June; Jefferson Co., Valley Station, June 21, S. Scholz; Meade Co., OCP, June 21, 25, R. A. Henderson. Previously reported from Illinois west to Nebraska (Forbes 1954).

Genus *Phyprosopus* Grt.

8525 ***P. callitrichoides*** Grt.
Curve-lined Owlet
PARKS AND RESERVES: BAN, FKN, OCP, RRG, SWR. **COUNTIES:** BAR, BOO, BUL, HEN, HND, JEF, LAU, MCC, MEA, MEN, NEL, OWS, WHI. **DATES:** Apr. 22, May 14–June 14, 16, 30, July 3, 9, 31, Aug. 17, Sept. 28. Moderately common.

Genus *Hypsoropha* Hbn.

8527 ***H. monilis*** (F.)
Large Necklace Moth
Jefferson Co., Louisville, Apr. 10, 18, 27, 28, May 15, 31, June 3, 7, July 8, 15, Aug. 8; Laurel Co., FS 615A/B, Apr. 11; Muhlenberg Co., Greenville, May 14. Locally common.

8528 ***H. hormos*** Hbn.
Small Necklace Moth
PARKS AND RESERVES: BAN, BEC, BMF, BUC, HWS, LBL, LCP, OCP, PNM, PYR, RGR. **COUNTIES:** ADA, BAR, BEL, BRE, BRK, BUL, CAD, FAY, GRY, HRD, JEF, MCC, MEA, NEL, OLD, OWS, PER, RUS, TRI, WHI. **DATES:** Apr. 25–Aug. 16. Common.

Genus *Plusiodonta* Gn.

8534 ***P. compressipalpis*** Gn.
Moonseed Moth
PARKS AND RESERVES: AJJ, BBL, CBB, COB, HWS, OCP, PYR, SWR. **COUNTIES:** BOO, BUL, CAM, CHR, FAY, HEN, HIC, HND, HRT, JEF, MCC, MEA, OLD, OWS, ROW, WOO. **DATES:** Apr. 11–May 23, June 7–14, 22, 26, July 13–Sept. 10. Common.

Genus *Calyptra* Ochs.

8536 ***C. canadensis*** (Bethune)
Canadian Owlet
Bullitt Co., BMF, June 1–6, Sept. 6; Madison Co., Oct. 17; Menifee Co., RRG, Aug. 31–Sept. 22; Oldham Co., HWS, June 14. Common.

Genus *Anomis* Hbn.

8545 ***A. erosa*** Hbn.
Yellow Scallop Moth
PARKS AND RESERVES: BAN, BBL, BOW, BUC, GBT, PNM, PYR, SWR. **COUNTIES:** BAR, BEL, BOO, BRE, CAD, CRL, FAY, FUL, HND, JEF, MCC, NEL, OWS, PER, ROW. **DATES:** May 31, June 8, Aug. 8, 21, 29–Oct. 22, Nov., 11. Common.

Genus *Alabama* Grt.

8554 ***A. argillacea*** (Hbn.)
Cotton Moth
PARKS AND RESERVES: BAN, BOW, CCR, LCP. **COUNTIES:** BAR, BOO, FAY, FUL, HIC, JEF, OWS, RUS, WAR. **DATES:** July 15, Aug. 10, Sept. 6–8, Oct. 1–8, 20, 23. **REMARKS:** Rare in recent years with the loss of cotton culture.

Genus *Scoliopteryx* Germ.

8555 ***S. libatrix*** (L.)
The Herald
PARKS AND RESERVES: BBL, BOW, LCP, OCP, PNM, RRG. **COUNTIES:** BEL, BOO, BRK, EDM, FAY, GAL, HIC, HRD, JEF, MCC, MEA, NEL, OLD, POW, ROC, RUS. **DATES:** Jan.–Mar. (adults overwintering in caves) Apr. 4, 5, May 27, June 9, July 18–21, 31, Aug. 4, 15, 22, Sept. 12, 16, Oct. 2, 18, 26–31, Nov. 9. Common.

Genus *Anticarsia* Hbn.

8574 ***A. gemmatalis*** Hbn.
Velvetbean Caterpillar Moth
PARKS AND RESERVES: BIP, BOW, BUC, GBT, LCP, OCP, PNM, RRG, SWR, WVP. **COUNTIES:** BEL, BOO, BRE, CRL, FUL, GUP, HND, JEF, LET, MCC, MEA, NEL, PER, PIK, POW, RUS, WHI. **DATES:** June 1, July 23, Aug. 19–Nov. 12. **REMARKS:** Common to abundant late in the season. A pest of soybean and related legume crops, especially in western counties.

Genus *Panopoda* Gn.

8587 ***P. rufimargo*** (Hbn.)
Red-lined Panopoda
PARKS AND RESERVES: BAN, BBM, BMF, BUC, CBF, KGC, KRC, LBL, LCP, LWO, NBP, PYR, RBF, RGR. **COUNTIES:** ADA, BAR, BRE, BRK, BUL, CAD, GRY, HLN, JEF, LET, MCC, MEA, MOR, PER, POW, ROW, RUS, TRI, WHI, WOL. **DATES:** May 4–Sept. 20. Common.

8588 ***P. carneicosta*** Gn.
Brown Panopoda
PARKS AND RESERVES: BAN, BBM, BUC, CBF, CMI, HWS, JCF, JWS, KGC, LCP, LWO, OCP, PMS, PNM, PYR, RGR, RRG. **COUNTIES:** BAR, BEL, BOO, BRE, BRK, BUL, CAD, EST, FLO, GRY, HLN, JEF, LET, MEA, NEL, OLD, OWS, PER, POW, ROW, RUS, WHI. **DATES:** May 9–Aug. 19, Sept. 1. Common.

8589 ***P. repanda*** (Wlk.)
Orange Panopoda
Forbes (1954) mentions the range of this southern species as extending into Kentucky.

Genus *Phoberia* Hbn.

8591 ***P. atomaris*** Hbn.
Common Oak Moth
PARKS AND RESERVES: BBL, BBM, BIP, BMF, CAC, CBB, CBF, GBT, HWS, IRO, JWS, KGC, LMP, MCV, MCY, NBP, OCP, PNM. **COUNTIES:** BEL, BOO, BUL, BRE, CAR, CRL, EDM, EST, FAY, FLO, HLN, JEF, LEE, MCC, MEA, MEN, MUH, OLD, OWS, PIK, POW, WHI, WOL. **DATES:** Mar. 11–16, 25–May 7, Aug. 14. **REMARKS:** Common to abundant. Sometimes becomes a serious defoliator of oaks, as revealed in an aerial survey of parts of Breathitt, Estill, Lee, Menifee, Powell, and Wolfe counties, June 2–5, 1981 (Carothers 1981b).

Genus *Cissusa* Wlk.

8592 ***C. spadix*** (Cram.)
Black-dotted Brown
PARKS AND RESERVES: BEC, BMF, CAC, KGC, KLK, LCP, LMP, MCV, OCP, PNM, RRG. **COUNTIES:** BEL, BUL, CAR, EDM, HLN, JEF, MCC, MEA, MEN, MRL, MUH, OWS, RUS, WHI. **DATES:** Mar. 29–May 4, July 16. Common.

Genus *Melipotis* Hbn.

8600 ***M. indomita*** (Wlk.)
Indomitable Melipotis
Edmonson Co., MCV, Apr. 14; Gallatin Co., Glencoe exit of I-71, Aug. 27; Russell Co., Alligator dock, July 4, and LCP, June 23. Uncommon.

8607 ***M. jucunda*** Hbn.
Merry Melipotis
Campbell Co., N side of SR 8, Melbourne, Aug. 22; Carlisle Co., Sept. 12; Fulton Co., Hickman, Aug. 20; Jefferson Co., FAO, Sept. 10, 18, 20. May be common at FAO in the daytime in sandy and exposed fossil rock habitats.

Genus *Synedoida* Hy. Edw.

8641 ***S. grandirena*** (Haw.)
Figure-seven Moth
Harlan Co., KGC, July 7; Knox Co., Beech Creek Hollow, Scalf, Aug. 1; Menifee Co., RRG, May 23, July 21; Owsley Co., 4.8 km (3 mi) NE of Booneville, May 7, July 14, 18; Rowan Co., Morehead, June 25–27, July 18, 19; Russell Co., Jamestown, July 19. Local and uncommon.

Genus *Hypocala* Gn.

8642 ***H. andremona*** (Cram.)
Rowan Co., SR 1274, 3.2 km (2 mi) W of SR 519, Oct 15, 1993, L. D. Gibson. Rare.

Genus *Thysania* Dalm.

8647 ***T. zenobia*** (Cram.)
Owl Moth
McCracken Co., Ray Black farm, 9.7 km (6 mi) SW of Paducah, at light, Aug. 23, 1974, W. R. Black, Jr. A rare stray from tropical America.

Genus *Ascalapha* Hbn.

8649 ***A. odorata*** (L.)
The Black Witch
Fayette Co., Lexington, July 13, 26; Henderson Co., AUD, Sept. 1; Jefferson Co., Anchorage (no date), and Valley Station, Aug. 14; Madison Co., June 24; McCracken Co., Paducah, July 12. The black witch is not a Kentucky resident, but often migrates northward from the neotropics and extreme southern United States.

Genus *Lesmone* Hbn.

8651 ***L. detrahens*** (Wlk.)
Detracted Owlet
PARKS AND RESERVES: BAN, BUC, HWS, KLK, LBL, LCP, PYR, SWR. **COUNTIES:** BAR, BRE, BUL, CHR, FAY, FUL, GAL, HND, JEF, MCC, MEA, MRL, NEL, OLD, OWS, PER, POW, ROW, RUS, TRI. **DATES:** Apr. 29–Sept. 9, 17. Common.

Genus *Matigramma* Grt.

8679 ***M. pulverilinea*** Grt.
Bullitt Co., PCP, July 22; Fulton Co., RNR, May 12; Hickman Co., Columbus, May 12. Very local.

Genus *Zale* Hbn.

8689 ***Z. lunata*** (Drury)
Lunate Zale
PARKS AND RESERVES: BAN, BEC, BIP, BMF, BUC, CAC, CBF, GBT, HWS, JWS, LBL, LCP, LMP, LWO, MCV, NBP, OCP, PNM, PYR, RGR, RRG, WVP. **COUNTIES:** BAR, BEL, BRE, BRK, BUL, CAD, CAR, CRL, EDM, FAY, FLO, FUL, GAL, GRY, HLN, HND, JEF, LET, MCC, MEA, MEN, MUH, NEL, OLD, OWS, PER, PIK, POW, ROW, RUS, SHE, TRI, WHI, WOL, WOO. **DATES:** Mar. 13–Nov. 11. Very common.

8692 ***Z. galbanata*** (Morr.)
Maple Zale
PARKS AND RESERVES: BBL, BBM, BEC, BMF, CAC, CBB, HWS, JWS, LCP, MCV, OCP, PNM, RGR, RRG. **COUNTIES:** BEL, BOO, BRA, BRK, BUL, CAR, EDM, FAY, FLO, FUL, GAL, GRY, HLN, HND, JEF, MEA, OLD, RUS, WOO. **DATES:** Apr. 6–Sept. 1, 28. Common.

8694 ***Z. aeruginosa*** (Gn.)
Green-dusted Zale
Bell Co., PNM, Apr. 17, 20, 28, 29–May 4, 12, July 19; Harlan Co., BBM, Apr. 20, and KGC, Apr. 20; Menifee Co., July 8; Powell Co., NBP, July 14, and RRG, July 8. Uncommon.

8695 ***Z. undularis*** (Drury)
Black Zale
PARKS AND RESERVES: BBM, BCN, BMF, BUC, PNM. **COUNTIES:** BEL, BOO, BRA, BRE, BUL, FAY, GAL, HLN, JEF, NEL, NIC, PER. **DATES:** Apr. 12, 22, May 3–Aug. 21. Locally common.

8697 ***Z. minerea*** (Gn.)
Colorful Zale
PARKS AND RESERVES: BIP, BMF, BUC, CAC, CBF, HNK, JWS, KGC, LCP, NBP, OCP, PMS, PNM, RRG. **COUNTIES:** BEL, BRE, BUL, CAL, CAR, FAY, FLO, HLN, JEF, MCC, MEA, PER, PIK, POW, ROW, RUS, WHI. **DATES:** Apr. 4–July 24, Aug. 14, 16, 29, Oct. 5. Common.

8698 ***Z. phaeocapna*** Franc.
Fayette Co., Elkhorn Creek S of Higbee Mill Rd., bait trap, June 19–20; Menifee Co., RRG, FS 9B along East Fork Indian Creek, Apr. 19; both J. S. Nordin. Rare.

8699 ***Z. obliqua*** (Gn.)
Oblique Zale
Owsley Co., 4.8 km (3 mi) NE of Booneville, June 22; Powell Co., RRG, Apr. 17; Whitley Co., CBF, Apr. 27, and Goldbug, June 30. Rare.

8700 ***Z. squamularis*** (Drury)
Bell Co., PNM, Apr. 28, May 2, 3, 12, 13, June 8; Owsley Co., 4.8 km (3 mi) NE of Booneville, July 1; Powell Co., RRG, Tunnel Ridge, Apr. 17, 18; Rowan Co., Morehead, June 28–30, July 1, 6. Uncommon.

8701 ***Z. confusa*** McD.
Bell Co., PNM, May 3; Jefferson Co., Valley Station, July 14; Powell Co., May 21; Whitley Co., CBF, Apr. 27, 28, and Goldbug, June 5. Uncommon.

8702 ***Z. submediana*** Strand
Bell Co., PNM, Apr. 14, 1978, C. Cornett. Rare.

8703 ***Z. duplicata*** (Bethune)
Powell Co., RRG, Tunnel Ridge, Apr. 17, 1989, L. D. Gibson. Rare.

8704 ***Z. helata*** (Smith)
Brown-spotted Zale
PARKS AND RESERVES: BIP, BMF, PNM, RRG. **COUNTIES:** BEL, BUL, FAY, JEF, LAU, MEN, OLD, OWS, PIK, POW, WHI. **DATES:** Apr. 12–May 30, July 25. Common.

8705 ***Z. bethunei*** (Smith)
Bethune's Zale
PARKS AND RESERVES: BMF, CAC, KLK, LMP, NBP, PYR, RRG. **COUNTIES:** BUL,

CAD, CAR, HLN, JEF, MEN, MRL, MUH, OWS, POW, PUL, ROW, WHI. **DATES:** Mar. 2, 11, 15, 29, Apr. 11–21, May 5, 21–Aug. 10. Common.

8707 ***Z. metatoides*** McD.
Washed-out Zale
PARKS AND RESERVES: BMF, BUC, CBF, JWS, LCP, NBP, PNM. **COUNTIES:** BEL, BRE, BUL, FAY, FLO, JEF, PER, POW, RUS, WHI. **DATES:** Apr. 27–May 3, 12–22, June 2–7, 22, 30, July 8, 20. Moderately common.

8708 ***Z. metata*** (Sm.)
PARKS AND RESERVES: BMF, CAC, CBF, JWS, LCP, NBP. **COUNTIES:** BUL, CAR, FLO, JEF, POW, RUS, WHI. **DATES:** Apr. 7–30, May 16, 22, July 5, 14. Moderately common.

8713 ***Z. lunifera*** (Hbn.)
Bold-based Zale
PARKS AND RESERVES: BBM, BIP, CAC, CBF, HWS, NBP, PNM, RRG. **COUNTIES:** BEL, CAR, HLN, JEF, LAU, MEN, OLD, OWS, PIK, POW, ROW, WHI, WOL. **DATES:** Apr. 12–May 27. Common.

8716 ***Z. unilineata*** (Grt.)
One-lined Zale
PARKS AND RESERVES: BBL, BBM, BIP, BMF, KGC, OCP, PNM. **COUNTIES:** BEL, BOO, BUL, HLN, JEF, MEA, NEL, PIK. **DATES:** Apr. 4–June 25. Common.

8717 ***Z. horrida*** Hbn.
Horrid Zale
PARKS AND RESERVES: BBL, BEC, BMF, CAC, CBF, COB, HNK, HWS, WVP. **COUNTIES:** BOO, BUL, CAL, CAR, HIC, JEF, MCC, NEL, OLD, WHI. **DATES:** Apr. 15–Sept. 7. Moderately common.

Genus *Euparthenos* Grt.

8719 ***E. nubilis*** (Hbn.)
Locust Underwing
PARKS AND RESERVES: BAN, BBM, BEC, BMF, BUC, CAC, GBT, JCF, JWS, LCP, PNM, RGR, WVP. **COUNTIES:** ADA, BAR, BEL, BRE, BRK, BUL, CAR, CLK, CRL, FAY, FLO, GRY, HLN, JEF, JES, LET, LEW, MAD, MCC, OWS, PER, PIK, ROW, RUS, WAR, WHI. **DATES:** Apr. 2–Oct. 11. **REMARKS:** Common. Two individuals with hindwing nearly entirely black were collected in Jefferson Co., Valley Station, S. Scholz on Aug. 10, 1974 and Aug. 2, 1983.

Genus *Allotria* Hbn.

8721 ***A. elonympha*** (Hbn.)
False Underwing
PARKS AND RESERVES: BAN, BMF, BUC, CAC, CBF, JWS, LBL, LCP, LMP, NBP, OCP, PMS, PNM, PYR, RRG, RGR, WVP. **COUNTIES:** BAR, BEL, BRE, BRK, BUL, CAD, CAR, FLO, GRY, HLN, JEF, KNO, LAR, MCC, MEA, MEN, MUH, NEL, OWS, PER, POW, ROW, RUS, TRI, WHI. **DATES:** Mar. 9–Sept. 7. **REMARKS:** Common. Forewing pattern is quite variable.

Genus *Dysgonia* Hbn.

8726 ***D. smithii*** (Gn.)
Smith's Darkwing
Graves Co., Bell Farm near Kaler, Sept. 13, 1980, 1 specimen in bait trap, L. C. Koehn and R. A. Henderson; Henderson Co., SWR, Sept. 10, 1983, R. A. Henderson. A southern species that is rare in Kentucky.

Genus *Parallelia* Hbn.

8727 ***P. bistriaris*** Hbn.
Maple Looper Moth
PARKS AND RESERVES: BAN, BEC, BMF, CAC, CBF, LCP, LVJ, MCV, OCP, PMS, PNM, PYR, RGR, RRG, WVP. **COUNTIES:** BAR, BEL, BRA, BRK, BUL, CAD, CAR, EDM, FAY, FUL, GRY, HLN, HND, JEF, LAU, MCC, MEA, OWS, POW, ROW, RUS, SPE, WHI. **DATES:** Apr. 8–Sept. 26. Common.

Genus *Cutina* Wlk.

8729 ***C. distincta*** (Grt.)
Fulton Co., SR 94, 4.0 km (2.5 mi) E of Cayce, Willingham Bottoms, at light, Sept. 8; Henderson Co., SWR, Aug. 22. Confined to cypress swamps.

8729.1 ***C. arcuata*** Pogue & Fgn.
Three paratypes from Kentucky are listed by Pogue and Ferguson (1998): Fulton Co., RNR, May 12, 1984, 2 males, and SR 94, 4.0 km (2.5 mi) E of Cayce, Willingham Bottoms, Aug. 3, 1986, 1 female; all collected by C.V. Covell, Jr., and in the NMNH.

Genus *Euclidia* Ochs.

8731 ***E. cuspidea*** (Hbn.)
Toothed Somberwing
PARKS AND RESERVES: BBL, HWS, PNM, RBF. **COUNTIES:** BEL, BOO, BRE, CRL, FAY, GAL, JEF, KNX, LEE, MCY, MEN, OLD, OWS, PUL, ROW. **DATES:** Apr. 8, 18–May 26, June 21–Aug. 1, 11, Sept. 3, 25. Locally common.

Genus *Caenurgia* Wlk.

8733 ***C. chloropha*** (Hbn.)
Vetch Looper Moth
PARKS AND RESERVES: BAN, OCP. **COUNTIES:** BAR, CHR, FAY, FUL, JEF, MCC, MCY,

MEA, OWS, WHI. **DATES:** Apr. 8, 10, 21, May 31, July 3, 23–Sept. 4, 13, 26, Oct. 5, 17, 24–28. Common.

Genus *Caenurgina* McD.

8738 ***C. crassiuscula*** (Haw.)
Clover Looper Moth
PARKS AND RESERVES: BAN, BBL, BMF, GBT, HWS, KRC, LBK, PYR, RGR, RRG. **COUNTIES:** BAR, BOO, BRA, BRK, BUL, CAD, CRL, FAY, GAL, GRY, HND, JEF, LAU, MCC, MCY, MEN, OLD, SHE, TRI, WOL. **DATES:** Mar. 9, 25–Apr. 26, June 2–7, 15, 27, July 15–28, Aug. 6–Sept. 8, 20, Oct. 20. **REMARKS:** Common to abundant, especially in clover and alfalfa fields.

8739 ***C. erechtea*** (Cram.)
Forage Looper Moth
PARKS AND RESERVES: BBL, BBM, BEC, BLA, BMF, BOW, CAV, CBB, HWS, IRO, OCP, RGR, SWR. **COUNTIES:** BAT, BOO, BRK, BUL, FAY, FUL, GRY, HLN, HND, JEF, JES, MEA, MER, MRL, OLD, ROW. **DATES:** Mar. 25, Apr. 10–Oct. 22. **REMARKS:** Common to abundant. Very similar to clover looper moth in appearance and habits.

Genus *Mocis* Hbn.

8743 ***M. latipes*** (Gn.)
Small Mocis
Bell Co., PNM, Oct. 23; Caldwell Co., PYR, Sept. 13; Fulton Co., Sept. 7; Jefferson Co., June 26, July 22, Aug. 30, Sept. 18, Oct. 13, Nov. 4; Kenton Co., Ft. Mitchell, Oct. 14; Letcher Co., Sept. 22. Common late in the season.

8745 ***M. texana*** (Morr.)
Texas Mocis
PARKS AND RESERVES: BMF, CAV, HWS, PNM. **COUNTIES:** BAT, BEL, BUL, FAY, FUL, GRV, JEF, MCC, MEA, OLD, OWS, ROW, WHI. **DATES:** Apr. 6–Aug. 28, Sept. 8–13. Common.

Genus *Celiptera* Gn.

8747 ***C. frustulum*** Gn.
Black Bit Moth
PARKS AND RESERVES: BMF, JWS, NBP, OCP, WVP. **COUNTIES:** BUL, FAY, FLO, FUL, HIC, JEF, MEA, MEN, OWS, POW, ROW, WOO. **DATES:** May 3–June 16, 30, July 13–30, Aug., Sept. 13. Common.

Genus *Ptichodis* Hbn.

8750 ***P. herbarum*** (Gn.)
Common Ptichodis
Fayette Co., Lexington, July 18; Jefferson Co., Valley Station, May 16, 19, June 1, 10, 25, Aug. 5, Sept. 6, 7, Oct. 10; Owsley Co., 4.8 km (3 mi) NE of Booneville, May 29; Powell Co., RRG, Tunnel Ridge, May 18; Trigg Co., LBL, July 13. Uncommon.

8751 ***P. bistrigata*** Hbn.
Southern Ptichodis
Bell Co., PNM, May 1; Carter Co., CAC, Apr. 29; Jefferson Co., Valley Station, Apr. 12, May 5, 28. Uncommon.

Genus *Argyrostrotis* Hbn.

8764 ***A. anilis*** (Drury)
Short-lined Chocolate
PARKS AND RESERVES: BBL, LBL, PNM, RBF, RRG. **COUNTIES:** BEL, BOO, BRE, FAY, JEF, MCC, OWS, POW, ROW, TRI. **DATES:** Apr. 4, May 15, 16, 25–June 15, 24–July 19, 28, Aug. 8–20, Sept. 5. Local and moderately common.

Genus *Spiloloma* Grt.

8769 ***S. lunilinea*** Grt.
Moon-lined Moth
PARKS AND RESERVES: BMF, CBB, CMI, HWS, OCP. **COUNTIES:** BOO, BUL, FAY, JEF, MEA, NEL, OLD. **DATES:** Apr. 5–Sept. 15. Common.

Genus *Catocala* Schr.

The Underwings

8770 ***C. innubens*** Gn.
The Betrothed
PARKS AND RESERVES: BCN, CER, HWS, LCP, MCV, OCP. **COUNTIES:** BOO, BUL, EDM, FAY, JEF, JES, MEA, OLD, ROW, RUS, WAY, WOO. **DATES:** July 2–Aug. 16, 25, Sept. 26. **REMARKS:** Records include form "*scintillans*" Grt., as well as typical *innubens*. Common.

8771 ***C. piatrix*** Grt.
The Penitent
PARKS AND RESERVES: CAV, HWS, LCP, OCP, PNM, RRG, SWR, WVP. **COUNTIES:** BAT, BEL, BOO, CUM, FAY, FUL, HND, JEF, MCC, MEA, MEN, OLD, OWS, ROW, RUS, UNI. **DATES:** July 16–Nov. 4. Common.

8773 ***C. epione*** (Drury)
Epione Underwing
PARKS AND RESERVES: BBM, BMF, BUC, KGC, LBL, LCP, MCV, NBP, OCP, PNM, RBF, RRG. **COUNTIES:** BEL, BRE, BRK, BUL, EDM, EST, FAY, HLN, HOP, JEF, KNX, LAU, MCC, MEA, OWS, PER, POW, ROW, RUS, TRI, WHI. **DATES:** June 12–Aug. 9. Common.

8778 ***C. habilis*** Grt.
Habilis Underwing
PARKS AND RESERVES: BMF, BOW, CAV, MCV, OCP, RRG. **COUNTIES:** BAT, BOO, BUL, EDM, JEF, MEA, ROW. **DATES:** July 7–Sept. 3, 12, Oct. 8. Moderately common.

8779 ***C. serena*** Edw.
Serene Underwing
PARKS AND RESERVES: BBM, BMF, CAV, CEW, KGC, LCP, OCP, PYR, RBF, WVP. **COUNTIES:** BAT, BRE, BUL, CHR, HLN, JEF, MCC, MEA, MET, OWS, RUS, WAY. **DATES:** June 21–Aug. 26. Locally common.

8780 ***C. robinsoni*** Grt.
Robinson's Underwing
PARKS AND RESERVES: AJJ, BBL, MCV, OCP, PNM, WVP. **COUNTIES:** BEL, BOO, CAM, EDM, FAY, JEF, MCC, MEA, OWS, ROW. **DATES:** July 3–Oct. 21, Nov. 9. Moderately common.

8781 ***C. judith*** Stkr.
Judith's Underwing
PARKS AND RESERVES: BBM, BMF, KGC, LCP, MCC, NBP, OCP, PMS, PYR, RRG. **COUNTIES:** BUL, CAD, HLN, JEF, MCC, MCY, MEA, MEN, POW, ROW, RUS, WHI. **DATES:** June 12, 14, 25–Aug. 16. Locally common.

8782 ***C. flebilis*** Grt.
Mourning Underwing
PARKS AND RESERVES: BBL, BMF, BOW, GBO, LCP, OCP, RBF, RRG. **COUNTIES:** BOO, BRE, BUL, FAY, GUP, JEF, LOG, MEA, OWS, POW, ROW, RUS, WAY, WHI. **DATES:** June 22–Sept. 13. Locally common.

8783 ***C. angusi*** Grt.
Angus's Underwing
PARKS AND RESERVES: BBL, BBM, BMF, BUC, MCV, OCP, PMS, PYR, RGR, WVP. **COUNTIES:** BOO, BRE, BRK, BUL, CAD, CHR, EDM, FAY, GRY, HLN, JEF, MEA, MCC, OLD, OWS, PER. **DATES:** July 9–Oct. 15. Common.

8784 ***C. obscura*** Stkr.
Obscure Underwing
PARKS AND RESERVES: KGC, LCP, MCV, OCP, SWR, WVP. **COUNTIES:** EDM, FAY, HEN, HLN, HND, JEF, MCC, MEA, OWS, ROW, RUS. **DATES:** July 18–Sept. 20. Common.

8785 ***C. residua*** Grt.
Residua Underwing
PARKS AND RESERVES: BMF, BOW, CAV, CEW, LBL, LCP, MCV, OCP, PYR, RGR, RRG, SWR, WVP. **COUNTIES:** BAT, BOO, BUL, BRK, CHR, EDM, FAY, GRY, HND, JEF, LYO, MEA, MEN, MET, OWS, ROW, RUS, WAY, WHI. **DATES:** July 7–22, Aug. 1–Sept. 11. Common.

8786 ***C. sappho*** Stkr.
Sappho Underwing
Harlan Co., PMS, Aug. 4 and 8, 1992, L. E. Smith; Wayne Co., CEW, July 22, 1978 (L. D. Gibson and R. A. Henderson collected one each during daytime). Very local in southern Kentucky.

8787 ***C. agrippina*** Stkr.
Agrippina Underwing
Fulton Co., SR 94, 4.0 km (2.5 mi) E of Cayce, Aug. 18, 1978, L. D. Gibson; Union County, Sullivan, July 19, 1952, J.A. Shields. Rare in southern Kentucky.

8788 ***C. retecta*** Grt.
Yellow-gray Underwing
PARKS AND RESERVES: AJJ, BBL, CAC, CAV, CER, CEW, KGC, KLK, LCP, MCV, OCP, RRG, WVP. **COUNTIES:** BAT, BOO, BUL, CAM, CAR, EDM, FAY, GAL, HLN, JEF, LOG, MEA, MRL, OWS, POW, ROW, RUS, UNI, WAY. **DATES:** July 7, 20–Oct. 13. Locally common.

8789 ***C. ulalume*** Stkr.
Ulalume Underwing
PARKS AND RESERVES: BUC, CEW, LBL, RRG. **COUNTIES:** FUL, LAU, LOG, PER, POW, TRI, WAY. **DATES:** July 14, 22, 31, Aug. 27. Uncommon. A southern species.

8790 ***C. dejecta*** Stkr.
Dejected Underwing
PARKS AND RESERVES: BBM, CEW, KGC, LCP, OCP, PNM, RRG. **COUNTIES:** BEL, BUL, HLN, JEF, MEA, MEN, MET, OWS, RUS, WAY. **DATES:** June 22, 23, July 2–22, Sept. 18. Moderately common.

8791 ***C. insolabilis*** Gn.
Inconsolable Underwing
PARKS AND RESERVES: BBL, LCP, NBP, OCP. **COUNTIES:** BOO, FAY, JEF, MCC, MEA, MEN, MET, OWS, POW, RUS. **DATES:** June 1, 15–July 23, Aug. 6, 8, 28. Uncommon.

8792 ***C. vidua*** (J. E. Smith)
Widow Underwing
PARKS AND RESERVES: BAN, BBM, BEC, BIP, CAV, GBO, KGC, LBL, LCP, MCV, OCP, PNM, PYR, RGR, SWR, WVP. **COUNTIES:** BAR, BAT, BEL, BRA, BRK, CAD, EDM, FAY, GRY, GUP, HLN, HND, JEF, LAU, LET, LYO, MCC, MEA, MEN, OWS, PIK, ROW, RUS.

DATES: July 7–Oct. 23. **REMARKS:** Very common – our most common *Catocala* with a black hindwing.

8793 ***C. maestosa*** Hulst
Sad Underwing
PARKS AND RESERVES: BBL, OCP, SWR. **COUNTIES:** BOO, CHR, FAY, FUL, HND, JEF, LYO, MEA. **DATES:** July 22, Aug. 8–Sept. 17, Oct. 10. Uncommon to rare.

8794 ***C. lacrymosa*** Gn.
Tearful Underwing
PARKS AND RESERVES: BIP, BUC, CAV, CEW, LCP, MCV, OCP, PNM, PYR, RGR, WVP. **COUNTIES:** ALL, BAT, BEL, BRE, BRK, BUL, CHR, EDM, FAY, GRY, GUP, JEF, LET, LOG, MAD, MCC, MEA, OWS, PER, PIK, ROW, RUS, WAY. **DATES:** July 2–Oct. 17. Locally common.

8795 ***C. palaeogama*** Gn.
Oldwife Underwing
PARKS AND RESERVES: BBL, BBM, BMF, BUC, CAC, CER, CEW, GBT, KGC, LCP, LMP, NBP, OCP, PMS, PYR, RGR, RRG. **COUNTIES:** BOO, BRE, BRK, BUL, CAD, CAR, CHR, CRL, FAY, GRY, HLN, JEF, LAU, LET, MCC, MEA, MEN, MET, MRL, MUH, PER, POW, ROW, RUS, WAY, WHI. **DATES:** June 12–Oct. 4. Common.

8796 ***C. nebulosa*** Edw.
Clouded Underwing
PARKS AND RESERVES: BBM, BCN, BIP, CAV, HWS, LCP, OCP, PCP, PYR, RGR, SWR. **COUNTIES:** BAT, BOO, BRK, BUL, CAD, FAY, GRY, HLN, HND, JEF, KNO, MEA, MEN, OLD, PIK, ROW, RUS, WHI. **DATES:** Apr. 30, May 1, July 6–19, 28–Sept. 12. Locally common.

8797 ***C. subnata*** Grt.
Youthful Underwing
PARKS AND RESERVES: BBM, BCN, CER, HWS, KGC, LCP, NBP, OCP, RGR, RRG. **COUNTIES:** BOD, BOO, BRK, GRY, HLN, JEF, MEA, MEN, OLD, OWS, POW, ROW, RUS. **DATES:** July 7–Sept. 4, 14. Moderately common.

8798 ***C. neogama*** (J. E. Smith)
The Bride
PARKS AND RESERVES: ASC, BAN, BBM, BCN, BMF, BUC, CAC, CAV, HWS, KGC, KLK, NBP, OCP, PMS, RGR, WVP. **COUNTIES:** BAR, BAT, BOO, BRA, BRE, BRK, BUL, CAR, FAY, GAL, GRY, HLN, JEF, MCC, MEA, MEN, MET, MRL, OLD, PER, POW, ROW, WAR. **DATES:** Apr. 14, June 15, 20, July 1–Sept. 22, Oct. 11. Common.

8801 ***C. ilia*** (Cram.)
Ilia Underwing
PARKS AND RESERVES: BBL, BBM, BMF, BUC, CAC, CER, HWS, KGC, LCP, LMP, LWO, MCV, OCP, PMS, PNM, RBF, RGR, RRG, WHI, WVP. **COUNTIES:** BEL, BOO, BRA, BRE, BRK, BUL, CAR, EDM, GAL, GRY, HLN, JEF, LAU, LET, MCC, MEA, MEN, MET, MUH, NEL, OLD, OWS, PER, ROW, RUS, WHI. **DATES:** May 27–Oct. 9. **REMARKS:** Forewing patterns quite variable. Our most common large underwing.

8802 ***C. cerogama*** Gn.
Yellow-banded Underwing
PARKS AND RESERVES: BCN, CAV, CER, PNM, RGR. **COUNTIES:** BAT, BEL, BOO, BRK, GRY, JEF, MEN, OWS, POW, ROW. **DATES:** July 19, Aug. 3–Sept. 3, 12, Oct. 5–23. Moderately common.

8803 ***C. relicta*** Wlk.
White Underwing
Sargent (1976) gives the range of *relicta* as "Hudson Bay area to Kentucky and west throughout area." This was our only "record" until L. D. Gibson took a fresh male specimen of form "*phrynia*" Hy. Edw. at a carwash in Rowan Co., Morehead, on Sept. 2, 1995. Rare this far south.

8804 ***C. marmorata*** Edw.
Marbled Underwing
Fayette Co., Lexington, no date; Jefferson Co., Louisville, Sept. 14; Lyon Co., LBL, Aug. 17; McCracken Co., Paducah, Sept. 8–18; Meade Co., OCP, June 26. Rare. It has been found in the wooded churchyard of Grace Episcopal Church, Paducah, on several occasions, most recently Sept. 11, 1994 by W. R. Black, Jr.

8805 ***C. unijuga*** Wlk.
Once-married Underwing
Fulton Co., Hickman, caught in daytime on Sept. 14, 1984, Jeffrey D. Hooper. Rare.

8806 ***C. parta*** Gn.
Mother Underwing
Jefferson Co., Louisville, Sept. 20, 1971, C. V. Covell, Jr. Rare.

8829 ***C. junctura*** Wlk.
Joined Underwing
Christian Co., PYR, Aug. 23; Fulton Co., Hickman, Aug. 8; Jefferson Co., July 1, 10, 11; Meade Co., OCP, Aug. 4, 24–27; Warren Co., July 4. Rare.

8832 ***C. cara*** Gn.
Darling Underwing
PARKS AND RESERVES: BCN, BUC, OCP, SWR. **COUNTIES:** BOO, BRE, FAY, FUL, GRV, HND, JEF, LEW, MCC, MEA, OWS, PER, ROW. **DATES:** July 19–Oct. 22. **REMARKS:** Moderately common; some specimens are form "*carissima*" Hulst.

8834 ***C. amatrix*** (Hbn.)
The Sweetheart
PARKS AND RESERVES: BOW, CER, GBT, LCP, OCP, SWR, WVP. **COUNTIES:** BOO, CRL, FAY, FUL, HND, JEF, MCC, MEA, PEN, ROW, RUS. **DATES:** July 24–Oct. 13. Common.

8835 ***C. delilah*** Stkr.
Delilah Underwing
Holland (1903) included Kentucky in its range in *The Moth Book*, but I have not found a precise record on which it was based. Questionably resident in Kentucky, very rare if so.

8840 ***C. illecta*** Wlk.
Magdalen Underwing
PARKS AND RESERVES: AJJ, BBL, BMF, HWS, OCP, WVP. **COUNTIES:** BOO, BRA, BUL, CAM, FAY, GAL, JEF, MEA, OLD, WAR, WOO. **DATES:** June 8–July 3, 11, 14. Moderately common.

8842 ***C. nuptialis*** Wlk.
Married Underwing
Henderson Co., SWR, Aug. 22; Jefferson Co., Valley Station, Aug. 25, 1975; Meade Co., OCP, no date. The Valley Station specimen is illustrated in Covell (1984), plate 33 (6). Rare.

8843 ***C. whitneyi*** Dodge
Whitney's Underwing
Jefferson Co., Valley Station, Aug. 25; Meade Co., OCP, Aug. 9. Rare.

8846 ***C. sordida*** Grt.
Sordid Underwing
PARKS AND RESERVES: BBM, KGC, LWO, RRG. **COUNTIES:** HLN, JEF, LET, MEN, POW, ROW. **DATES:** June 21, 29, July 6–26, Aug. 7, 25. Uncommon.

8847 ***C. gracilis*** Edw.
Graceful Underwing
Laurel Co., SR 192, 7.3 km (6 mi) SW of London, July 21; Menifee Co., RRG, July 21; Powell Co., NBP, July 14, RRG, Tunnel Ridge, July 8–13, 23. *C. sordida* is possibly a variation of this species; both were taken in RRG the same night, and are very similar except for the basal dash present in *gracilis*.

8849 ***C. andromedae*** Gn.
Andromeda Underwing
PARKS AND RESERVES: BBM, KGC, KRC, LWO, PMS, PNM, RRG. **COUNTIES:** BEL, HLN, JEF, LET, OWS, POW, ROW, WOL. **DATES:** June 30, July 1–18, 27, Aug. 7. Locally common.

8851 ***C. coccinata*** Grt.
Scarlet Underwing
PARKS AND RESERVES: BBM, BMF, HWS, KGC, LCP, LWO, NBP, OCP, RRG. **COUNTIES:** BOD, BUL, HLN, JEF, LET, MCC, MEA, MEN, OLD, OWS, POW, ROW, RUS. **DATES:** May 30–Aug. 7, mostly June and July. Sometimes locally common.

8857 ***C. ultronia*** (Hbn.)
Plum Tree Underwing
PARKS AND RESERVES: BBL, BCN, BMF, CER, HWS, KGC, LBL, OCP, PMS. **COUNTIES:** BOO, BRA, BUL, FAY, HLN, JEF, LAR, MCC, MEA, OLD, ROW, RUS, TRI, WAR. **DATES:** June 8–Aug. 12. Common.

8858 ***C. crataegi*** Saund.
McCracken Co., Paducah at light, June 15, 1996, Philip. M. Sisto. Rare

8863 ***C. mira*** Grt.
Wonderful Underwing
Boone Co., BBL, June 20–July 3, 11, 13, 27; Fayette Co., June 25, 29, July 1; Harlan Co., BBM, July 8, and PMS, July 8; Menifee Co., RRG, July 13, 21; Oldham Co., HWS, July 9. Local and uncommon.

8864 ***C. grynea*** (Cram.)
Woody Underwing
PARKS AND RESERVES: AJJ, BBL, BMF, MCV, MUR, NBP, RRG. **COUNTIES:** BOO, BUL, CAM, EDM, FAY, HIC, JEF, MCC, MEN, OWS, POW, ROW, WHI. **DATES:** June 4–Aug. 30. Common.

8865 ***C. praeclara*** G. & R.
Laurel Co., DBNF, W side of Marsh Branch Rd., July 25 1997, L. D. Gibson. Rare.

8867 ***C. blandula*** Hulst
Charming Underwing
Boone Co., BBL, July 2, 27; Menifee Co., RRG, July 7, 8, 21. Barnes and McDunnough (1918) stated: "We have a single specimen before us labelled 'Kentucky'." Rare.

8869 ***C. alabamae*** Grt.
Alabama Underwing
Henderson Co., AUD, Aug. 15, 1968, J. B. Wood. Rare.

8871 ***C. dulciola*** Grt.
Sweet Underwing
Boone Co., BBL, June 9, 26, 27, and CER, July 2; Caldwell Co., PYR, June 18. Uncommon to rare.

8872 ***C. clintoni*** Grt.
Clinton's Underwing
PARKS AND RESERVES: BBL, CMI, HWS, OCP, RRG. **COUNTIES:** BOO, JEF, MEA, OLD, OWS, ROW, WHI. **DATES:** June 8–July 2, Aug. 13. **REMARKS:** Locally common some years.

8873 ***C. similis*** Edw.
Similar Underwing
PARKS AND RESERVES: KGC, NBP, PNM, RRG. **COUNTIES:** BEL, FAY, HLN, MEN, OWS, POW, ROW. **DATES:** June 28–July 28. Uncommon.

8874 ***C. minuta*** Edw.
Little Underwing
PARKS AND RESERVES: BBL, BMF, HWS, OCP. **COUNTIES:** BOO, BUL, FAY, JEF, MEA, OLD. **DATES:** June 16–July 9, Aug. 14, Sept. 12. Uncommon to rare.

8876 ***C. micronympha*** Gn.
Little Nymph
PARKS AND RESERVES: AJJ, BBL, BBM, BMF, BUC, CCR, HWS, KGC, LBL, LCP, LWO, NBP, OCP, PCP, PMS, PNM, RRG, WVP. **COUNTIES:** BEL, BOO, BRE, BUL, CAM, HLN, JEF, LET, MCC, MEA, MEN, OLD, OWS, PER, POW, RUS, TRI. **DATES:** June 5–Sept. 6. **REMARKS:** Very common at times, and variable – especially in forewing pattern. A form with totally black hindwings is known only from three Kentucky specimens, and was described as form "*sargenti*" by Covell (1978). This form has been collected in Bullitt, Meade, and Oldham counties, and has not been reported anywhere else in the world.

8877 ***C. connubialis*** Gn.
Connubial Underwing
PARKS AND RESERVES: BBL, BMF, CCR, NBP, OCP, RRG. **COUNTIES:** BOO, BUL, HLN, JEF, MEA, MEN, POW, ROW. **DATES:** Apr. 17, June 12–Aug. 22. **REMARKS:** Fairly common but distribution is very local. Several forms represented.

8878 ***C. amica*** (Hbn.)
Girlfriend Underwing
PARKS AND RESERVES: BBL, BBM, BMF, BUC, HWS, KGC, LBL, LCP, OCP, PMS, PNM, RBF, RRG. **COUNTIES:** BEL, BOO, BRE, BUL, EST, FAY, HLN, JEF, LAU, MCC, MEA, OLD, OWS, PER, POW, ROW, RUS, TRI, WHI. **DATES:** May 5–Sept. 28. **REMARKS:** Common. Forewing pattern is quite variable.

8878.1 ***C. lineella*** Grt.
PARKS AND RESERVES: BBL, CER, LCP, RRG. **COUNTIES:** BOO, JEF, OWS, POW, ROW, RUS. **DATES:** July 2, 11–19, 28–Aug. 7. **REMARKS:** Common. Until recently thought to be a subspecies of *C. amica*, but it is now considered to be a distinct species (Gall 1990).

Subfamily **Plusiinae**

Genus ***Abrostola*** Ochs.

8880 ***A. ovalis*** Gn.
Menifee Co., RRG, July 19, Aug. 29; Powell Co., NBP, July 7, 9; all in light traps. Rare.

8881 ***A. urentis*** Gn.
Boone Co., BBL, Aug. 25; Bracken Co., Meldahl Dam, May 9, Aug. 9, Sept. 16. Reported from Kentucky by Eichlin and Cunningham (1978). Rare.

Genus ***Mouralia*** Wlk.

8884 ***M. tinctoides*** (Gn.)
McCracken Co., Paducah, Sept. 27, 1979, W. R. Black, Jr. It is a fresh specimen, and far from its known range of Florida to Texas. Rare.

Genus ***Argyrogramma*** Hbn.

8885 ***A. verruca*** (F.)
Golden Looper Moth
Jefferson Co., Sept. 21; Russell Co., LCP, Oct. 11. Rare.

Genus ***Enigmogramma*** Lafontaine & Poole

8886 ***E. basigera*** (Wlk.)
Pink-washed Looper Moth
PARKS AND RESERVES: LCP, OCP, PYR. **COUNTIES:** CHR, FUL, JEF, MEA, NEL, OLD, ROW, RUS. **DATES:** June 1, 4, 20, July 1, 16, Aug., Sept. 1, 10–12, Oct. 10. Common.

Genus ***Trichoplusia*** McD.

8887 ***T. ni*** (Hbn.)
Cabbage Looper Moth
PARKS AND RESERVES: ASC, KGC, KLK, LMP, RGR. **COUNTIES:** BRA, BRK, FAY, FUL, GRY, HLN, JEF, KNO, LAU, MAD, MCC, MRL, MUH, OWS, ROW. **DATES:** Apr. 29–Oct. 21, Dec. 1. **REMARKS:** Common, and a pest of cole crops and other vegetables.

Genus ***Ctenplusia*** Hbn.

8889 ***C. oxygramma*** (Gey.)
Sharp-stigma Looper Moth
PARKS AND RESERVES: BOW, CAC, LCP,

PNM, PYR. **COUNTIES:** BEL, BOO, BRK, CAR, CHR, FUL, HND, JEF, RUS, WOO. **DATES:** May 31, July 25–Sept. 1, 10–19, Oct. 1, 10–25. Common.

Genus ***Pseudoplusia*** McD.

8890 ***P. includens*** (Wlk.)
Soybean Looper Moth
PARKS AND RESERVES: BEC, BIP, BMF, BOW, BUC, CAC, CBB, GBO, GBT, LCP, PNM, SWR. **COUNTIES:** BEL, BOO, BUL, CAR, CRL, FAY, FUL, GAL, GUP, HND, JEF, LAU, OLD, PER, PIK, RUS. **DATES:** Mar. 27–Apr. 3, June 19, July 16, Sept. 1–19, 29–Oct. 18. Common.

Genus ***Rachiplusia*** Hamp.

8895 ***R. ou*** (Gn.)
Gray Looper Moth
PARKS AND RESERVES: BAN, BMF, BOW, BUC, JWS, LBL, OCP, RGR. **COUNTIES:** BAR, BOO, BRE, BRK, BUL, FAY, FLO, GRY, JEF, MEA, OWS, PER, ROW, TRI. **DATES:** May 16, June 7–July 1, 15, 28–Aug. 12, Oct, 13, 15. Common.

Genus ***Diachrysia*** Hbn.

8896 ***D. aeroides*** (Grt.)
Dark-spotted Looper Moth
Green Co., Crailhope, May 22; Harlan Co., BBM, July 13, 14. Rare.

8897 ***D. balluca*** Gey.
Harlan Co., BBM, July 13; Oldham Co., Sept. 29. Rare.

Genus ***Allagrapha*** Franc.

8898 ***A. aerea*** (Hbn.)
Unspotted Looper Moth
PARKS AND RESERVES: BBM, BMF, BOW, BUC, GBO, GBT, HWS, JCF, LCP, OCP, RRG, SWR. **COUNTIES:** BOO, BRA, BRE, BUL, CRL, FAY, GUP, HLN, HND, JEF, KNX, MCC, MEA, MEN, NEL, NIC, OLD, PER, ROW, RUS, WHI. **DATES:** Apr. 26–Oct. 18. **REMARKS:** Common. L. D. Gibson reported this species nectaring on *Gaura biennis* blossoms.

Genus ***Pseudeva*** Hamp.

8899 ***P. purpurigera*** (Wlk.)
Straight-lined Looper Moth
Menifee Co., RRG, FS 9B along Kelley Branch, July 8; Oldham Co., HWS, June, 14; Owsley Co., 4.8 km (3 mi) NE of Booneville, June 4, Woodford Co., 1.6 km (1 mi) SW of SR 68, along Kentucky River, May 25, June 8. Uncommon.

Genus ***Polychrysia*** Hbn.

8902 ***P. morigera*** (Hy. Edw.)
PARKS AND RESERVES: BBL, BCN, BMF, OCP. **COUNTIES:** BOO, BRA, BUL, GAL, MEA, MER, OLD, OWS, WOO. **DATES:** May 4, 20–25, June 2–8, 20–23 (Malaise trap), Sept. 27. **REMARKS:** Local and uncommon. A western species, known in the east only from Pennsylvania and Ohio before the first Kentucky specimens were taken by A. J. Brownell in 1976 at BMF. Covell and Medley (1986) reported it from Tennessee (two specimens trapped in flowers of southern lady's-slipper).

Genus ***Chrysanympha*** Grt.

8904 ***C. formosa*** (Grt.)
Formosa Looper Moth
Bell Co., PNM, June 8; Bullitt Co., BMF, May 22; Laurel Co., DBNF, FS 121 near Bolton Branch, June 17; Owsley Co., 4.8 km (3 mi) NE of Booneville, July 3. Rare.

Genus ***Eosphoropteryx*** Dyar

8905 ***E. thyatyroides*** (Gn.)
Pink-patched Looper Moth
PARKS AND RESERVES: BBL, BBM, PYR, RRG. **COUNTIES:** BOO, CHR, FAY, GAL, GUP, HLN, JEF, LEE, MEN, NEL, OWS. **DATES:** June 4–14, 22, 28 (found dead), July 8–24, Aug. 20–30, Sept. 19, Oct. 12. Uncommon.

Genus ***Megalographa*** Lafontaine & Poole

8907 ***M. biloba*** (Steph.)
Bilobed Looper Moth
PARKS AND RESERVES: BBL, BMF, BUC, CAC, JWS, LCP, OCP, PNM. **COUNTIES:** BEL, BOO, BRE, BUL, CAR, FAY, FLO, JEF, LEE, MEA, MEN, NEL, PER, ROW, RUS, WAR, WHI. **DATES:** Apr. 13–Sept. 28, Nov. 21. Common.

Genus ***Autographa*** Hbn.

8908 ***A. precationis*** (Gn.)
Common Looper Moth
PARKS AND RESERVES: BLA, BMF, BUC, CAC, CBB, FKN, GBT, HWS, JWS, LCP, NBP, PNM, PYR, RGR. **COUNTIES:** BEL, BRA, BRE, BRK, BUL, CAD, CAR, CRL, FAY, FLO, FUL, GAL, GRY, HLN, HND, HRD, JEF, MCC, MEA, NEL, OLD, OWS, PER, POW, ROW, RUS. **DATES:** Mar. 27–Oct. 23. Common.

8913 ***A. pseudogamma*** (Grt.)
Meade Co., OCP, Apr. 12, 1985, R. A. Henderson. Rare.

8923 *A. ampla* (Wlk.)
Large Looper Moth
Harlan Co., BBM, July 9, 12, 14, KGC, July 6. Rare.

Genus *Anagrapha* McD.

8924 *A. falcifera* (Kby.)
Celery Looper Moth
PARKS AND RESERVES: BAN, BBL, BEC, BMF, BOW, FKN, GBT, LCP, OCP, RGR. **COUNTIES:** BAR, BOO, BRA, BRK, BUL, CRL, FAY, GRY, JEF, KNX, LAU, MEA, ROW, RUS, SHE. **DATES:** Mar. 13–Nov. 1. Common.

Subfamily **Euteliinae**

Genus *Marathyssa* Wlk.

8955 *M. inficita* (Wlk.)
Dark Marathyssa
PARKS AND RESERVES: BAN, HWS, LBL, LCP, NBP, RBF. **COUNTIES:** BAR, BRE, GAL, HRT, JEF, MCC, MEA, OLD, OWS, POW, RUS, TRI, WHI. **DATES:** May 16–Sept. 23. Common.

8956 *M. basalis* Wlk.
Light Marathyssa
PARKS AND RESERVES: BBL, CAC, GBT, LCP, OCP, PNM, RRG. **COUNTIES:** BEL, BOO, CAR, CRL, FAY, JEF, MEA, MEN, RUS. **DATES:** Mar. 20, Apr. 9, 22–30, May 3–16, June 21. Common.

Genus *Paectes* Hbn.

8957 *P. oculatrix* (Gn.)
Eyed Paectes
PARKS AND RESERVES: BBL, BMF, BUC, CAC, CBB, LBL, LCP, LMP, OCP, PNM, PYR. **COUNTIES:** BEL, BOO, BRA, BRE, BUL, CAD, CAR, FAY, JEF, MCC, MEA, MEN, MUH, OLD, OWS, PER, RUS, TRI, WOO. **DATES:** Apr. 23–Aug. 22. Common.

8959 *P. pygmaea* Hbn.
Pygmy Paectes
PARKS AND RESERVES: BAN, BMF, JWS, LBK, LCP, LMP, OCP, PNM, PYR, RGR. **COUNTIES:** BAR, BEL, BRA, BRK, BUL, CAD, FLO, GAL, GRY, HLN, JEF, MCC, MEA, MUH, OWS, ROW, RUS, TRI, WHI, WOO. **DATES:** May 14, 22–Aug. 23. Common.

8959.1 *P. abrostolella* (Wlk.)
Bullitt Co., PCP, July 22; Hardin Co., Elizabethtown, June; Jefferson Co., Valley Station, July 4; Meade Co., Doe Run Inn area, July 21. This uncommon species was recently recognized as separate from *P. pygmaea* by Metzler and Franclemont (1991).

8962 *P. abrostoloides* (Gn.)
Large Paectes
PARKS AND RESERVES: BBL, BBM, BMF, CBF, GBT, HWS, JWS, KLK, LCP, OCP, PNM, PYR, RGR. **COUNTIES:** BEL, BOO, BRK, BUL, CAD, CHR, CRL, FAY, FLO, FUL, GRV, GRY, HLN, JEF, MCC, MEA, MRL, OLD, ROW, RUS, WHI. **DATES:** Apr. 15, May 15, 16, June 8–18, July 23–Oct. 31. Common.

Genus *Eutelia* Hbn.

8968 *E. pulcherrima* (Grt.)
Beautiful Eutelia
PARKS AND RESERVES: BBL, BOW, COB. **COUNTIES:** BOO, GAL, HIC, JEF, LAR, MCC. **DATES:** May 5–June 8, July 5, 25. Moderately common.

Subfamily **Sarrothripinae**

Genus *Baileya* Grt.

8969 *B. doubledayi* (Gn.)
Doubleday's Baileya
Laurel Co., DBNF (2 sites), May 17; Menifee Co., RRG, July 8; both L. D. Gibson. Uncommon.

8970 *B. ophthalmica* (Gn.)
Eyed Baileya
PARKS AND RESERVES: BAN, BIP, BMF, CAC, CBB, CBF, JWS, KRC, LBL, LCP, LMP, OCP, PNM, RRG, WVP. **COUNTIES:** BAR, BEL, BOO, BUL, CAR, FLO, JEF, MCC, MEA, MEN, MUH, OLD, OWS, PIK, POW, RUS, TRI, WHI, WOL. **DATES:** Apr. 12–Aug. 21. Common.

8971 *B. dormitans* (Gn.)
Sleeping Baileya
PARKS AND RESERVES: BBM, LCP, LMP, NBP, PMS, PNM, RRG. **COUNTIES:** BEL, GAL, HLN, JEF, MEN, MUH, OWS, POW, RUS. **DATES:** Apr. 19, 21, May 3, 6, 21–June 2, 25–30, July 8, Aug. 11, 14, Sept. 8. Common.

8972 *B. levitans* (Smith)
Pale Baileya
PARKS AND RESERVES: BBL, BIP, BMF, BUC, CAC, HWS, KRC, LCP, LWO, NBP, OCP, PNM, RRG. **COUNTIES:** BEL, BOO, BRE, BUL, CAR, EST, JEF, LET, MEA, MEN, OLD, OWS, PER, PIK, POW, RUS, WOL. **DATES:** Mar. 20–Aug. 21. Common.

8973 *B. australis* (Grote)
Small Baileya
PARKS AND RESERVES: BBL, BMF, BUC, HWS, LCP, LMP, PYR. **COUNTIES:** BOO, BRE, BUL, CAD, FAY, GAL, GUP, JEF, MCC, MEA, MEN, MUH, NEL, OLD, PER, RUS. **DATES:** Apr. 6–Sept. 23. Common.

Genus ***Characoma*** Wlk.

8974 ***C. nilotica*** (Rogenhofer)
Boone Co., BBL, May 29, June 28–July 11, Aug. 3; Jefferson Co., Valley Station, July 11; Owsley Co., 4.8 km (3 mi) NE of Booneville, July 5. Rare.

Genus ***Nycteola*** Hbn.

8975 ***N. frigidana*** (Wlk.)
Frigid Owlet
PARKS AND RESERVES: BMF, HWS, RRG. **COUNTIES:** BUL, FAY, JEF, MEN, OLD, WHI. **DATES:** Apr. 26, May 10–30, Aug. 9–19, 30, Sept. 6, 10. Common.

Subfamily **Nolinae**

Genus ***Meganola*** Dyar

8983 ***M. minuscula*** (Zell.)
Confused Meganola
PARKS AND RESERVES: BMF, CAC, CBB, CBF, HWS, KGC, KRC, LBL, LCP, NBP, OCP, PNM, RRG. **COUNTIES:** BEL, BUL, CAR, HLN, JEF, LAU, MCC, MEA, OLD, OWS, POW, RUS, TRI, WHI, WOL, WOO. **DATES:** Apr. 12–Sept. 8. Common.

8983.1 ***M. phylla*** (Dyar)
Carroll Co., GBT, May 6; Meade Co., OCP, Aug. 10; both L. D. Gibson. It is not clear if this is a valid species or merely a color form of the confused meganola.

8983.2 ***M. spodia*** Franc.
A specimen from Green Co., Crailhope, collected by C. Cook, was recorded in the original description (Franclemont 1985). Bell Co., PNM, May 12, July 18, 19; Carter Co., CAC, Apr. 23; Owsley Co., 4.8 km (3 mi) NE of Booneville, June 16; Russell Co., LCP, Aug. 14, 16. Uncommon.

Genus ***Nola*** Leach

8989 ***N. pustulata*** (Wlk.)
Sharp-blotched Nola
Edmonson Co., Park City, Park Mammoth Resort, May 16; Jefferson Co., Valley Station, May 7; Muhlenberg Co., LMP, May 13, 14. Rare.

8990 ***N. cilicoides*** (Grt.)
Meade Co., OCP, June 13; Oldham Co., HWS, June 22; Owsley Co., 4.8 km (3 mi) NE of Booneville, June 14, July 14; Rowan Co., Morehead, June 20, 28. Uncommon.

8991 ***N. cereella*** (Bosc)
Sorghum Webworm Moth
PARKS AND RESERVES: BMF, CBF, LBL. **COUNTIES:** BUL, FAY, FUL, JEF, MCC, TRI, WHI. **DATES:** Apr. 26, June 19, July 15, Aug. 6, 17–25, Sept. 3–19, Oct. 6–13. **REMARKS:** Common. This species was known as *N. sorghiella* Riley until the above name was resurrected by Miller and Becker (1989).

8992 ***N. triquetrana*** (Fitch)
Three-spotted Nola
PARKS AND RESERVES: BIP, CBF, NBP, OCP. **COUNTIES:** BUL, LAU, MEA, OWS, PIK, POW, WHI. **DATES:** Mar. 25, Apr. 16, 18, 30, May 28, June 7, Sept. 8, 19. **REMARKS:** Uncommon, but it may be locally common.

Subfamily **Acontiinae**

Genus ***Tripuda*** Grt.

9003 ***T. quadrifera*** (Zell.)
Boone Co., BBL, May 31, Aug. 25; Christian Co., PYR, Aug. 22; Owsley Co., 4.8 km (3 mi) NE of Booneville, June 10, Aug. 22; all L. D. Gibson. These specimens may represent an undescribed species closely related to *T. quadrifera.*

Genus ***Oruza*** Wlk.

9025 ***O. albocostaliata*** (Pack.)
The White-edge
PARKS AND RESERVES: ASC, BBL, BMF, BUC, HWS, LBK, LBL, LCP, NBP, OCP, PNM, RRG. **COUNTIES:** BEL, BOO, BRE, BUL, GRY, JEF, MEA, NEL, OLD, OWS, PER, POW, RUS, TRI. **DATES:** Apr. 25–Sept. 12. Common.

Genus ***Ozarba*** Wlk.

9030 ***O. aeria*** (Grt.)
Aerial Brown
PARKS AND RESERVES: BMF, CBB, OCP, RRG. **COUNTIES:** BUL, FUL, JEF, LIV, MEA, MEN, OLD. **DATES:** May 14, 30, July 6, Aug. 10, 17, Sept. 1–8. Local.

Genus ***Hyperstrotia*** Hamp.

9037 ***H. pervertens*** (B. & McD.)
Dotted Graylet
Jefferson Co., Valley Station, May 28, June 13, July 9, 11; Oldham Co., Goshen, June 14, July 29, Aug. 13; Trigg Co., LBL, July 14. Uncommon.

9038 ***H. villificans*** (B. & McD.)
White-lined Graylet
PARKS AND RESERVES: BMF, CBF, LBK, LCP, PNM. **COUNTIES:** BEL, BUL, JEF, NEL, OLD, RUS, TRI, WHI. **DATES:** Apr. 19–30, May 16–25, June 1–23, July, Aug. 2–19, Sept. 1. Common.

9039 ***H. flaviguttata*** (Grt.)
Yellow-spotted Graylet
Christian Co., PYR, in light trap, Aug. 22, 1986, L. D. Gibson. Rare.

9040 ***H. secta*** (Grt.)
Black-patched Graylet
PARKS AND RESERVES: BBL, BMF, CAC, HWS, LMP, PNM, RRG. **COUNTIES:** BEL, BOO, BUL, CAR, JEF, MEN, MUH, OLD, OWS, POW. **DATES:** Apr. 23, May 10–June 13, 26, 27, July 8–19, Aug. 22, 25, Sept. 8. Common.

Genus ***Thioptera*** Franc.

9044 ***T. nigrofimbria*** (Gn.)
Black-bordered Lemon Moth
PARKS AND RESERVES: BEC, BMF, CAH, GBT, HWS, LBK, LBL, LCP, OCP, PMS, RGR, SWR, WKM. **COUNTIES:** BRK, BUL, CRL, FAY, FUL, GAL, GRY, HLN, HND, HOP, HRT, JEF, KNX, MCC, MCY, MEA, NEL, OLD, OWS, ROW, RUS, SPE, TRI. **DATES:** Apr. 26–Sept. 22, Oct. 2, 9, Nov. 14. Common.

Genus ***Lithacodia*** Hbn.

9047 ***L. muscosula*** (Gn.)
Large Mossy Lithacodia
PARKS AND RESERVES: BBL, BMF, CAH, CBB, HWS, LCP, LMP, OCP, PNM, PYR, RGR. **COUNTIES:** BEL, BOO, BRK, BUL, CAD, FAY, FUL, GAL, GRY, JEF, MEA, MEN, MUH, OLD, RUS, WHI, WOO. **DATES:** Apr. 25–Sept. 5. Common.

9050 ***L. concinnimacula*** (Gn.)
Red-spotted Lithacodia
Meade Co., OCP, 2 specimens, May 3, 1980; Powell Co., RRG, Tunnel Ridge, May 3, 1991; L. D. Gibson. Rare.

9051 ***L. musta*** (G. & R.)
Small Mossy Lithacodia
PARKS AND RESERVES: BBL, BEC, BMF, BUC, CER, HWS, KGC, LBK, LBL, LCP, OCP, PNM, PYR, RGR, RRG. **COUNTIES:** BEL, BOO, BRE, BRK, BUL, CHR, FAY, GAL, GRY, HLN, JEF, MEA, MEN, OLD, OWS, PER, RUS, TRI, WOO. **DATES:** May 14–Sept. 8. Common.

9054 ***L. indeterminata*** B. & McD.
Indeterminate Lithacodia
Fulton Co., SR 94, 4.0 km (2.5 mi) E Cayce, Sept. 8, 1990, C. V. Covell, Jr. Rare.

Genus ***Pseudeustrotia*** Warr.

9053 ***P. carneola*** (Gn.)
Pink-barred Lithacodia
PARKS AND RESERVES: AJJ, BEC, BMF, BUC, CBB, GBT, HWS, KRC, LBL, LMP, PNM, PYR. **COUNTIES:** BEL, BRA, BRE, BUL, CAD, CAM, CRL, FAY, GRV, HRT, JEF, MUH, OLD, OWS, PER, ROW, TRI, WOL. **DATES:** May 6–Sept. 28. **REMARKS:** Common. This species was transferred from *Lithacodia* by Poole (1989).

Genus ***Maliattha*** Wlk.

9055.1 ***M. synochitis*** (G. & R.)
Black-dotted Lithacodia
PARKS AND RESERVES: BBL, BMF, HWS, PMS, WVP. **COUNTIES:** BOO, BRA, BUL, HLN, JEF, OLD. **DATES:** May 6, 11, 23–June 1, 10–22, July 8. **REMARKS:** Formerly in *Lithacodia* (Hodges et al. 1983, number 9049).

Genus ***Anterastria*** Sugi

9055.3 ***A. teratophora*** (H.-S.)
The Gray Marvel
PARKS AND RESERVES: BBL, HWS, LBK, LBL, LMP, PYR. **COUNTIES:** BOO, BUL, CAD, FAY, JEF, MCC, MUH, OLD, OWS, TRI. **DATES:** May 15, June, July 8, 28, Sept. 27. **REMARKS:** Formerly in *Agriopodes* (Hodges et al. 1983, number 9243).

Genus ***Homophoberia*** Morr.

9056 ***H. cristata*** Morr.
Water-lily Moth
Caldwell Co., PYR, June 12, 1973, C. Cornett. Rare.

9057 ***H. apicosa*** (Haw.)
Black Wedge-spot
PARKS AND RESERVES: BBL, BMF, LBL, LMP, OCP, PYR. **COUNTIES:** BOO, BUL, CAD, FAY, FUL, GRV, HRT, JEF, MCC, MEA, MUH, NEL, OWS, ROW, TRI. **DATES:** May 6–18, 31, June 12, 25–Sept. 5–8, 22. Common.

Genus ***Argillophora*** Grt.

9060 ***A. furcilla*** Grt.
Fulton Co., SR 94, 4.0 km (2.5 mi) E of Cayce, Sept. 9; Jefferson Co., May 5, June 20, Sept. 12; Letcher Co., LWO, July 11; Menifee Co., RRG, May 26, June 21, 28, July 13. Uncommon.

Genus ***Cerma*** Hbn.

9061 ***C. cora*** Hbn.
Jefferson Co., WVP, May 9, June 4–12, July 8–12, 20, 30; Nelson Co., Nelsonville, May 18. A rare and very local moth in Kentucky.

9062 ***C. cerintha*** (Tr.)
Tufted Bird-dropping Moth
PARKS AND RESERVES: BAN, BBL, BBM, BMF, HWS, LBK, PYR, RBF, RRG. **COUN-**

TIES: BAR, BOO, BRA, BRE, BUL, CAD, FAY, HLN, JEF, MCC, MEN, NEL, OLD, OWS, ROW, TRI. **DATES:** Apr. 2–Sept. 7. Locally common.

Genus ***Leuconycta*** Hamp.

9065 ***L. diphteroides*** (Gn.)
Green Leuconycta
PARKS AND RESERVES: AJJ, BAN, BBL, BCN, BMF, BUC, CMI, HWS, LBL, PNM, PYR, RBF, RGR, WVP. **COUNTIES:** BAR, BEL, BOO, BRE, BRK, BUL, CAD, CAM, EST, FAY, GRY, JEF, MCC, MEA, MEN, OLD, OWS, PER, ROW, TRI, WHI. **DATES:** Apr. 5–Nov. 18. Common.

9066 ***L. lepidula*** (Grt.)
PARKS AND RESERVES: BBL, BMF, HWS, KGC, PMS, PNM, RRG. **COUNTIES:** BEL, BOO, BUL, GAL, HLN, MEN, OLD, OWS, ROW. **DATES:** Apr. 12, May 6, 19, June 7–10, 20, July 7, 13, 24, 26, Aug. 1, 23. Moderately and locally common.

Genus ***Amyna*** Gn.

9069 ***A. bullula*** (Grt.)
Hook-tipped Amyna
Henderson Co., SWR, Sept. 4, 11; Jefferson Co., Valley Station, Aug. 22; McCracken Co., Paducah, Oct. 4; Menifee Co., RRG, FS 9A, Sept. 21. Rare.

9070 ***A. octo*** (Gn.)
Eight-spot
PARKS AND RESERVES: GBT, HWS, LCP. **COUNTIES:** CRL, FUL, JEF, MCC, MRL, OLD, RUS. **DATES:** June 9, Aug. 6, 12, 22, Sept. 10–28, Oct. 13. Uncommon.

Genus ***Tarachidia*** Hamp.

9085 ***T. semiflava*** (Gn.)
The Half-yellow
Laurel Co., FS 121 near Bolton Bridge, June 11; McCreary Co., Yahoo Ridge, June 14. Uncommon.

9090 ***T. candefacta*** (Hbn.)
Olive-shaded Bird-dropping Moth
PARKS AND RESERVES: BBL, BMF, CBB, CER, HWS, LBL, RGR. **COUNTIES:** BOO, BRA, BRK, BUL, FAY, GRY, HRD, HRT, JEF, MCC, OLD, OWS, ROW, TRI, WOO. **DATES:** Apr. 1, May 4, 17–28, June 1, 3, 14–24, July 1–Aug. 2–25, Sept. 6–17. Common.

9095 ***T. erastrioides*** (Gn.)
Small Bird-dropping Moth
PARKS AND RESERVES: BAN, BBL, HWS, LBK, LBL, LMP, NBP. **COUNTIES:** BAR, BOO, FAY, HRT, JEF, MUH, OLD, OWS, POW, ROW, TRI. **DATES:** Apr. 30, May 7, 10, 19–Sept. 1, 19–24. Common.

Genus ***Spragueia*** Grt.

9122 ***S. dama*** (Gn.)
Southern Spragueia
PARKS AND RESERVES: LBK, PYR. **COUNTIES:** CHR, FUL, HOP, HRT, MCC, MRL, TRI. **DATES:** June 17, 18, 26, Aug. 22, 23, Sept. 4–13. **REMARKS:** Apparently restricted to western counties. Locally common.

9127 ***S. leo*** (Gn.)
Common Spragueia
PARKS AND RESERVES: BBL, BMF, CBB, HWS, LBL, LCP, PNM, PYR, SWR. **COUNTIES:** BEL, BOO, BRA, BUL, CHR, FAY, FUL, HND, HRT, JEF, LAU, MCC, MEA, MRL, OLD, OWS, ROW, RUS, TRI, WHI. **DATES:** May 27–Sept. 23. Common.

9131 ***S. apicalis*** (H.-S.)
Yellow Spragueia
PARKS AND RESERVES: BBL, HWS, LCP, SWR. **COUNTIES:** BOO, BRA, BUL, FAY, FUL, GAL, GRY, HND, HRD, JEF, LIV, MCC, OLD, OWS, ROW, RUS, WAR, WOO. **DATES:** June 2, 14, 15, 27, July 19, 22–Sept. 7. **REMARKS:** Our subspecies is *apicella* (Grt.). Locally common.

Genus ***Acontia*** Ochs.

9136 ***A. aprica*** (Hbn.)
Exposed Bird-dropping Moth
PARKS AND RESERVES: AJJ, BBL, CER, HWS, LBL, RNR, SWR. **COUNTIES:** BOO, BUL, CAD, CAM, FAY, FUL, GRV, HND, JEF, MCC, NEL, OLD, ROC, ROW, TRI, WAR. **DATES:** Apr. 10, May 9–June 6, 14–Sept. 1, 10, 13. Locally common.

9137 ***A. lactipennis*** (Harv.)
Jefferson Co., July 22, 1964, J. R. Merritt. Rare.

9138 ***A. abdominalis*** (Grt.)
Jefferson Co., July 22, 1964, J. R. Merritt. Rare.

9146 ***A. delecta*** Wlk.
Delightful Bird-dropping Moth
PARKS AND RESERVES: BLD, RNR, SWR. **COUNTIES:** BAL, FUL, GRV, HND, JEF, TRI. **DATES:** May 27, 28, June 7, July 15, Sept. 10, 13. **REMARKS:** Common in western counties.

Genus ***Bagisara*** Wlk.

9169 ***B. rectifascia*** (Grt.)
Bell Co., PNM, July 18, 1975, C. V. Covell, Jr. Rare.

9175 ***B. gulnare*** (Stkr.)
Gallatin Co., Markland Dam, 1 female, June 24, 1989, L. D. Gibson. Rare.

Subfamily **Pantheinae**

Genus ***Panthea*** Hbn.

9177 ***P. acronyctoides*** (Wlk.)
Black Zigzag
Harlan Co., PMS, July 8, 13, and PNM, May 12; Menifee Co., RRG, May 12, June 29, July 13, 20, 21; Powell Co., NBP, July 13, and RRG, July 23; Rowan Co., DBNF, W side of SR 1274, 5.3 km (3.3 mi) E of SR 519, July 23. Uncommon.

9182 ***P. furcilla*** (Pack.)
Eastern Panthea
PARKS AND RESERVES: BMF, CCR, NBP, OCP, PNM, RRG. **COUNTIES:** BEL, BUL, JEF, MEA, MEN, MOR, OWS, POW, ROW, WHI. **DATES:** Mar. 20, May 3–16, 25, June 5–12, 23, July 21, 23, Aug. 4–Sept. 4, 27. Uncommon.

Genus ***Colocasia*** Ochs.

9184 ***C. flavicornis*** (Sm.)
Yellowhorn
PARKS AND RESERVES: BBM, BIP, BMF, CAC, CBB, HWS, LCP, NBP, OCP, PMS, PNM, RRG. **COUNTIES:** BEL, BUL, CAR, HLN, JEF, MEA, MEN, OLD, PIK, POW, ROW, RUS. **DATES:** Mar. 19–Sept. 11. Common.

9185 ***C. propinquilinea*** (Grt.)
Closebanded Yellowhorn
PARKS AND RESERVES: BBM, BIP, BMF, CAC, CAH, CBF, HWS, LCP, LMP, NBP, OCP, PNM, RBF, RGR, RRG. **COUNTIES:** BEL, BRE, BRK, BUL, CAR, EST, GRY, HLN, JEF, MEA, MEN, MUH, NEL, OLD, OWS, PIK, POW, ROW, RUS, WHI. **DATES:** Mar. 20–Sept. 2, 8. Common.

Genus ***Charadra*** Wlk.

9189 ***C. deridens*** (Gn.)
The Laugher
PARKS AND RESERVES: BAN, BBL, BIP, BMF, BUC, CAC, CBF, HWS, JWS, LBL, LCP, LWO, NBP, OCP, PNM, RGR, RRG. **COUNTIES:** BAR, BEL, BOO, BRE, BRK, BUL, CAR, FAY, FLO, GAL, GRY, JEF, LET, MEA, MEN, NEL, OLD, OWS, PER, PIK, POW, RUS, TRI, WHI. **DATES:** Mar. 3–Sept. 12. Very common.

Genus ***Raphia*** Hbn.

9193 ***R. frater*** Grt.
The Brother
PARKS AND RESERVES: LCP, OCP, RNR. **COUNTIES:** FAY, FUL, GAL, JEF, MCC, MEA, OLD, RUS, WOO. **DATES:** Apr. 20–May 2, 12, 20, 29, June 2, 15–29, July 5, 21–27, Aug. 5–22. Moderately common.

Subfamily **Acronictinae**

Genus ***Acronicta*** Ochs.

9199 ***A. rubricoma*** Gn.
Ruddy Dagger Moth
PARKS AND RESERVES: AJJ, BAN, BBL, BMF, HWS, LBL, NBP, OCP, PNM, RGR. **COUNTIES:** BAR, BEL, BOO, BRA, BRK, BUL, CAM, FAY, GRY, JEF, MEA, MEN, OLD, POW, TRI. **DATES:** Apr. 27–Aug. 18. Common.

9200 ***A. americana*** (Harr.)
American Dagger Moth
PARKS AND RESERVES: BAN, BBM, BMF, BUC, CBB, HWS, JWS, KRC, LBL, LCP, LWO, MCV, NBP, OCP, PMS, PNM, RGR, RRG. **COUNTIES:** BAR, BEL, BRE, BRK, BUL, EDM, EST, FAY, FLO, FUL, GAL, GRY, HLN, JEF, LAR, LET, MCC, MEA, MEN, NEL, OLD, OWS, PER, POW, ROW, RUS, TRI, WOL, WOO. **DATES:** Apr. 6–Oct. 18. Common.

9203 ***A. dactylina*** Grt.
Fingered Dagger Moth
Rowan Co., Morehead, June 20, 30, July 1–8, 25; Whitley Co., Goldbug, Aug. 2. Uncommon.

9205 ***A. lepusculina*** Gn.
Cottonwood Dagger Moth
Gallatin Co., Markland Dam, June 21; Jefferson Co., Valley Station, Aug. 17; Kenton Co., Latonia, reared on cottonwood, Aug. 11; Meade Co., OCP, Apr. 13, 15, June 12, Aug. 10, 17. Uncommon.

9207 ***A. innotata*** Gn.
Unmarked Dagger Moth
PARKS AND RESERVES: BBM, NBP, PNM, RRG. **COUNTIES:** BEL, HLN, LAU, MEN, OWS, POW. **DATES:** May 21, July 8–24, Aug. 26. Uncommon.

9208 ***A. betulae*** Riley
Birch Dagger Moth
Hopkins Co., SR 502, 16.1 km (10 mi) W of Madisonville, in black-light trap, June 26; Jefferson Co., Louisville, May 11; Perry Co., BUC, May 2; Rowan Co., Morehead, Apr. 17. Uncommon.

9209 ***A. radcliffei*** (Harv.)
Radcliffe's Dagger Moth
PARKS AND RESERVES: AJJ, BBL, BBM, HNK, LCP, NBP, OCP, PMS. **COUNTIES:** BOO, CAL, CAM, HLN, JEF, MEA, MEN, POW, RUS. **DATES:** Apr. 16–22, May 4, 12–24, June 9, 16, July 8, 14, Aug. 1, 6, 15, 16. Locally common.

9211 ***A. tritona*** (Hbn.)
Laurel Co., FS 121, SW of London, July 24, and FS 4176 at ridge top, SW of Ned Branch, June 6; Powell Co., Tunnel Ridge, RRG, July 13. Rare.

9216 ***A. albarufa*** Grt.
Hardin Co., Elizabethtown, Aug. 1992; Jefferson Co., Mar. 30, 1974. Rare.

9219 ***A. connecta*** Grt.
Connected Dagger Moth
PARKS AND RESERVES: FKN, HWS. **COUNTIES:** FAY, FUL, MCC, MEA, JEF, OLD. **DATES:** Apr. 10, 18–24, May 25, 29, June 14, July 16–23, 31, Aug. 2, Sept. 5–13. Locally common.

9221 ***A. funeralis*** G. & R.
Funerary Dagger Moth
PARKS AND RESERVES: AJJ, BAN, BBL, BBM, BMF, HWS, KGC, LCP, LMP, OCP, PNM, PYR, RRG. **COUNTIES:** BAR, BEL, BOO, BUL, CAM, CHR, HLN, JEF, MEA, MUH, OLD, OWS, POW, RUS. **DATES:** Apr. 20–June 27, July 6, 9, 20–28, Aug. 1–21, Sept. 4. **REMARKS:** Common in both normal and melanic forms.

9225 ***A. vinnula*** (Grt.)
Delightful Dagger Moth
PARKS AND RESERVES: BMF, BUC, CAC, HWS, LBL, LCP, PNM, RRG, WVP. **COUNTIES:** BEL, BOO, BRA, BRE, BUL, CAR, FAY, HOP, JEF, MCC, MEA, MEN, OLD, OWS, PER, RUS, TRI, WAR. **DATES:** Apr. 12–Sept. 18. Locally common.

9226 ***A. superans*** Gn.
Splendid Dagger Moth
Harlan Co., BBM, July 14; Menifee Co., RRG, July 8, 10, 21; Powell Co., RRG, Tunnel Ridge, July 10. Rare.

9227 ***A. laetifica*** Smith
Pleasant Dagger Moth
PARKS AND RESERVES: BBL, BMF, HWS, LCP, OCP, PNM, RGR. **COUNTIES:** BEL, BOO, BRK, BUL, GRY, JEF, MEA, MEN, OLD, POW, RUS, WAR, WOO. **DATES:** Apr. 12–Sept. 18. Locally common.

9229 ***A. hasta*** Gn.
Speared Dagger Moth
PARKS AND RESERVES: BAN, BBL, BBM, BMF, CAC, GBT, HWS, KGC, LCP, LWO, OCP, PNM, PYR. **COUNTIES:** BAR, BEL, BOO, BUL, CAD, CAR, CRL, FAY, HLN, JEF, LET, MCC, MEA, OLD, OWS, ROW, RUS. **DATES:** Apr. 17–Sept. 17. Common.

9235 ***A. spinigera*** Gn.
Nondescript Dagger Moth
PARKS AND RESERVES: CAC, CAV, CBB, HWS, OCP, RBF, RRG. **COUNTIES:** BAT, BRE, CAR, FAY, JEF, MEA, OLD, POW, PUL. **DATES:** Apr. 18, 24, May 21, June 15, 16, July, Aug. 10, 21, 26. **REMARKS:** Locally common. At Shopville, Pulaski Co., a larva was found on apple foliage on Sept. 10. It pupated Sept. 13, and emerged as an adult on Nov. 19.

9236 ***A. morula*** G. & R.
Ochre Dagger Moth
PARKS AND RESERVES: BBL, BBM, BUC, LCP, LMP, NBP, OCP, PNM. **COUNTIES:** BEL, BOO, BRE, BUL, GAL, HLN, JEF, MEA, MEN, MUH, PER, POW, RUS, WHI, WOL. **DATES:** Apr. 23–30, May 11, 23, 28–June 9, 20, July 9, Aug. 5–22, Sept. 23. Locally common.

9237 ***A. interrupta*** Gn.
Interrupted Dagger Moth
PARKS AND RESERVES: BAN, BBL, BMF, BUC, CAC, CBF, HWS, JWS, LCP, NBP, OCP, PNM, PYR, RGR, RRG. **COUNTIES:** BAR, BEL, BOO, BRE, BRK, BUL, CAD, CAR, FAY, FLO, GAL, GRY, JEF, MEA, MEN, OLD, OWS, PER, POW, PUL, ROW, RUS, WHI. **DATES:** Apr. 11–June 5, 14–July 7, 23, Aug. 2–22, Sept. 2, 4, 12, 17. Common.

9238 ***A. lobeliae*** Gn.
Lobelia Dagger Moth
PARKS AND RESERVES: AJJ, BAN, BBL, BBM, BUC, CAC, JWS, KGC, LBL, LCP, LWO, NBP, OCP, PNM, RGR, RRG, SWR. **COUNTIES:** BAR, BEL, BOO, BRE, BRK, CAM, CAR, FAY, FLO, GAL, GRY, HLN, HND, JEF, LET, MEA, MRL, NEL, PER, POW, ROW, RUS, TRI. **DATES:** Apr. 11–Sept. 10. Common.

9241 ***A. fragilis*** (Gn.)
Fragile Dagger Moth
PARKS AND RESERVES: BBM, BIP, HWS, KGC, NBP, PNM, RRG. **COUNTIES:** BEL, HLN, MEA, MEN, OLD, PIK, POW, WOO. **DATES:** Apr. 26, 30, May 3, 11, 12, 22, June 2, 9, July 8–14, 23, Aug. 8, 14, 22. Locally common.

9241.1 ***A. heitzmani*** Covell & Metzler
Heitzman's Dagger Moth
Paratypes of this recently described species (Covell and Metzler 1992) are as follows: Boone Co., BBL, May 2 and 19, 1979, May 5 and 18, 1980, all L. D. Gibson; Jefferson Co., Valley Station, May 3, 1987, B. S. Nichols; Meade Co., FKN, Aug. 9, 1975, S. Scholz, and OCP, May 10, 1987, B. S. Nichols; Oldham Co., HWS, Aug. 14, 1967, C. V.

Covell, Jr.; Powell Co., RRG, Tunnel Ridge, May 14, 1988 and May 3, 1991, L. D. Gibson. Locally common.

9242 ***A. exilis*** Grt.
Exiled Dagger Moth
PARKS AND RESERVES: BAN, CAC, CBB, HWS, KLK, LBL, LCP, OCP, PNM. **COUNTIES:** BAR, BEL, BUL, CAR, EST, FAY, JEF, MCC, MEA, MRL, OLD, RUS, TRI. **DATES:** Apr. 21, May 13–15, 28–June 14, July 3–28, Aug. 6–16, Sept. 5. Common.

9243 ***A. ovata*** Grt.
Ovate Dagger Moth
PARKS AND RESERVES: BBM, KGC, KRC, MCV, NBP, PMS, RBF, RRG. **COUNTIES:** BRE, EDM, HLN, JEF, MCC, MEN, OWS, POW, ROW, WHI, WOL. **DATES:** Apr. 22, May 19, June 25, July, Aug. 3, 9, 26. **REMARKS:** Common. See Rings et al. (1992) for comments on this and the following species.

9244 ***A. modica*** Wlk.
Medium Dagger Moth
PARKS AND RESERVES: AJJ, BAN, BBM, BMF, HWS, JWS, LCP, NBP, OCP, PNM, PYR, RGR, RRG. **COUNTIES:** BAR, BEL, BRK, BUL, CAM, CHR, FAY, FLO, GRY, HLN, MEA, OLD, POW, ROW, RUS. **DATES:** May 5, 12, 28–July 14, Aug. 8, 9, 17, 28, 31. Common.

9245 ***A. haesitata*** (Grt.)
Hesitant Dagger Moth
PARKS AND RESERVES: BAN, BBM, BMF, BUC, CBF, HWS, JWS, KRC, LBL, LCP, LWO, NBP, OCP, PNM, PYR, RBF, RGR. **COUNTIES:** BAR, BEL, BRE, BRK, BUL, CAD, FAY, FLO, GRY, HLN, JEF, LET, MCC, MEA, MEN, NEL, OLD, PER, POW, RUS, TRI, WHI, WOL. **DATES:** Apr. 10–Oct. 13. Common.

9246 ***A. clarescens*** Gn.
Bracken Co., 0.8 km (0.5 mi) W of Foster, May 14; Grayson Co., RGR, Aug. 9; Jefferson Co., WVP, melanic specimen, June 30; Russell Co., LCP, May 28. Rare. Includes what were identified as *A. pruni* Harr., a synonym.

9247 ***A. tristis*** Sm.
Caldwell Co., PYR, June 5; Jefferson Co., CCR, Aug. 25, and Valley Station, June 3, 12–18, Aug. 4, 15, 29; Meade Co., OCP, June 2, Aug. 10; Menifee Co., RRG, June 8, 15, Aug. 9. Uncommon.

9248 ***A. hamamelis*** Gn.
Bell Co., PNM, Apr. 30; Estill Co., Irvine, July 27; Harlan Co., PMS, July 8; Rowan Co., Morehead, July 4, 6. Uncommon.

9249 ***A. increta*** Morr.
PARKS AND RESERVES: BBL, BBM, BMF, HWS, KGC, KRC, LBK, LBL, LCP, LWO, NBP, OCP, PNM, RBF, RGR, RRG, WVP. **COUNTIES:** BEL, BOO, BRE, BRK, BUL, GRY, HLN, JEF, LET, MCC, MEA, MEN, OLD, POW, ROW, RUS, TRI, WOL. **DATES:** May 1–Sept. 1. **REMARKS:** Common. These records include those for *A. inclara* Smith, which is a synonym.

9251 ***A. retardata*** (Wlk.)
Retarded Dagger Moth
PARKS AND RESERVES: BAN, BBM, BMF, CBF, HWS, KGC, LBL, LCP, NBP, OCP, PNM, PYR, RGR, RRG. **COUNTIES:** BAR, BEL, BRK, BUL, CAD, FAY, GRY, HLN, JEF, MCC, MEA, MEN, OLD, POW, RUS, TRI, WHI. **DATES:** Apr. 17–30, May 9–31, June 10–14, July 9–17, 24–Aug. 19. **REMARKS:** Locally common. These records include those for *A. caesarea* Sm., a synonym.

9254 ***A. afflicta*** Grt.
Afflicted Dagger Moth
PARKS AND RESERVES: BAN, BIP, BMF, BUC, CAV, LBK, LCP, LMP, NBP, OCP, PNM, RGR, RRG. **COUNTIES:** BAR, BAT, BEL, BRE, BRK, BUL, FAY, GAL, GRV, GRY, HLN, JEF, MCC, MEA, MEN, MUH, OLD, OWS, PER, PIK, POW, ROW, RUS, TRI, WAR. **DATES:** Apr. 4–Sept. 26. Common.

9257 ***A. impleta*** Wlk.
Yellow-haired Dagger Moth
PARKS AND RESERVES: BAN, BBL, BIP, CBF, HWS, KGC, LCP, NBP, OCP, PNM, RGR, RRG. **COUNTIES:** BAR, BEL, BOO, BRA, FAY, GRY, HLN, JEF, MCC, MEA, MEN, OLD, OWS, PIK, POW, ROW, RUS, WHI. **DATES:** Apr. 1, 17–Aug. 30. Common.

9259 ***A. noctivaga*** Grt.
Night-wandering Dagger Moth
PARKS AND RESERVES: BBL, RRG. **COUNTIES:** BOO, BUL, FAY, JAC, JEF, LAU, OWS, POW. **DATES:** Apr. 9, 12, 21–25, May 2, 4, 29, June 14, July 15–23. Uncommon.

9261 ***A. impressa*** Wlk.
Boone Co., BBL, July 2, 1979, L. D. Gibson. Rare.

9264 ***A. longa*** Gn.
Long-winged Dagger Moth
PARKS AND RESERVES: BAN, BBL, BMF, HWS. **COUNTIES:** BAR, BOO, BUL, FAY, JEF, MOR, OLD, OWS. **DATES:** Apr. 13, 22, May 11–22, July 13, Aug. 7–15, Sept. 3. Common.

9266 ***A. lithospila*** Grt.
Streaked Dagger Moth
PARKS AND RESERVES: BAN, LCP, OCP, PYR, RGR. **COUNTIES:** BAR, BRK, CAD, CHR, FAY, GRY, JEF, MCC, MEA, OWS, ROW, RUS. **DATES:** Apr. 11, 23, May 23–31, June 20–28, July 25, 31, Aug. 2–25. Common.

9272 ***A. oblinita*** (J. E. Smith)
Smeared Dagger Moth
PARKS AND RESERVES: BAN, BMF, BOW, OCP, PYR, SWR. **COUNTIES:** BAR, BOO, BRA, BUL, CAD, CHR, CLK, FAY, HND, JEF, LET, MEA, OLD, ROW, WAR, WOO. **DATES:** Apr. 11, 12, 22–May 10, 22, 29, June 2, 10, 16–July 23, Aug. 1, 12, 31, Sept. 10, 12, 21. **REMARKS:** Common. The Woodford County specimen fed on oak, pupated October 15, and emerged in January 1929.

Genus ***Meroloncha*** Grt.

9277 ***M. dolli*** B. & McD.
Laurel Co., DBNF, FS 615A/B, Apr. 6, 1991, L. D. Gibson and D. J. Wright. A rare species; this record constitutes a substantial southward extension of its known range.

Genus ***Simyra***

9280 ***S. henrici*** (Grt.)
Henry's Marsh Moth
Boone Co., BCN, July 5; Fulton Co., Willingham Bottoms, 4.0 km (2.5 mi) E of Cayce, Sept. 11; Henderson Co., SWR, Sept. 9–12; Woodford Co., June 29. Very local in marsh and cypress swamp habitats.

Genus ***Agriopodes*** Hamp.

9281 ***A. fallax*** (H.-S.)
The Green Marvel
PARKS AND RESERVES: BAN, BMF, CAC, LCP, NBP, OCP. **COUNTIES:** BAR, BOO, BRA, BUL, CAR, JEF, MEA, NEL, POW, RUS. **DATES:** Apr. 19–May 7, 16, 29, June 1, 12, July 3, 4, 19, Aug. 9, 10, 30, Sept. 5, 6. Common.

Genus ***Polygrammate*** Hbn.

9285 ***P. hebraeicum*** Hbn.
The Hebrew
PARKS AND RESERVES: BBM, BMF, BUC, FKN, HWS, JCF, LBL, LCP, LWO, NBP, OCP, PNM, RBF, RGR, WVP. **COUNTIES:** BEL, BRE, BRK, BUL, GRV, GRY, HLN, JEF, LAR, LET, MEA, NEL, OLD, OWS, PER, POW, ROW, RUS, TRI, WHI. **DATES:** Apr. 20–Aug. 30. **REMARKS:** Common. Two melanic specimens with uniformly dark gray forewings have been collected at PNM, Bell Co.

Genus ***Harrisimemna*** Grt.

9286 ***H. trisignata*** (Wlk.)
Harris's Three-spot
PARKS AND RESERVES: BBL, BBN, CAV, HWS, LBK, LBL, OCP, RGR. **COUNTIES:** BAT, BOO, BRA, BRK, GAL, GRY, JEF, MEA, OLD, ROW, TRI, WOO. **DATES:** May 18, June 27–July 18, Aug. 7–21, Sept. 17. Locally common.

Subfamily **Agaristinae**

Genus ***Eudryas*** Bdv.

9299 ***E. unio*** (Hbn.)
Pearly Wood-nymph
PARKS AND RESERVES: BBL, HWS, LBL, SWR. **COUNTIES:** BOO, BRA, FAY, FUL, GAL, GRV, HND, JEF, NEL, OLD, ROW, TRI. **DATES:** Apr. 23–Sept. 10. Common.

9301 ***E. grata*** (F.)
Beautiful Wood-nymph
PARKS AND RESERVES: AJJ, BEC, BMF, BUC, CAC, CBB, HWS, JCF, KGC, LCP, LWO, NBP, OCP, PNM, PYR, RGR, WVP. **COUNTIES:** BEL, BRA, BRE, BRK, BUL, CAD, CAM, CAR, CLI, FAY, GRY, HLN, JEF, LAR, LET, MCC, MEA, NEL, NIC, OLD, OWS, PER, PIK, POW, RUS. **DATES:** Apr. 7–Sept. 30. Common.

Genus ***Psychomorpha*** Harr.

9309 ***P. epimenis*** (Drury)
Grapevine Epimenis
PARKS AND RESERVES: BBM, BEC, BMF, CAC, CBB, RRG. **COUNTIES:** BOO, BUL, CAR, FAY, JEF, HLN, MAD, MEA, MEN, NEL, OLD, OWS, PER. **DATES:** Mar. 25–May 4, 13. **REMARKS:** Common. Found flying in woodlands on warm spring days.

Genus ***Alypia*** Hbn.

9314 ***A. octomaculata*** (F.)
Eight-spotted Forester
PARKS AND RESERVES: BBL, BBM, BMF, SGS, WVP. **COUNTIES:** BOO, BRA, BUL, FAY, HLN, JEF, MAD, MAR, MEA, MER, NEL, OWS. **DATES:** Apr. 12–Aug. 12. Moderately common and active in the daytime.

Subfamily **Amphipyrinae**

Genus ***Apamea*** Ochs.

9326 ***A. verbascoides*** (Gn.)
Harlan Co., BBM, July 13, 1984, R. A. Henderson. Rare.

9329 ***A. cariosa*** (Gn.)
PARKS AND RESERVES: BBL, BMF, CCR, OCP. **COUNTIES:** BOO, BUL, GAL, JEF, MEA,

RRG. **DATES:** June 7, 9, 19, 22–30, July 4, Aug. 1. Uncommon.

9331 ***A. cristata*** (Grt.)
Harlan Co., KGC, July 9; Jefferson Co., Valley Station, May 12, July 1; Letcher Co., LWO, July 11; Meade Co., OCP, July l. Rare.

9332 ***A. vulgaris*** (G. & R.)
Common Apamea
Bullitt Co., BMF, May 20–3l, June 1–15; Fayette Co., Lexington, May 28, June 2; Jefferson Co., May 20, 26, July 5, Aug. 13; Meade Co., FKN, May 16–18. Uncommon.

9333 ***A. lignicolora*** (Gn.)
Jefferson Co., Valley Station, July 5; Meade Co., OCP, June 16. Rare.

9344 ***A. plutonia*** (Grt.)
Bullitt Co., BMF, in Malaise trap, May 23–31, A. J. Brownell. Rare.

9348 ***A. amputatrix*** (Fitch)
Yellow-headed Cutworm Moth
Jefferson Co., May 28, 1941, collected by "Bisig." Rare.

9361 ***A. mixta*** (Grt.)
Menifee Co, RRG, Tunnel Ridge, June 29, 1986, D. J. Wright. Rare.

9362 ***A. remissa*** (Hbn.)
Ignorant Apamea
Bullitt Co., BMF, June 22–30; Harlan Co., BBM, June 8; Owsley Co., 4.8 km (3 mi) NE of Booneville, a very dark specimen, June 22. Our subspecies is *indocilis* (Wlk.). Uncommon.

9373 ***A. helva*** (Grt.)
Yellow Three-spot
PARKS AND RESERVES: BMF, CAC, RRG, SWR. **COUNTIES:** BRA, BUL, CAR, FAY, HND, JEF, MEA, MEN, OLD, ROW. **DATES:** Aug. 23–Sept. 19, Oct. 13. Common.

Genus *Luperina* Bdv.

9391 ***L. passer*** (Gn.)
Dock Rustic
PARKS AND RESERVES: HWS. **COUNTIES:** BRA, FAY, JEF, MEN, OLD, ROW. **DATES:** May 13, 26, June 3, July 8, 22, Aug. 18–Sept. 4, 16, 19, Oct. 29. Common.

Genus *Oligia* Hbn.

9402 ***O. chlorostigma*** (Harv.)
Boone Co., BBL, June 19, 26; Gallatin Co., Ambrose Rd., June 23; Harlan Co., BBM, July 8; Powell Co., RRG, Tunnel Ridge, June 19. L. D. Gibson (*pers. comm.*) has found it common on baited trees.

9404 ***O. modica*** (Gn.)
Black-banded Brocade
PARKS AND RESERVES: BBL, BBM, BMF, CCR, COB, KGC, PYR, SWR. **COUNTIES:** BOO, BRA, BUL, CAD, FAY, HIC, HLN, HND, HRD, JEF. **DATES:** Apr. 29, June 3, 12–30, July 15, Aug. 18, 19, 30–Sept. 11, Oct. 2. Common.

9406 ***O. fractilinea*** (Grt.)
Broken-lined Brocade
Bath Co., CAV, FS 918, Aug. 28; Jefferson Co., Valley Station, (dark form "*misera*" Grt.), Aug. 20, Sept. 12; Laurel Co., DBNF, Craig Creek Rd., Sept. 19; Rowan Co., DBNF, E side of SR 1274, 3.2 km (2 mi) W of SR 519, Aug. 23, and Morehead (dark form), Aug. 30. Uncommon.

9408 ***O. exhausta*** (Smith)
Boone Co., BBL, July 10, 18, both L. D. Gibson. Rare.

9410 ***O. crytora*** (Franc.)
Harlan Co., BBM, July 14; Laurel Co., DBNF, ridge top SW of Ned Branch, June 19; Russell Co., LCP, 1 female, June 9. Uncommon.

9411 ***O. semicana*** (Wlk.)
Owsley Co., 4.8 km (3 mi) NE of Booneville, May 16, 23, L. D. Gibson. Rare.

9420 ***O. illocata*** (Wlk.)
Wandering Brocade
Bell Co., PNM, Oct. 9–23; Fayette Co., Aug. 20; Jefferson Co., Valley Station, Sept. 26, 29, Oct. 2, 3, 22–25; Menifee Co., RRG, Oct. 11–26; Powell Co., RRG, Tunnel Ridge, Oct. 6. Uncommon.

Genus *Meropleon* Dyar

9427 ***M. diversicolor*** (Morr.)
Fulton Co., Sept. 19, 1975, C. Cornett. L. D. Gibson collected what is possibly this species from Rowan Co., DBNF, E side of SR 1274, 3.2 km (2 mi) E of SR 519, Aug. 26, 28, and Sept. 2. Rare.

9428 ***M. ambifusca*** (Newman)
Rowan Co., DBNF, E side of SR 1274, 3.2 km (2 mi) E of SR 519, Sept. 2; both L. D. Gibson.

Genus *Archanara* Wlk.

9449 ***A. oblonga*** (Grt.)
Oblong Sedge Borer Moth
Henderson Co., SWR, at light, Sept. 11; Jefferson Co., Valley Station, June 15. Local and uncommon.

Genus *Macronoctua* Grt.

9452 ***M. onusta*** Grt.
Iris Borer Moth
Fayette Co., Lexington, emerged from iris roots, Sept. 24–29; Jefferson Co., Louisville, Sept. 12,

29, Oct. 2–10, 25, 31; Powell Co., NBP, July 7. Common. A pest of iris.

Genus *Amphipoea* Billberg

9454 ***A. velata*** (Wlk.)
Veiled Ear Moth
PARKS AND RESERVES: BBL, BBM, BMF, HWS, NBP, OCP. **COUNTIES:** BOO, BUL, FAY, HLN, JEF, MEA, OLD, POW, ROW, WHI. **DATES:** Apr. 27, June 5–July 9, 25, Aug. 4, Oct. 10. Locally common.

9456 ***A. interoceanica*** (Sm.)
Harlan Co., BBM, July 9, 1977 at light, C. V. Covell, Jr. Very local and uncommon.

9457 ***A. americana*** (Speyer)
American Ear Moth
Breathitt Co., RBF, in Malaise trap, July 22–28; Harlan Co., BBM, July 8–15, and KGC, July 6, 13; Menifee Co., RRG, July 13; Rowan Co., Morehead, June 25, 30, July 7, 8; Whitley Co., Goldbug, June 23. Locally common in eastern counties at high elevations.

Genus *Parapamea* Bird

9463 ***P. buffaloensis*** (Grt.)
Buffalo Moth
Fulton Co., Willingham Bottoms, SR 94, 4.0 km (2.5 mi) E of Cayce, Sept. 5, 7, and Bayou de Chien, Sept.; Graves Co., Mayfield, Sept. 21; Henderson Co., SWR, Sept. 9–11; McCracken Co., Paducah, Sept. 24; Menifee Co., RRG, Leatherwood Fork of Indian Creek, Sept. 14. Locally common.

Genus *Papaipema* Sm.

9466 ***P. cataphracta*** (Grt.)
Burdock Borer Moth
PARKS AND RESERVES: BBL, BBM, CBF, OCP, RRG. **COUNTIES:** BOO, CHR, HLN, JEF, MEA, MEN, OWS, WHI, WOO. **DATES:** Sept. 18–Oct. 20. Locally common.

9469 ***P. polymniae*** Bird
Bath Co., CAV, FS 918, Sept. 11; Greenup Co., GBO, Sept. 13; Jefferson Co., Oct. 5, 6, 12; Meade Co., OCP, Oct. 12; Menifee Co., RRG, Oct. 3. Uncommon.

9470 ***P. araliae*** Bird & Jones
Christian Co., Crofton, Sept. 18; Jefferson Co., Valley Station, Aug.18, 22, Sept. 12, 24, Oct. 15; Livingston Co., Carrsville, Aug. Rare.

9471 ***P. arctivorens*** Hamp.
Northern Burdock Borer Moth
PARKS AND RESERVES: AJJ, BBL, BOW, CAV, CER, HWS. **COUNTIES:** BAT, BOO, BRA, CAM, GAL, JEF, LAU, MCC, MEN, OLD, ROW. **DATES:** July 10–Oct. 16. Locally common.

9473 ***P. impecuniosa*** (Grt.)
Fayette Co., Oct. 15; Jefferson Co., Louisville, Oct. 1–30, and Prospect, Oct. 17, 19; McCracken Co., Ray Black farm, 9.7 km (6 mi) SW of Paducah, Oct. 27; Menifee Co., RRG, Oct. 3; Owsley Co., 4.8 km (3 mi) NE of Booneville, Oct. 8, 10. Uncommon.

9477 ***P. astuta*** Bird
Bath Co., CAV, FS 918, Sept. 5–11; Greenup Co., Aug. 27. Rare.

9478 ***P. leucostigma*** (Harr.)
Meade Co., OCP, Aug. 12, 18; Menifee Co., RRG, Sept. 1, 6; Jefferson Co., Valley Station, Sept. 15, Oct. 8. Uncommon.

9479 ***P. lysimachiae*** Bird
PARKS AND RESERVES: OCP, RRG. **COUNTIES:** JEF, LAU, MEA, MEN, OWS, ROW. **DATES:** Aug. 18, Sept. 8, 9, 19, 29–Oct. 22. Uncommon.

9483 ***P. inquaesita*** (G. & R.)
Sensitive Fern Borer Moth
Greenup Co., Aug. 27; Hickman Co., MUR, Oct. 2; Jefferson Co., Sept. 15, 29, Oct. 23; Menifee Co., RRG, Sept. 14–22; Powell Co., 1.6 km (1 mi) SW of Clay City, Oct. 7. Locally common.

9484 ***P. rutila*** (Gn.)
PARKS AND RESERVES: BBL, OCP, RRG. **COUNTIES:** BOO, BRA, JEF, KEN, LAU, MEA, MEN, OWS. **DATES:** Aug. 14, 17, Sept. 13–Oct. 14. Locally common.

9485 ***P. baptisiae*** (Bird)
PARKS AND RESERVES: AJJ, CCR, GBT, KLK, OCP, SWR. **COUNTIES:** BAT, CAM, CRL, FAY, FUL, HND, JEF, KEN, MCC, MEA, MEN, MRL, OLD, OWS, ROW. **DATES:** Aug. 8–Oct. 20. Common.

9486 ***P. birdi*** (Dyar)
PARKS AND RESERVES: RRG, SWR. **COUNTIES:** FUL, GUP, HEN, HIC, HND, JEF, MEN, POW. **DATES:** Aug. 27, Sept. 8–14, 22, Oct. 5–11. Uncommon.

9490 ***P. nepheleptena*** (Dyar)
Bell Co., PNM, Oct. 16, 1973, C. Cornett. Rare.

9492 ***P. marginidens*** (Gn.)
PARKS AND RESERVES: CAV, CBF, LCP, OCP, PCP, PNM, RRG. **COUNTIES:** BAT, BEL, BUL, FAY, JEF, LAU, MEA, MEN, OWS, ROW, RUS, WHI, WOL. **DATES:** Sept. 5–Oct. 28, Nov. 7. Locally common.

9494 **P. eryngii** Bird
Christian Co., PYR, sight record, R. A. Henderson; Hardin Co., larvae in *Eryngium yuccafolium*, 1991–98, J. A. Bess. Rare. Listed as an endangered species in Kentucky (KSNPC 1996).

9495 **P. furcata** (Sm.)
Ash Tip Borer Moth
PARKS AND RESERVES: BBL, BOW, CCR, GBT, LCP, OCP, PNM, RRG, WVP. **COUNTIES:** BEL, BOO, BRA, CRL, GAL, HIC, JEF, MEA, MEN, OLD, RUS, WOO. **DATES:** Aug. 25–Oct. 14. Common.

9496 **P. nebris** (Gn.)
Stalk Borer Moth
PARKS AND RESERVES: AJJ, ASC, BBL, HWS, SWR. **COUNTIES:** BOO, BRA, CAL, CAM, FAY, FUL, GAL, GRY, HND, JEF, KEN, LAR, MCC, MEN, OLD, OWS, POW, ROW, TOD. **DATES:** Aug. 1, 8, 22–Oct. 23. **REMARKS:** Common. "Caused some damage to corn in Calloway and Larue counties" (*Kentucky Cooperative Economic Insect Report* 1975). "Larvae damaging corn, Todd County" (*Kentucky Cooperative Plant Pest Report* 1978).

9497 **P. necopina** (Grt.)
Sunflower Borer Moth
PARKS AND RESERVES: BBL, BOW, SWR. **COUNTIES:** BOO, BRA, FAY, HND, JEF, MCC, WOO. **DATES:** Aug. 12, 24–Oct. 20. Common.

9499 **P. duplicata** Bird
Jefferson Co., Valley Station, Oct. 2, 15; Meade Co., OCP, Oct. 18. Rare.

9500 **P. maritima** Bird
McCracken Co., Ray Black farm, 9.7 km (6 mi) SW of Paducah, Oct. 17, 27, W. R. Black, Jr. Rare.

9501 **P. eupatorii** (Lyman)
Jefferson Co., Valley Station, Oct. 5–29, Nov. 5; Meade Co., OCP, Oct. 12, 18, 30; Menifee Co., RRG, Oct 15; Owsley Co., 4.8 km (3 mi) NE of Booneville, Oct. 4. Locally common.

9502 **P. nelita** (Stkr.)
Greenup Co., Aug. 27; Jefferson Co., Aug. 24; Menifee Co., RRG, Aug. 26, 31, Sept. 7. Uncommon.

9503 **P. rigida** (Grt.)
Rigid Sunflower Borer Moth
Jefferson Co., Camp Dan Beard, Sept. 12; Menifee Co., RRG, Sept. 12–Oct. 11; Oldham Co., HWS, Sept. 6. Locally common.

9505 **P. cerussata** (Grt.)
Ironweed Borer Moth
PARKS AND RESERVES: PNM, RRG. **COUNTIES:** BEL, BOO, FAY, FUL, JEF, KEN, MCC, MEN, OWS, POW, ROW. **DATES:** Sept. 21–Oct. 31. Common.

9507 **P. limpida** (Gn.)
Gallatin Co., Ambrose Rd., Sept. 29; Jefferson Co., Louisville, Sept. 8–28, Oct. 4; Oldham Co., Pewee Valley, Sept. 12. Uncommon.

9509 **P. unimoda** (Sm.)
Bullitt Co., E side of SR 1442, 12.9 km (8 mi) E of SR 61, reared from larva in *Thalictrum*, emerged Sept. 26; Menifee Co., RRG, FS 9B, East Fork Indian Creek, Sept. 14–Oct.15; Rowan Co., SR 1274, 3.2 km (2 mi) W of SR 519, Oct. 15. Uncommon.

9509.1 **P.** undescribed species
A new species which bores in *Arundinaria* was collected by L. D. Gibson in Owsley Co., 4.8 km (3 mi) NE of Booneville, Oct. 8, 1982.

Genus *Hydraecia* Gn.

9516 **H. stramentosa** Gn.
PARKS AND RESERVES: CAV, CCR, OCP, WVP. **COUNTIES:** BAT, BRA, BRK, JEF, MEA, OLD. **DATES:** Aug. 7, 28–Oct. 2. Local and uncommon.

Genus *Achatodes* Gn.

9520 **A. zeae** (Harr.)
Elder Shoot Borer Moth
PARKS AND RESERVES: BBL, OCP, RRG. **COUNTIES:** BOO, FAY, JEF, MEA, MEN, OWS, ROW. **DATES:** June 3–July 4, 13, 14. Moderately common.

Genus *Iodopepla* Franc.

9522 **I. u-album** (Gn.)
PARKS AND RESERVES: BBL, OCP, PCP, PYR. **COUNTIES:** BOO, BUL, CHR, FUL, HRD, MEA. **DATES:** Mar. 25, Apr. 16, May 9, 31, Aug. 14, 22, Sept. 6. Uncommon.

Genus *Bellura* Wlk.

9523 **B. gortynoides** Wlk.
White-tailed Diver
Henderson Co., SWR, Aug. 22, Sept. 10, 11; Trigg Co., LBL, July 12. Uncommon. Very local in swampy areas. *Bellura* larvae are aquatic, boring in emergent plants.

9525 **B. obliqua** (Wlk.)
Cattail Borer Moth
Boone Co., BBL, June 1; Menifee Co., RRG, May 12. Uncommon and local in swampy areas.

9526 ***B. densa*** (Wlk.)
Pickerelweed Borer Moth
Boone Co., BBL, July 18; Jefferson Co., June 18, July 13, Aug. 2. Local and uncommon in swampy habitats.

Genus ***Euplexia*** Steph.

9545 ***E. benesimilis*** McD.
American Angle Shades
PARKS AND RESERVES: BBM, BUC, CAV, LCP, OCP, PNM, PYR, RRG, WVP. **COUNTIES:** BAT, BEL, BRE, CHR, FAY, HLN, JEF, MEA, MEN, MRL, OWS, PER, ROW, RUS. **DATES:** May 10–Sept. 6. Common.

Genus ***Phlogophora*** Tr.

9546 ***P. iris*** Gn.
Olive Angle Shades
Harlan Co., BBM, June 8, 1985, one each collected by R. A. Henderson and D. J. Wright. Rare.

9547 ***P. periculosa*** Gn.
Brown Angle Shades
PARKS AND RESERVES: BBM, CAC, CAV, RRG. **COUNTIES:** BAT, CAR, HLN, LET, MEN, OWS, ROW, WHI. **DATES:** Aug. 8–Oct. 18. Uncommon.

Genus ***Ipimorpha*** Hbn.

9555 ***I. pleonectusa*** Grt.
Even-lined Sallow
Gallatin Co., Markland Dam, June 22, 29; Jefferson Co., May 1, June 14, 21, Aug. 8, 12, Sept. 2; Meade Co., OCP, June 11; Powell Co., July 4. Uncommon.

Genus ***Chytonix*** Grt.

9556 ***C. palliatricula*** (Gn.)
Cloaked Marvel
PARKS AND RESERVES: BCN, BMF, BUC, CBF, HWS, JWS, LBL, NBP, PNM, PYR, RGR. **COUNTIES:** BEL, BOO, BRE, BRK, BUL, CAD, FLO, GRY, JEF, MCC, OLD, OWS, PER, POW, ROW, TRI, WHI. **DATES:** Apr. 17–Oct. 3. Common.

Genus ***Dypterygia*** Steph.

9560 ***D. rozmani*** Berio
American Bird's Wing Moth
PARKS AND RESERVES: AJJ, BBL, OCP. **COUNTIES:** BOO, CAM, FAY, FUL, GAL, JEF, MEA, OWS. **DATES:** May 6, 22–July 3, 11–20, 28–Sept. 1, 14. Common.

9561 ***D. patina*** (Harv.)
Bullitt Co., PCP, Sept. 6; Harlan Co., BBM, Apr. 21; Jefferson Co., Valley Station, Sept. 20; Meade Co., OCP, Sept. 5. Rare.

Genus ***Hyppa*** Dup.

9578 ***H. xylinoides*** (Gn.)
Common Hyppa
Powell Co., RRG, Rough Trail parking lot, one male at light, June 9, 1989, L. D. Gibson. Not "common" in Kentucky.

Genus ***Nedra*** Clarke

9582 ***N. ramosula*** (Gn.)
Gray Half-spot
PARKS AND RESERVES: BAN, BIP, BMF, CAV, HWS, OCP, PYR. **COUNTIES:** BAR, BAT, BUL, CAD, FAY, GRV, JEF, LAU, MCC, MEA, MEN, MRL, OLD, OWS, PIK, ROW. **DATES:** Apr. 6–23, May 17, June 4, 20–Oct. 15. Common.

Genus ***Properigea*** B. & Benj.

9589 ***P. costa*** (B. & Benj.)
Woodford Co., 1.6 km (1 mi) W of highway 68, along Kentucky River, in light trap, May 25, 1985, J. S. Nordin. This is a western species recently appearing widely in the east (J. Rawlins, *pers. comm.*).

Genus ***Phosphila*** Hbn.

9618 ***P. turbulenta*** Hbn.
Turbulent Phosphila
PARKS AND RESERVES: BBM, BCN, BMF, KGC, LCP, OCP, PMS, PNM, RGR, RRG, WVP. **COUNTIES:** BEL, BOO, BRK, BUL, GRY, HLN, JEF, MEA, OLD, OWS, POW, RUS, WHI. **DATES:** May 23–Aug. 26. Common.

9619 ***P. miselioides*** (Gn.)
Spotted Phosphila
PARKS AND RESERVES: BAN, BBL, BBM, BIP, HWS, LBK, LCP, LWO, OCP, PNM, PYR, RGR, WVP. **COUNTIES:** BAR, BEL, BOO, BRK, BUL, CAD, FAY, GRY, HLN, JEF, LET, MEA, MEN, OLD, OWS, PIK, ROW, RUS, TRI, WAR, WOO. **DATES:** Apr. 25–Oct. 15. Common.

Genus ***Trachea*** Ochs.

9626 ***T. delicata*** (Grt.)
Bullitt Co., BMF, June 5, A. J. Brownell. Kentucky is included in this species range (Forbes 1954). Rare.

Genus ***Callopistria*** Hbn.

9631 ***C. mollissima*** (Gn.)
Pink-shaded Fern Moth
PARKS AND RESERVES: BBM, HWS, LCP, NBP, PNM, PYR, RRG. **COUNTIES:** BEL,

CHR, GUP, HLN, JEF, MCC, MEN, OLD, OWS, POW, RUS. **DATES:** Apr. 26, May 6, 16–28, June 9–July 19, Aug. 8, 14, 22, 23, 27. Common.

9633 ***C. cordata*** (Ljungh)
Silver-spotted Fern Moth
Bell Co., PNM, July 19; Meade Co., OCP, June 21; Owsley Co., 4.8 km (3 mi) NE of Booneville, May 28; Powell Co., RRG, Tunnel Ridge, July 23. Uncommon.

Genus ***Magusa*** Wlk.

9637 ***M. orbifera*** (Wlk.)
Orbed Narrow-wing
PARKS AND RESERVES: BBL, CAV, LCP, NBP, WVP. **COUNTIES:** BAT, BOO, GAL, JEF, POW, ROW, RUS. **DATES:** July 2–Nov. 23. Moderately common.

Genus ***Amphipyra*** Ochs.

9638 ***A. pyramidoides*** Gn.
Copper Underwing
PARKS AND RESERVES: BEC, BIP, BMF, BUC, CAC, GBT, HLN, HWS, KGC, LBL, LCP, LWO, MCV, NBP, OCP, PNM, PYR, WVP. **COUNTIES:** BEL, BRE, BUL, CAD, CAR, CRL, EDM, FAY, GRV, HLN, JEF, LET, MCC, MEA, OLD, PER, PIK, POW, ROW, RUS, TRI. **DATES:** June 6–Oct. 31, Nov. 23. **REMARKS:** Common. Attracted to sugar bait as well as to lights.

9640 ***A. glabella*** (Morr.)
Jefferson Co., WVP, June 10, July 2, 6, S. Scholz. Local and uncommon.

Genus ***Micrathetis*** Hamp.

9644 ***M. triplex*** (Wlk.)
Jefferson Co., WVP, Valley Station, Sept. 5, 1975, S. Scholz. Rare.

Genus ***Athetis*** Hbn.

9647 ***A. miranda*** (Grt.)
Miranda Moth
PARKS AND RESERVES: BBL, BBM, BMF, FKN, KLK, LBL, RGR. **COUNTIES:** BOO, BRA, BRK, BUL, EST, FAY, FUL, GRY, HLN, HRD, HRT, JEF, MCC, MEA, MRL, ROW, TRI. **DATES:** Apr. 20–Sept. 16. **REMARKS:** Common. Genus changed by Poole (1989), who synonymized *Proxenus* H.-S. to *Athetis*.

Genus ***Anorthodes*** Sm.

9650 ***A. tarda*** (Gn.)
The Slowpoke
PARKS AND RESERVES: BIP, BMF, BUC, CBB, CBF, HNK, HWS, JWS, KLK, LCP, LMP, NBP, OCP, PNM, PYR, RBF. **COUNTIES:** BEL, BRE, BUL, CAD, CAL, FAY, FLO, HRD, JEF, MCC, MEA, MRL, MUH, OLD, PER, PIK, POW, RUS, WHI. **DATES:** Apr. 20–Oct. 18. Abundant.

Genus ***Crambodes*** Gn.

9661 ***C. talidiformis*** Gn.
Verbena Moth
PARKS AND RESERVES: BBL, CBB, HWS, SWR. **COUNTIES:** BOO, BRA, FAY, GAL, HND, HRT, JEF, MCC, OLD, ROW. **DATES:** May 3, 14–Aug. 18. Common.

Genus ***Balsa*** Wlk.

9662 ***B. malana*** (Fitch)
Many-dotted Appleworm Moth
PARKS AND RESERVES: AJJ, BBL, KLK. **COUNTIES:** BOO, CAM, FAY, JEF, LYO, MAD, MRL, ROC, ROW. **DATES:** Apr. 8, 20–May 5, 19, 20, June 2, 16, 25, July 9, 12, 30–Aug. 6, Sept. 1, 7. Common.

9663 ***B. tristrigella*** (Wlk.)
Three-lined Balsa
Bracken Co., 0.8 km (0.5 mi) W of Foster, May 14; Gallatin Co., Markland Dam, May 27; Jefferson Co., Valley Station, June 4, 8; Oldham Co., CBB, May 14. More common than our records indicate.

9664 ***B. labecula*** (Grt.)
White-blotched Balsa
PARKS AND RESERVES: BBL, BBM, CBF, PMS, PNM, PYR, RRG. **COUNTIES:** BEL, BOO, CHR, HLN, MEN, OWS, WHI. **DATES:** Apr. 26–May 3, 12, June 7, 9, 16, 29, July 8–28. Common.

Genus ***Spodoptera*** Gn.

9665 ***S. exigua*** (Hbn.)
Beet Armyworm Moth
Jefferson Co., Valley Station, Sept. 24, 30, S. Scholz. Uncommon.

9666 ***S. frugiperda*** (J. E. Smith)
Fall Armyworm Moth
PARKS AND RESERVES: BEC, BIP, BOW, CAV, GBT, LCP, OCP, PNM, RGR. **COUNTIES:** BAT, BEL, BOO, BRA, BRK, CAM, CRL, FAY, FUL, GRV, GRY, HND, JEF, MCC, MEA, MRL, OWS, PIK, RUS. **DATES:** Apr. 12, June 4, 15, July 16, 28–Oct. 21, 30, Nov. 23. **REMARKS:** Common to abundant. A pest of corn and other herbaceous plants.

9669 ***S. ornithogalli*** (Gn.)
Yellow-striped Armyworm Moth
PARKS AND RESERVES: BAN, BBL, BBM,

BEC, BMF, BUC, CAC, GBO, GBT, HWS, KLK, LBL, LCP, LMP, OCP, PNM, PYR, RGR, WVP. **COUNTIES:** BAR, BAT, BEL, BOO, BRA, BRE, BRK, BUL, CAD, CAR, CRL, FAY, GAL, GRY, GUP, HLN, JEF, MCC, MEA, MRL, MUH, OLD, PER, ROW, RUS, SPE, TRI. **DATES:** Mar. 17, Apr. 13, 21–Nov. 5, 14, Dec. 7. **REMARKS:** Common to abundant. A pest on cotton and other crops.

9671 ***S. dolichos*** (F.)
Dolichos Armyworm Moth
Fayette Co., Aug. 5, 7, 1890, and Aug. 6, 12, 1891; all UK.

9672 ***S. eridania*** (Cram.)
Southern Armyworm Moth
Jefferson Co., Valley Station, Aug. 27, Sept. 22, S. Scholz. Uncommon.

Genus ***Elaphria*** Hbn.

9677 ***E. agrotina*** (Gn.)
Meade Co., OCP, at light, Oct. 11, 1986, B. S. Nichols. Rare.

9678 ***E. versicolor*** (Grt.)
Variegated Midget
PARKS AND RESERVES: AJJ, BBL, BMF, BUC, CBF, GBT, HWS, LBK, NBP, OCP, PNM, RRG. **COUNTIES:** BEL, BOO, BRE, BUL, CAM, CAR, CRL, EST, JEF, MEA, MEN, OLD, OWS, PER, POW, TRI, WHI. **DATES:** Apr. 13–Sept. 9. Common.

9679 ***E. chalcedonia*** (Hbn.)
Chalcedony Midget
Fulton Co., Fulton, Sept. 17, 19; Jefferson Co., Valley Station, Sept. 29. Local and uncommon.

9680 ***E. georgei*** (Moore & Rawson)
PARKS AND RESERVES: BBL, BMF, MCV, RRG. **COUNTIES:** BOO, BUL, EDM, JEF, MEN, ROW. **DATES:** Apr. 12–17, May 26, 31, June 8, Oct. 22, 31. Uncommon.

9681 ***E. festivoides*** (Gn.)
Festive Midget
PARKS AND RESERVES: BBM, BMF, CAC, CBF, KGC, KLK, LBL, LCP, OCP, PNM, PYR, RBF, RRG. **COUNTIES:** BEL, BRE, BUL, CAR, CHR, EST, FAY, HLN, JEF, MCC, MEA, MRL, OWS, POW, RUS, TRI, WHI. **DATES:** Apr. 17–May 3, 13–26, June 14, 22, July 10–28, Aug. 4–9, 19–31, Sept. 12, 16. Common.

9684 ***E. grata*** Hbn.
Grateful Midget
PARKS AND RESERVES: BBL, BEC, BMF, CAC, CBF, GBT, HWS, KLK, LBK, LBL, LCP, NBP, OCP, PYR, RGR, RRG. **COUNTIES:** BOO, BRA, BRK, BUL, CAD, CAR, CRL, FAY, FUL, GRY, GUP, HRT, JEF, MCC, MEA, MEN, MRL, NEL, OLD, OWS, POW, ROW, RUS, TRI, WHI, WOO. **DATES:** Apr. 12–Nov. 7. Common to abundant.

Genus ***Galgula*** Gn.

9688 ***G. partita*** Gn.
The Wedgling
PARKS AND RESERVES: BAN, BEC, BLA, BMF, CAC, FKN, GBT, HWS, LBK, LBL, LCP, NBP, PNM, RGR. **COUNTIES:** BAR, BEL, BOO, BRA, BRK, BUL, CAR, CRL, FAY, FUL, GRY, HRD, JEF, MCC, MEA, MRL, NEL, OLD, OWS, POW, ROW, RUS, TRI, WOO. **DATES:** Mar. 28–Oct. 15. **REMARKS:** Very common throughout the state.

Genus ***Perigea*** Gn.

9689 ***P. xanthioides*** Gn.
Red Groundling
PARKS AND RESERVES: BBL, BBM, BMF, CBB, HWS. **COUNTIES:** BOO, BRA, BUL, FAY, HLN, JEF, MCC, MEN, NEL, OLD, ROW. **DATES:** Apr. 4–Sept. 15. Common.

Genus ***Condica*** Wlk.

9690 ***C. videns*** (Gn.)
White-dotted Groundling
PARKS AND RESERVES: BBL, BCN, HWS, LBL, PYR, SWR. **COUNTIES:** BOO, BRA, CAD, CHR, FUL, HND, JEF, OLD, OWS, TRI. **DATES:** Apr. 24, May 16–20, June 3–6, 15, July 6–10, 24–Sept. 12. **REMARKS:** Common. Listed from Kentucky in Crumb (1956). Poole (1989) made *Platysenta* Grt. a junior synonym of *Condica*, the genus name formerly used.

9693 ***C. mobilis*** (Wlk.)
Mobile Groundling
Bath Co., CAV, FS 918, Aug. 28, Sept. 5; Fayette Co., Lexington, July 21, Sept. 26, Nov. 13; Jefferson Co., Aug. 3, Sept. 8, Oct. 7; Meade Co., OCP, Sept. 9; Russell Co., LCP, Oct. 11, 14. Uncommon.

9696 ***C. vecors*** (Gn.)
Dusky Groundling
PARKS AND RESERVES: BAN, BBL, BMF, BUC, CBB, GBT, HWS, JWS, LCP, LMP, NBP, OCP, PNM, PYR, RBF, RGR. **COUNTIES:** BAR, BEL, BOO, BRE, BRK, BUL, CAD, CRL, FAY, FLO, GRY, JEF, MEA, MRL, MUH, OLD, OWS, PER, POW, ROW, RUS. **DATES:** Apr. 8–Oct. 13. Abundant.

9699 ***C. sutor*** (Gn.)
The Cobbler
PARKS AND RESERVES: FKN, GBO, GBT, HWS, LCP, OCP, PYR. **COUNTIES:** CRL, FAY, FLO, GAL, GUP, HRD, JEF, MCC, MEA, MEN, OLD, OWS, ROW, RUS, WHI. **DATES:** Apr. 27–Oct. 15, 28. Common.

9714 ***C. confederata*** (Grt.)
The Confederate
Boone Co., BOW, Aug. 13, Sept. 19; Gallatin Co., Sept. 29; Jefferson Co., Valley Station, Sept. 28, Oct. 8; Meade Co., OCP, Aug. 8; Rowan Co., Morehead, Sept. 8. Uncommon.

Genus ***Ogdoconta*** Butler

9720 ***O. cinereola*** (Gn.)
Common Pinkband
PARKS AND RESERVES: BBL, BMF, CBB, GBT, HWS, LCP, LMP, NBP, PYR. **COUNTIES:** BOO, BUL, CAD, CRL, FAY, FUL, GAL, HRT, JEF, MCC, MEN, MRL, MUH, OLD, OWS, POW, ROW, RUS. **DATES:** Apr. 18, May 6, 25–Oct. 14. Abundant.

Genus ***Stiriodes*** Hamp.

9725 ***S. obtusa*** (H.-S.)
Obtuse Yellow
PARKS AND RESERVES: BBL, BMF, CCR, CER, HWS, LBK, LBL, LMP, OCP, PYR, RGR. **COUNTIES:** BOO, BRK, BUL, CAD, FAY, FUL, GRY, HND, JEF, MCC, MEA, MUH, NEL, OLD, OWS, TRI, WHI. **DATES:** May 12–Sept. 28. Common.

Genus ***Plagiomimicus*** Grt.

NOTE: The genus formerly used, *Stibadium*, was synonymized to *Plagiomimicus* by Poole (1989).

9748 ***P. spumosum*** (Grt.)
Frothy Moth
Fayette Co., Lexington, Aug. 12, 14, 1928, H. Garman, and Lexington, larvae collected in 1974, P. H. Freytag; Henderson Co., SWR, Sept. 10. Uncommon.

9754 ***P. pityochromus*** Grt.
Black-barred Brown
PARKS AND RESERVES: BBL, CCR, RGR, SWR, WVP. **COUNTIES:** BOO, BRA, BRK, FAY, GRY, HND, JEF, MCC, OWS, ROW. **DATES:** June 13, 24, Aug. 3–25, Sept. 2–10. Locally common.

Genus ***Cirrhophanus*** Grt.

9766 ***C. triangulifer*** Grt.
Goldenrod Stowaway
PARKS AND RESERVES: ASC, CAC, OCP, PYR, RRG, SWR, WVP. **COUNTIES:** CAR, CHR, FRA, FUL, GRY, HND, JEF, MCC, MEA, MEN, OHI, OWS, ROW. **DATES:** Aug. 1, 15–Sept. 23. **REMARKS:** Found at light, or well camouflaged among goldenrod flowers during the day. Common.

Genus ***Basilodes*** Gn.

9781 ***B. pepita*** Gn.
Gold Moth
PARKS AND RESERVES: BBL, BMF, OCP, RRG, WVP. **COUNTIES:** BOO, BRA, BUL, FAY, JEF, KEN, MEA, MEN, OWS, WOO. **DATES:** July 23–Sept. 3. Uncommon.

Genus ***Stiria*** Grt.

9785 ***S. rugifrons*** Grt.
Christian Co., PYR, Aug. 21; Fayette Co., Lexington, Aug. 17, "Sept."; Jefferson Co., Valley Station, Aug. 15; Muhlenberg Co., LMP, Aug. 10; Owsley Co., 4.8 km (3 mi) NE of Booneville, Aug. 10. Uncommon.

Genus ***Cosmia*** Ochs.

9815 ***C. calami*** (Harv.)
American Dun-bar
PARKS AND RESERVES: AJJ, BBM, BMF, CBF, KLK, LBK, NBP, OCP, PYR, WVP. **COUNTIES:** BUL, CAD, CAM, HLN, JEF, MCC, MEA, MRL, OWS, POW, ROW, TRI, WHI. **DATES:** Apr. 14–Sept. 16. Common.

Genus ***Amolita*** Grt.

9818 ***A. fessa*** Grt.
Feeble Grass Moth
PARKS AND RESERVES: BBL, BCN, HWS, KGC, RRG, WVP. **COUNTIES:** BOO, FUL, HLN, HOP, JEF, MEN, OLD, PUL. **DATES:** Apr. 27, May 17–31, June 15–July 13, Aug., Sept. 3. Common.

Subfamily **Cuculliinae**

Genus ***Lithophane*** Hbn.

NOTE: Members of this genus may be found in late fall to late spring, and are best seen at sugar bait on warm, late winter evenings.

9886 ***L. patefacta*** (Wlk.)
Bell Co., PNM, Oct. 10, 23; Harlan Co., KGC, at sugar bait, Oct. 28; Menifee Co., RRG, Oct. 22–28. Uncommon.

9887 ***L. bethunei*** (G. & R.)
Bethune's Pinion
PARKS AND RESERVES: BBL, BCN, BMF, GBT, OCP. **COUNTIES:** BOO, BUL, CRL, FAY, JEF, MEA, MEN, WOO. **DATES:** Feb. 12, Mar.

1– Apr. 1, 19, Oct. 10, 24, Nov. 5–11, 22–30, Dec. 11, 14, 24, 27. Locally common.

9888 **L. innominata** (Sm.)
Nameless Pinion
Bell Co., PNM, Apr. 13; Harlan Co., KGC, at sugar bait, Oct. 28; Powell Co., NBP, Mar. 26, and RRG, Tunnel Ridge, Nov. 1, 15; other Powell Co. sites, at bait or in bait traps, Oct. 25–31, Nov. 1–9. Uncommon.

9889 **L. petulca** Grt.
Wanton Pinion
Harlan Co., KGC, at sugar bait, Oct. 28; Menifee Co., RRG, Mar. 7; Powell Co., RRG, Nov. 1–15, Tunnel Ridge, Nov. 1, 9. Uncommon.

9889.1 **L. joannis** Metzler & Covell
Kentucky paratypes of this recently described species (Covell and Metzler 1992) have data as follows: Boone Co., BBL, Feb. 20 and Mar. 25, 1986, Mar. 20, 1979, Mar. 23, 1980, L. D. Gibson and D. J. Wright, and BCN, Mar. 17 and 24, 1982, L. D. Gibson; Carroll Co., GBT, Mar. 29, 1980, L. D. Gibson; Jefferson Co., CCR, Feb. 15, 1981, R. A. Henderson, and Louisville, Mar. 9, 1988, C. V. Covell, Jr.; Menifee Co., RRG, north of Slade Rd., Edward Branch, 274 m (900 ft) elevation, Mar. 3–8, 1980, J. S. Nordin; Meade Co., OCP, Mar. 17, 1979, R. A. Henderson. Locally common.

9892 **L. disposita** Morr.
Jefferson Co., Valley Station, Mar. 17; Powell Co., RRG, 1.3 km (0.8 mi) E of Nada Tunnel, at sugar bait, Oct. 25, Nov., 4, 7, 15. Uncommon.

9893 **L. hemina** Grt.
Hemina Pinion
PARKS AND RESERVES: BBL, CAC, KGC, LMP, NBP, RRG. **COUNTIES:** BOO, CAR, JEF, HLN, MEN, MUH, POW, WOO. **DATES:** Mar. 3, 17–20, 29, Apr. 4, 10, 20, 21, Oct. 2–Nov. 15, Dec. 27. Locally common.

9895 **L. signosa** (Wlk.)
PARKS AND RESERVES: BBL, BCN, BMF, GBT, OCP, RRG. **COUNTIES:** BOO, BUL, CAR, CRL, FAY, JEF, MEA, MEN, POW. **DATES:** Jan. 3–Apr. 3, Oct. 12, Nov., Dec. 14, 28. Common.

9902 **L. baileyi** Grt.
Bailey's Pinion
Harlan Co., KGC, at sugar bait, Oct. 28; Menifee Co., RRG, in bait traps, Oct. 14–20, Nov. 9–16; Powell Co., RRG, Apr. 5, Oct. 25, 28, Nov. 4–15. Moderately common.

9904 **L. querquera** Grt.
Shivering Pinion
Harlan Co., KGC, Apr. 20, Oct. 28; Menifee Co., RRG, in bait traps, Mar. 11–16, Apr. 26, Oct. 13–Nov. 16; Owsley Co., 4.8 km (3 mi) NE of Booneville, Oct. 8; Powell Co., RRG, Mar. 11, Apr. 26, Oct. 19–31, Nov. 1–15; Rowan Co., DBNF, 3.2 km (2 mi) W of SR 519 on SR 1274, Oct. 15. Locally common.

9909 **L. tepida** Grt.
Harlan Co., KGC, at sugar bait, Oct. 28; Menifee Co., RRG near Edward Branch, 213 m (700 ft) elevation, N of Slade, Oct. 22–28, Nov. 2–8; Powell Co., RRG, Nov. 7, 15, Tunnel Ridge, Nov. 21. Our subspecies is *atincta* (Smith). Rare.

9910 **L. antennata** (Wlk.)
Ashen Pinion
PARKS AND RESERVES: BAN, BBL, BCN, BMF, BOW, GBT, KGC, LMP, NBP, OCP, PNM, RRG. **COUNTIES:** BAR, BEL, BOO, BUL, CRL, FAY, HLN, JEF, MEA, MEN, MUH, OLD, POW. **DATES:** Jan. 3–Apr. 28, Oct. 21–Dec. 21. Locally common to abundant.

9914 **L. laticinerea** Grt.
PARKS AND RESERVES: BBL, BCN, CCR, GBT, RRG. **COUNTIES:** BOO, CRL, FAY, JEF, POW, WOO. **DATES:** Feb. 12, Mar. 3–5, 19, Apr. 5, Oct. 30–Dec. 12. Moderately common.

9915 **L. grotei** Riley
Grote's Pinion
PARKS AND RESERVES: BBL, BBM, BCN, KGC, LMP, PNM, RRG. **COUNTIES:** BEL, BOO, HLN, JEF, MEN, MUH, OLD, PER, POW, WOO. **DATES:** Jan. 18, Feb. 12–18, 28, Mar. 1–14, 28, Apr. 8, Oct. 28, 31, Nov., Dec. 28. Locally common.

9916 **L. unimoda** (Lint.)
Dowdy Pinion
PARKS AND RESERVES: BBL, BCN, CAH, CCR, KGC, OCP. **COUNTIES:** BOO, FAY, HLN, JEF, MEA, WOO. **DATES:** Oct. 1–Apr. 22. Locally common.

9925 **L. lepida** Grt.
One specimen of subspecies *adipel* (Benjamin) was collected by C. Cornett at Powell Co., NBP, Apr. 11, 1980. D. F. Schweitzer identified it and said that this record represents a considerable westward range extension for this species.

Genus *Pyreferra* Franc.

9929 **P. hesperidago** (Gn.)
Mustard Sallow
Jefferson Co., Louisville, Feb. 1, Mar. 9, 19–27; Meade Co., OCP, Mar. 28; Menifee Co., RRG, Mar. 11–28; Powell Co., NBP, Apr. 11, 28, Nov. 4, RRG, Tunnel Ridge, Apr. 17, near Nada Tunnel, Nov. 4, 7. Uncommon.

9930 ***P. citrombra*** Franc.
Boone Co., BCN, Mar. 17; Pike Co., BIP, Oct. 16; Powell Co., RRG, Mar. 11, Apr. 5, Nov. 7, 9. Uncommon. Tentative identification; it is possibly an undescribed species (D. F. Schweitzer, *pers. comm.*).

9932 ***P. pettiti*** (Grt.)
Boone Co., BBL, Mar. 20; Jefferson Co., Louisville, Feb. 1, Mar. 20–27, Apr. 1; Meade Co., OCP, Mar. 17; Menifee Co., Mar. 23–28. Uncommon. These may represent an undescribed species as they are larger and deeper orange than typical *pettiti* from northward (D. F. Schweitzer, *pers. comm.*).

Genus ***Eupsilia*** Hbn.

9933 ***E. vinulenta*** (Grt.)
Straight-toothed Sallow
PARKS AND RESERVES: BBL, BCN, BOW, CCR, GBT, LCP, OCP, RRG. **COUNTIES:** BOO, BRE, CRL, JEF, MEA, MEN, RUS, WOO. **DATES:** Jan. 28, Feb. 13–22, Mar., Apr. 3, Oct. 27, 28, Nov. 14–Dec. 28. Our most common *Eupsilia* species.

9933.1 ***E. sidus*** (Gn.)
Sidus Sallow
Boone Co., BBL, Mar. 20, and Florence, Mar. 19; Jefferson Co., Valley Station, Jan. 28, Feb. 13, Nov. 18, Dec. 6, 10, 23; Menifee Co., RRG, Mar. 23–28; Powell Co., RRG, Nov. 4–21. Uncommon.

9934 ***E. cirripalea*** Franc.
Franclemont's Sallow
PARKS AND RESERVES: BMF, CAC, CCR, GBT, KGC, LCP, LMP, NBP. **COUNTIES:** BUL, CAR, CRL, FAY, HLN, JEF, MEN, MUH, POW, RUS. **DATES:** Jan. 28, Feb. 21–23, Mar. 13–Apr. 24, Oct. 28, Nov. 2–23, Dec. 7–16, 28. More common than *E. sidus*, but less so than *E. vinulenta*.

9935 ***E. tristigmata*** (Grt.)
Three-spotted Sallow
Jefferson Co., Jan. 21, Feb. 10, Oct. 20, Dec. 3; Powell Co., RRG, 1.3 km (0.8 mi) E of Nada Tunnel, at sugar bait, Nov. 4, 9. Uncommon.

9936 ***E. morrisoni*** (Grt.)
Morrison's Sallow
PARKS AND RESERVES: BBL, BBM, GBT, KGC, RRG. **COUNTIES:** BOO, CRL, FAY, HLN, JEF, MEN, POW. **DATES:** Mar. 7–14, Apr. 1–20, Oct. 28, Nov. 4, 14, 15, Dec. 3–19. Common.

9939 ***E. devia*** (Grt.)
Lost Sallow
Boone Co., BBL, Mar. 20, 23, and BCN, Mar. 17, and Florence, Mar. 19; Fayette Co., Feb. 23; Jefferson Co., Louisville, Feb. 1, Nov. 4, Mar. 9; Menifee Co., RRG, Mar. 9–15. Uncommon.

Genus ***Sericaglaea*** Franc

9941 ***S. signata*** (French)
Variable Sallow
PARKS AND RESERVES: BBL, BMF, CAC, KLK, LCP, LMP, MCV, OCP, RRG, WVP. **COUNTIES:** BOO, BUL, CAR, EDM, JEF, MCC, MEA, MRL, MUH, POW, RUS. **DATES:** Jan. 2–May 4, June 19, Sept. 26, Oct. 3–Dec. 28. Common to abundant.

Genus ***Xystopeplus*** Franc.

9942 ***X. rufago*** (Hbn.)
Red-winged Sallow
PARKS AND RESERVES: BMF, CAC, NBP, PNM, RRG. **COUNTIES:** BEL, BUL, CAR, JEF, MEN, OWS, POW. **DATES:** Jan. 3, Feb. 6–15, Mar., Apr. 13–29, Nov. 2, Dec. 5–16, 28. Common.

Genus ***Metaxaglaea*** Franc.

9943 ***M. inulta*** (Grt.)
Unsated Sallow
Boone Co., BBL, Oct. 4, 23, 31; Jefferson Co., CCR, Oct. 21; McCracken Co., Ray Black farm, 9.7 km (6 mi) SW of Paducah, "early spring." Uncommon.

9944 ***M. viatica*** (Grt.)
Roadside Sallow
PARKS AND RESERVES: GBT, PNM, RRG. **COUNTIES:** BEL, CRL, JEF, MCC, MEN, OLD, POW. **DATES:** Feb. 2, Sept. 16, 20, Oct. 12, 22–Dec. 9, 19. Common.

9945 ***M. semitaria*** Franc.
Footpath Sallow
PARKS AND RESERVES: BBM, BIP, KGC, PNM, RRG. **COUNTIES:** BEL, HLN, MEN, PIK, POW, WOL. **DATES:** Oct. 9–28, Nov. 5–15. Locally common.

9945.2 ***M. violacea*** Schweitzer
Bell Co., PNM, Oct. 23; Boone Co., BBL, Oct. 31; Menifee Co., RRG, in bait traps, Oct. 26–31, Nov. 1; Powell Co., RRG, at sugar bait, Nov. 4–15; Whitley Co., Goldbug, Oct. 20. Locally common.

Genus ***Epiglaea*** Grt.

9946 ***E. decliva*** (Grt.)
Sloping Sallow
PARKS AND RESERVES: AJJ, BBL, BCN, KGC, PNM. **COUNTIES:** AND, BEL, BOO, CAM, HLN, JEF. **DATES:** Oct. 15–Nov. 22, Dec. 5, 19. Local and uncommon.

Genus ***Cheataglaea*** Franc.

9949 ***C. tremula*** (Harv.)
Trembling Sallow
Jefferson Co., Valley Station, Nov. 1; Powell Co., RRG, at sugar bait, Oct. 25, Nov. 4–15. Uncommon.

9950 ***C. sericea*** (Morr.)
Silky Sallow
Bell Co., PNM, Oct. 8, 13, 15; Harlan Co., KGC, at sugar bait, Oct. 28; Powell Co., RRG, Tunnel Ridge, Nov. 9, 15; Wolfe Co., DBNF, Rock Bridge Trail, Nov. 5. Uncommon.

Genus ***Eucirroedia*** Grt.

9952 ***E. pampina*** (Gn.)
Scalloped Sallow
PARKS AND RESERVES: BBL, BOW, GBT, KGC, LCP, PNM, RRG, WVP. **COUNTIES:** BEL, BOO, CRL, HLN, JEF, MEN, OLD, OWS, POW, RUS. **DATES:** June 4, Sept. 10, 19–Nov. 15. Locally common.

Genus ***Sunira*** Franc.

9957 ***S. bicolorago*** (Gn.)
Bicolored Sallow
PARKS AND RESERVES: BBL, BBM, BIP, BMF, CBF, GBT, KGC, LCP, NBP, OCP, PNM, RRG. **COUNTIES:** BEL, BOO, BUL, CRL, FAY, HLN, JEF, MEA, MEN, OLD, OWS, PIK, POW, RUS, SHE, WHI. **DATES:** Aug. 4–Dec. 7, 16. **REMARKS:** Abundant. Seems well adapted to city neighborhoods.

Genus ***Anathix*** Franc.

9961 ***A. ralla*** (G. & R.)
Dotted Sallow
PARKS AND RESERVES: BMF, CAV, CBF, CCR, FKN, GBT, LBK, MCV, PNM, RRG, WVP. **COUNTIES:** BAT, BEL, BUL, CRL, EDM, HRD, JEF, LAU, MEA, MEN, TRI, WHI. **DATES:** June 7, Aug. 4, 28–Oct. 31. Common.

Genus ***Platypolia*** Grt.

9976 ***P. anceps*** (Steph.)
Fayette Co., Lexington, Trail's End Camp, Oct., 1924, M. Didlake (UK); Powell Co., RRG, 1.3 km (0.8 mi) E of Nada Tunnel, Nov. 4, 1988, L. D. Gibson. Uncommon to rare.

Genus ***Xylotype*** Hamp.

9979 ***X. capax*** (G. & R.)
Broad Sallow
Bell Co., PNM, Oct. 23; Harlan Co., KGC, at lights, Oct. 28; Powell Co., NBP, Nov. 22. Local in eastern highlands.

Genus ***Pachypolia*** Grt.

9992 ***P. atricornis*** Grt.
PARKS AND RESERVES: BBL, BBM, BCN, OCP, RRG. **COUNTIES:** BOO, HLN, MEA, MEN, OLD, POW. **DATES:** Oct. 1–Nov. 17. **REMARKS:** Common.

Genus ***Feralia*** Grt.

10005 ***F. jocosa*** (Gn.)
Jocose Sallow
Laurel Co., DBNF, SR 1193 and Twin Branch Trail, Mar. 15; Menifee Co., RRG, Mar. 7–19, 29, Apr. 3; Morgan Co., Mar. 21; Powell Co., NBP, Apr. 11, and RRG, Mar. 4. Local and uncommon.

10007 ***F. major*** Sm.
Major Sallow
PARKS AND RESERVES: RRG. **COUNTIES:** LAU, MEN, MOR, PER, POW, ROW. **DATES:** Mar. 4–21, Apr. 11, 20–26, May 2, 5. Local and uncommon.

10008 ***F. comstocki*** (Grt.)
Comstock's Sallow
PARKS AND RESERVES: KGC, NBP, PNM, RRG. **COUNTIES:** BEL, HLN, LAU, MCY, MEN, POW, WOL. **DATES:** Apr. 11–May 2. **REMARKS:** Apparently our most common *Feralia* species.

Genus ***Psaphida*** Wlk.

NOTE: Poole (1989) synonymized *Eutolype* Grt. and *Copipanolis* Grt. to *Psaphida*.

10012 ***P. electilis*** (Morr.)
Chosen Sallow
PARKS AND RESERVES: BBL, BBM, BMF, CAC, GBT, KGC, LCP, LMP, NBP, OCP, PNM, RRG. **COUNTIES:** BAT, BEL, BOO, BUL, CAR, CRL, FAY, GAL, HLN, JEF, MCC, MEA, MEN, MUH, OWS, POW, RUS. **DATES:** Mar. 19–May 3. **REMARKS:** Records include those for *P. depilis* Grt., now considered a form of *P. electilis*. Common.

10013 ***P. grandis*** (Sm.)
Gray Sallow
PARKS AND RESERVES: BMF, LCP, LMP, OCP, RRG. **COUNTIES:** BUL, JEF, LAU, MEA,

MEN, MUH, POW, RUS. **DATES:** Mar. 4–19, Apr. 4, 13, 26. Common.

10014 ***P. rolandi*** (Grt.)
Roland's Sallow
PARKS AND RESERVES: BBL, BBM, BMF, CAC, HWS, LMP, NBP, OCP, RRG. **COUNTIES:** BOO, BUL, CAR, HLN, JEF, MCC, MEA, MEN, MUH, OLD, POW. **DATES:** Mar. 3–Apr. 29. Common.

10016 ***P. styracis*** (Gn.)
Fawn Sallow
PARKS AND RESERVES: BMF, CBF, OCP, RRG. **COUNTIES:** BUL, MEA, MEN, NEL, PER, POW, WHI. **DATES:** Mar. 4–Apr. 1, May 22. **REMARKS:** Locally common. L. D. Gibson found it abundant on Mar. 4, 1989, in Powell Co., RRG.

10019 ***P. resumens*** Wlk.
Figure-eight Sallow
PARKS AND RESERVES: BMF, CAC, GBT, KGC, LCP, LMP, MCV, NBP, OCP, RRG. **COUNTIES:** BUL, CAR, CRL, EDM, HLN, JEF, MEA, MEN, MUH, OWS, POW, RUS. **DATES:** Mar. 19–Apr. 22. Common.

10020 ***P. thaxteriana*** (Grt.)
Bullitt Co., BMF, Mar. 30; Jefferson Co., CAH, Mar. 5, and Louisville, Mar. 10; Menifee Co., RRG, Mar. 7, 12; Powell Co., RRG, Tunnel Ridge, Mar. 4, 11. Locally common.

Genus ***Copivaleria*** Grt.

10021 ***C. grotei*** (Morr.)
Grote's Sallow
PARKS AND RESERVES: BBL, BBM, BIP, BMF, CAC, CBB, GBT, HWS, LMP, NBP, OCP, RRG. **COUNTIES:** BOO, BUL, CAR, CRL, FAY, HLN, JEF, MEA, MEN, MUH, OLD, OWS, PIK, POW, ROW. **DATES:** Mar. 9, 28–Apr. 2, 11–26. Common.

Genus ***Homohadena*** Grt.

10065 ***H. infixa*** (Wlk.)
Broad-lined Sallow
PARKS AND RESERVES: BAN, BBL, BMF, FKN, HWS, LBK, OCP, PYR, WVP. **COUNTIES:** BAR, BOO, BRA, BUL, CAD, HRD, JEF, MEA, NEL, OLD, TRI, WOO. **DATES:** May 23–June 23, July 9, Aug. 16. Locally common.

Genus ***Adita*** Grt.

10067 ***A. chionanthi*** (J. E. Smith)
Bath Co., CAV, FS 918, Aug. 28; Menifee Co., RRG, N of Slade, East Fork Indian Creek, in black-light trap, Sept. 21, 22. Local and uncommon.

Genus ***Cucullia*** Schr.

10197 ***C. florea*** Gn.
Harlan Co., BBM, July 14, 1984, D. J. Wright. Rare.

10199 ***C. omissa*** Dod
Omitted Cucullia
Harlan Co., BBM, July 14, Aug. 8; Jefferson Co., Valley Station, Sept. 5. Uncommon.

10200 ***C. asteroides*** Gn.
The Asteroid
PARKS AND RESERVES: BAN, BCN, BMF, BOW, BUC, KGC, LCP, RGR, RRG. **COUNTIES:** BAR, BOO, BRA, BRE, BRK, BUL, FAY, FUL, GRY, HLN, JEF, MAS, MEA, MEN, OWS, PER, ROW, RUS. **DATES:** Apr. 1, May 10–June 8, 18, July 5, 14, Aug. 4–25, Sept. 5–7, 22. Common.

10202 ***C. convexipennis*** G. & R.
Brown-hooded Owlet; Brown-bordered Cucullia
PARKS AND RESERVES: BBM, BMF, BOW. **COUNTIES:** BOO, BRA, BRK, BUL, HLN, JEF, MEA, NEL. **DATES:** May 29, June 4, July 9, 29, Aug. 7–21, Sept. 1, 26. Local and uncommon.

Subfamily **Hadeninae**

Genus ***Discestra*** Hamp.

10223 ***D. trifolii*** (Hufn.)
The Nutmeg
Boone Co., BBL, May 24, and BCN, Aug. 26, L. D. Gibson; Fayette Co., Lexington, Aug. 18, 1891, Aug. 6, 1896, Aug. 14, 1923; also a specimen from Lexington reared from cabbage, pupated Sept. 1, 1903, emerged Apr. 8, 1904; all (except Boone Co., records) H. Garman (UK). Uncommon.

Genus ***Sideridis*** Hbn.

10266 ***S. congermana*** (Morr.)
The German Cousin
Hardin Co., Elizabethtown, June; Harlan Co., BBM, July 9, Aug. 9. Local and rare.

Genus ***Polia*** Ochs.

10275 ***P. nimbosa*** (Gn.)
Stormy Arches
Harlan Co., BBM, July 8–14. Common at light on summit.

10276 ***P. imbrifera*** (Gn.)
Cloudy Arches
Harlan Co., BBM, July 9–14. Less common than stormy arches at lights.

10288 ***P. detracta*** (Wlk.)
Disparaged Arches
PARKS AND RESERVES: BBL, BBM, BMF, BUC, HWS, JWS, LCP, NBP, PNM, PYR. **COUNTIES:** BEL, BOO, BRE, BUL, CAD, FLO, HLN, JEF, MEN, OLD, OWS, PER, POW, PUL, RUS, WOO. **DATES:** May 23–Aug. 8, 9, Sept. 1. **REMARKS:** Common. McCabe (1980) indicated this does not belong in *Polia*, but it is retained here until its generic placement is established.

Genus ***Melanchra*** Hbn.

10292 ***M. adjuncta*** (Gn.)
Hitched Arches
PARKS AND RESERVES: BBL, HWS, OCP, RRG. **COUNTIES:** BOO, FAY, JEF, JES, MEA, MEN, MER, OLD, OWS, ROW. **DATES:** May 9–Sept. 2. Common to abundant.

10293 ***M. picta*** (Harr.)
Zebra Caterpillar Moth
Boone Co., BBL, May 9, and BCN, June 15; Fayette Co., larva collected at Lexington, Sept. 22, pupated Oct. 20, and emerged in May (also collected May 18, Aug. 26); Jefferson Co., Prospect, Sept. 12. Uncommon.

Genus ***Lacanobia*** Billberg

10299 ***L. subjuncta*** (G. & R.)
Boone Co., BBL, May 20, 23, Aug. 1; Harlan Co., BBM, Aug. 8; Meade Co., Doe Valley, July 29, and OCP, Aug. 12. Uncommon.

Genus ***Trichordestra*** McCabe

10304 ***T. legitima*** (Grt.)
Striped Garden Caterpillar Moth
PARKS AND RESERVES: BBM, BOW, FKN, OCP, RRG. **COUNTIES:** BOO, BRA, BRK, FAY, GUP, HLN, JEF, MEA, MEN, OWS, ROW. **DATES:** May 19, June 30, July 8, 14, 29–Sept. 15. **REMARKS:** Common. This species was listed in *Lacanobia* in Hodges et al. (1983), but was placed in this genus by McCabe (1980).

Genus ***Hadena*** Schr.

10316 ***H. ectypa*** (Morr.)
Boone Co., BBL, June 20; Powell Co., RRG, July 11; Woodford Co., 1.6 km (1 mi) W of SR 68, along Kentucky River, May 25. Uncommon. This species was noted in Forbes (1954) as "very rare."

Genus ***Anepia*** Hamp.

10317 ***A. capsularis*** (Gn.)
Capsule Moth
Jefferson Co., Valley Station, May 17, 20, 22. Local and uncommon.

Genus ***Lacinipolia*** McD.

10368 ***L. meditata*** (Grt.)
The Thinker
PARKS AND RESERVES: ASC, BBL, GBO, GBT, RRG, WVP. **COUNTIES:** BOO, BUL, CRL, FAY, GRY, GUP, JEF, LAU, MEN, OWS, POW. **DATES:** Apr. 16–Oct. 7, Nov. 8–18. Very common.

10372 ***L. anguina*** (Grt.)
Snaky Arches
Bell Co., PNM, May 4, 12, 13; Caldwell Co., PYR, May 10; Carter Co., CAC, Apr. 21; Owsley Co., 4.8 km (3 mi) NE of Booneville, May 23; Powell Co., RRG, Tunnel Ridge, Apr. 28, May 14–21. Local and uncommon.

10393 ***L. teligera*** (Morr.)
Woodford Co., 1.6 km (1 mi) W of Highway 68 on Kentucky River, Oct. 9, 1980, J. S. Nordin. Rare.

10397 ***L. renigera*** (Steph.)
Bristly Cutworm Moth
PARKS AND RESERVES: AJJ, BAN, BMF, BUC, CBB, FKN, GBT, HWS, KLK, LBK, LMP, NBP, PNM, PYR, RRG. **COUNTIES:** BAR, BEL, BOO, BRA, BRE, BUL, CAD, CAM, CRL, FAY, FUL, HLN, HRD, HRT, JEF, MCC, MEA, MEN, MRL, MUH, NIC, OLD, OWS, PER, POW, ROW, SHE, TRI. **DATES:** Apr. 30–Oct. 27. **REMARKS:** Very common. Males of this species respond to pheromones released by the bolas spider, *Mastophora hutchinsoni*, which then preys on the moths (Yeargan 1988).

10405 ***L. lorea*** (Gn.)
Bridled Arches
PARKS AND RESERVES: BBM, HWS, JWS, WVP. **COUNTIES:** FLO, GAL, HLN, JEF, OLD, OWS, WOO. **DATES:** May 9, 26–June 20, July 11. Uncommon.

10406 ***L. olivacea*** (Morr.)
Olive Arches
Harlan Co., BBM, July 8–11. Common at one site on the summit.

10413 ***L. explicata*** McD.
Explicit Arches
In his original description, McDunnough (1937) lists as paratypes 1 male and 2 females labelled "Kentucky" in the Barnes Collection, the male dated Sept. 1, 1899, and the females Sept. 5, 1899, and no date. These specimens are now in the NMNH. Rare; no recent records.

10414 ***L. implicata*** McD.
Implicit Arches
PARKS AND RESERVES: AJJ, ASC, BAN, BBL, BUC, CAC, CAV, CCR, GBT, HWS, KLK, LCP, PYR, RRG. **COUNTIES:** BAR, BAT, BOO, BRA, BRE, BUL, CAD, CAM, CAR, CRL, GRY, HND, JEF, LAU, MRL, OLD, OWS, PER, POW, RUS, WAR, WHI, WOO. **DATES:** Apr. 18, 23, May 3–8, 17, June 3, 12, 24, 30, July 7, 18–29, Aug. 23–Oct. 9. **REMARKS:** Locally common. Most records are in September.

Genus *Faronta* Sm.

10431 ***F. diffusa*** (Wlk.)
Wheat Head Armyworm Moth
PARKS AND RESERVES: BBL, BBM, RGR. **COUNTIES:** BOO, BRK, FAY, GRY, HLN, JEF, MEN, MOR, NEL. **DATES:** Apr. 11, 17, 26–May 5, 13–17, July 23, Aug. 8–23, Sept. 2, 3, 16, Oct. 1. Common.

10434 ***F. rubripennis*** (G. & R.)
The Pink-streak
Henderson Co., SWR, one fresh male specimen taken in light trap, Sept. 9, 1983, L. D. Gibson. A rare southern species. Rings et al. (1992) considered it threatened in Ohio prairie remnants.

Genus *Aletia* Hbn.

10436 ***A. oxygala*** (Grt.)
Bullitt Co., BMF, in Malaise trap, May 11–20, 1976, A. J. Brownell. Rare.

Genus *Pseudaletia* Franc.

10438 ***P. unipuncta*** (Haw.)
Armyworm Moth
PARKS AND RESERVES: ASC, BAN, BEC, BMF, BUC, CAC, CBB, CMI, FKN, GBO, GBT, HWS, JWS, KLK, KRC, LCP, MCV, NBP, OCP, PNM, PYR, RGR, SWR. **COUNTIES:** BAR, BEL, BOO, BOU, BRA, BRE, BRK, BUL, CAD, CAR, CRL, EDM, FAY, FLO, FUL, GAL, GRY, GUP, HND, JEF, JOH, MAD, MCC, MEA, MRL, MOR, OLD, PER, POW, ROW, RUS, SHE, WOL. **DATES:** Mar. 30–Dec. 8. Common to abundant and a major pest of corn and other grain crops.

Genus *Leucania* Ochs.

10444 ***L. phragmitidicola*** Gn.
Phragmites Wainscot
PARKS AND RESERVES: BMF, HWS, LBK, PYR, RGR. **COUNTIES:** BRK, BUL, CAD, FAY, GRY, JEF, JOH, OLD, OWS, ROW, TRI. **DATES:** Apr. 30–Oct. 26. **REMARKS:** Common. Misspelled in Hodges et al. (1983).

10445 ***L. linda*** Franc.
Linda's Wainscot
PARKS AND RESERVES: ASC, CAV, GBT, HNK, HWS, LCP. **COUNTIES:** BAT, BRA, CAL, CRL, GAL, GRY, JEF, MCC, OLD, OWS, ROW, RUS. **DATES:** May 4, 19, 20, June 9–27, Aug. 21–Sept. 7, 16, 29. Common.

10455 ***L. scirpicola*** Gn.
Scirpus Wainscot
PARKS AND RESERVES: BMF, HWS, KLK. **COUNTIES:** BUL, FAY, FUL, JEF, MRL, NIC, OLD, SHE, TRI, WOO. **DATES:** Apr. 28, May 11–23, June 2–21, Aug. 8–Oct. 8. Locally common.

10456 ***L. adjuta*** (Grt.)
Bath Co., CAV, Oct. 15; Jefferson Co., Valley Station, Mar. 21, June 8, 12, 23, 29, July 8, Aug. 10, 21, 26, Sept. 11, 21, Oct. 4, 6, Nov. 5; Powell Co., DBNF, 1.3 km (0.8 mi) E of Nada Tunnel along Martin Fork, Oct 15, 25; Russell Co., LCP, Oct. 13. Locally common.

10459 ***L. inermis*** (Fbs.)
Unarmed Wainscot
PARKS AND RESERVES: ASC, BMF, HWS, PCP, WVP. **COUNTIES:** BOO, BUL, GRY, HRD, JEF, LIV, MEN, OLD, OWS. **DATES:** May 11–30, June 1–16, Aug. 8, 10, Sept. 2–7. Common.

10461 ***L. ursula*** (Fbs.)
Ursula Wainscot
PARKS AND RESERVES: ASC, BAN, BBL, BMF, CBF, HWS, OCP, RGR, RRG. **COUNTIES:** BAR, BOO, BRK, BUL, FAY, GRY, JEF, MEA, OLD, POW, ROW, RUS, WHI. **DATES:** May 10, 19–Oct. 5. Common.

10462 ***L. pseudargyria*** Gn.
False Wainscot
PARKS AND RESERVES: KRC, PMS, PNM. **COUNTIES:** BEL, FAY, HLN, MEN, NIC, OWS, WOL. **DATES:** May 25, June 30–July 25, Aug. 6, 9, 27. Common.

Genus *Orthosia* Ochs.

10487 ***O. rubescens*** (Wlk.)
Ruby Quaker
PARKS AND RESERVES: BBL, BBM, BIP, BMF, CAC, GBT, KGC, LCP, LMP, NBP, OCP, PNM, RRG. **COUNTIES:** BEL, BOO, BUL, CAR, CRL, HLN, JEF, MEA, MUH, PIK, POW, RUS. **DATES:** Mar. 4, 28–May 20. Locally common to abundant.

10488 **O. garmani** (Grt.)
Garman's Quaker
PARKS AND RESERVES: BBL, CAC, HWS, KLK, LMP, NBP, RRG. **COUNTIES:** BOO, CAR, JEF, MEN, MRL, MUH, OLD, OWS, POW. **DATES:** Mar. 25–Apr. 20. Common.

10490 **O. revicta** (Morr.)
Subdued Quaker
Harlan Co., BBM, in light trap, Apr. 20, and KGC, Apr. 20; Powell Co., RRG, Tunnel Ridge, Apr. 28. Uncommon.

10491 **O. alurina** (Sm.)
Gray Quaker
PARKS AND RESERVES: BBM, BMF, GBT, HWS, LMP, NBP, OCP. **COUNTIES:** BUL, CRL, HLN, JEF, MEA, MUH, OLD, POW. **DATES:** Feb. 29, Mar. 11–20, 29, Apr. 1–11, 20. Common.

10495 **O. hibisci** (Gn.)
Freckled Green Fruitworm Moth
PARKS AND RESERVES: BBM, BMF, CAC, GBT, HWS, LMP, NBP, OCP, RRG. **COUNTIES:** BOO, BUL, CAR, CRL, FAY, HLN, JEF, MEA, MUH, NEL, OLD, OWS, POW, SHE, TRM, WHI. **DATES:** Mar. 1, 9–Apr. 21. Common to abundant.

Genus *Crocigrapha* Grt.

10501 **C. normani** (Grt.)
Norman's Quaker
PARKS AND RESERVES: BBL, BBM, BMF, CAC, KGC, OCP, PNM, RRG. **COUNTIES:** BEL, BOO, BUL, CAR, HLN, JEF, MCC, MEA, MEN, OLD, OWS, POW. **DATES:** Mar. 4, 11, Apr. 8–May 7. Common to abundant.

Genus *Himella* Grt.

10502 **H. intractata** (Morr.)
Intractable Quaker
PARKS AND RESERVES: BMF, BUC, CAC, KGC, LMP, MCV, NBP, OCP. **COUNTIES:** BRE, BUL, CAR, EDM, HLN, JEF, MCC, MEA, MUH, OLD, PER, POW, ROW. **DATES:** Mar. 20–May 3, June 1. Common.

Genus *Egira* Dup.

10517 **E. alternans** (Wlk.)
Alternate Woodling
PARKS AND RESERVES: BIP, BMF, CAC, KGC, NBP, OCP, PNM, RRG. **COUNTIES:** BEL, BUL, CAR, HLN, JEF, MEA, MEN, OWS, PIK, POW, ROW. **DATES:** Apr. 6–May 2. Common.

Genus *Achatia* Hbn.

10518 **A. distincta** Hbn.
Distinct Quaker
PARKS AND RESERVES: BBL, BIP, BMF, BUC, CAC, CBF, JWS, KGC, LCP, LMP, MCV, NBP, OCP, PNM. **COUNTIES:** BAR, BEL, BOO, BRE, BUL, CAR, EDM, FLO, FUL, HLN, JEF, MCC, MEA, MUH, OLD, OWS, PER, PIK, POW, ROW, RUS, WHI. **DATES:** Mar. 19, 29, Apr., May 21, July 27, Sept. 7, Oct. 1, Nov. 5. Common.

Genus *Morrisonia* Grt.

10520 **M. evicta** (Grt.)
Bicolored Woodgrain
PARKS AND RESERVES: BBL, BIP, BUC, KGC, NBP, PNM, RRG. **COUNTIES:** BEL, BOO, BRE, HLN, MEN, PER, PIK, POW. **DATES:** Apr. 12–May 4, 12. Common.

10521 **M. confusa** (Hbn.)
Confused Woodgrain
PARKS AND RESERVES: BBL, BIP, BMF, CAC, CBF, HNK, JWS, KGC, LCP, MCV, NBP, OCP, PNM, RBF, RGR, RRG. **COUNTIES:** BEL, BOO, BRE, BRK, BUL, CAL, CAR, EDM, FAY, FLO, GRY, HLN, JEF, MEA, MEN, NEL, PIK, POW, ROW, RUS, WHI. **DATES:** Apr. 11–May 28, July 25, Aug. 10, 28. Common to abundant.

10521.1 **M. latex** (Gn.)
Fluid Arches
PARKS AND RESERVES: AJJ, BAN, BBL, BBM, BMF, BUC, CBF, FKN, HWS, JWS, KGC, LCP, LMP, OCP, PNM, RRG. **COUNTIES:** BAR, BEL, BOO, BRE, BUL, CAM, FLO, HLN, HRD, JEF, MEA, MEN, MUH, NEL, OLD, OWS, PER, ROW, RUS, WHI. **DATES:** Apr. 19–July 18, Sept. 22, Nov. 17–23. **REMARKS:** Common. This species was transferred from *Polia* (Hodges et al. 1983, checklist number 10291) by Poole (1989).

Genus *Nephelodes* Gn.

10524 **N. minians** Gn.
Bronzed Cutworm Moth
PARKS AND RESERVES: BAN, BLA, BOW, CAC, CCR, GBT, KLK, LCP. **COUNTIES:** BAR, BOO, CAR, CRL, FAY, JEF, MEN, MRL, ROW, RUS, SHE. **DATES:** Aug. 4, 31–Oct. 13. **REMARKS:** Common. This species is among four preyed upon by the bolas spider, *Mastophora hutchinsoni*, which attracts male moths with its scent (Yeargan 1988).

Genus *Homorthodes* McD.

10532 ***H. furfurata*** (Grt.)
Scurfy Quaker
PARKS AND RESERVES: BMF, BOW, BUC, CBF, JWS, LCP, OCP, PNM, PYR. **COUNTIES:** BEL, BOO, BRE, BUL, CHR, FLO, JEF, MEA, OWS, PER, RUS, WHI. **DATES:** Apr. 22–June 18, Aug. 6, 16, 17–25. **REMARKS:** Our subspecies is *lindseyi* (Benj.). Common.

Genus *Ulolonche* Smith

10567 ***U. culea*** (Gn.)
Sheathed Quaker
Calloway Co., HNK, May 4; Christian Co., PYR, May 10; McCracken Co., Ray Black farm, 9.7 km (6 mi) SW of Paducah, Apr. 30; Russell Co., LCP, May 16. Local and uncommon.

Genus *Pseudorthodes* McD.

10578 ***P. vecors*** (Gn.)
Small Brown Quaker
PARKS AND RESERVES: BBL, BBM, BMF, CBB, OCP, RRG. **COUNTIES:** BOO, BUL, FAY, HLN, JEF, MEA, MEN, OLD, ROW. **DATES:** Apr. 3, May 13–27, June 10, 13, 21–23, July 7–24, Aug. 6–31, Oct. 16, 21. Locally common.

Genus *Orthodes* Gn.

10585 ***O. crenulata*** (Butler)
Rustic Quaker
PARKS AND RESERVES: BBL, BBM, BMF, HWS, LCP, NBP, OCP, PNM, RGR, WVP. **COUNTIES:** BEL, BOO, BRA, BRK, BUL, FAY, GRY, HLN, JEF, LAU, MEA, OLD, POW, ROW, RUS, WHI. **DATES:** Mar. 13–Sept. 3, 12, 19. Common.

10587 ***O. cynica*** Gn.
Cynical Quaker
PARKS AND RESERVES: BBM, BUC, COB, GBT, HWS, PNM. **COUNTIES:** BEL, BRA, BRE, CRL, GAL, HIC, HLN, JEF, MEN, OLD, OWS, PER, WOO. **DATES:** May 6, 14, 16, 27–July 11, Sept. 16. Common.

10589.1 ***O. goodelli*** (Grt.)
Goodell's Arches
PARKS AND RESERVES: BBM, HNK, OCP, RGR, RRG. **COUNTIES:** BEL, BRK, CAL, FUL, GRY, HLN, MEA, MEN, ROW. **DATES:** May 4, 15, June 21, July 8, 14, 27, Aug. 12–17, Sept. 5, 6. **REMARKS:** Common. McCabe (1980) indicates this species seems related to *Orthodes cynica*; but it is retained here until final placement is published. This species was moved from *Polia* (Hodges et al. 1983, checklist number 10289) by Poole (1989).

Genus *Tricholita* Grt.

10627 ***T. signata*** (Wlk.)
Signate Quaker
PARKS AND RESERVES: BBL, NBP, RRG. **COUNTIES:** BOO, BRA, BRK, JEF, LAU, MEN, OLD, OWS, POW. **DATES:** May 29, June 18, July 31, Aug. 2–Oct. 10, 28. Common.

Genus *Xanthopastis* Hbn.

10640 ***X. timais*** (Cram.)
Spanish Moth
Livingston Co., Carrsville, June 1998; Meade Co., probably collected in 1966–1968 at OCP; Warren Co., Bowling Green, no date (an old specimen in UK Collection). Rare. A migrant from the deep south, where it is common.

Subfamily **Noctuinae**

NOTE: Changes in nomenclature and order of checklist numbers in this subfamily follow recent publications such as Lafontaine (1987, 1998).

Genus *Agrotis* Ochs.

10648 ***A. gladiaria*** Morr.
Swordsman Dart
PARKS AND RESERVES: CCR, GBT, LCP, PNM, WVP. **COUNTIES:** BEL, CRL, FAY, JEF, KEN, OLD, OWS, RUS. **DATES:** Sept. 1–Oct. 21. Common.

10651 ***A. venerabilis*** Wlk.
Venerable Dart
Jefferson Co., Valley Station, Oct. 12, 15; Owsley Co., 4.8 km (3 mi) NE of Booneville, Oct. 8; Powell Co., 1.6 km (1 mi) SW of Clay City, Oct. 7. Local and uncommon.

10661 ***A. malefida*** Gn.
Rascal Dart
Bullitt Co., Mt. Washington, July 28, 1965, C. V. Covell, Jr. Rare.

10663 ***A. ipsilon*** (Hufn.)
Black Cutworm Moth
PARKS AND RESERVES: BEC, BMF, BOW, BUC, CMI, GBT, HWS, JWS, KLK, LBK, LBL, LCP, NBP, OCP, PNM, PYR, RGR, RRG, WVP. **COUNTIES:** BEL, BOO, BRE, BRK, BUL, CAD, CRL, FAY, FLO, FUL, GRY, HOP, JEF, MCC, MEA, MRL, OLD, OWS, PER, POW, RUS, SHE, SPE, TRI. **DATES:** Apr. 9–Nov. 24. **REMARKS:** Common. A pest of many plants, including lawn grasses and garden vegetables.

10664 ***A. subterranea*** (F.)
Subterranean Dart
PARKS AND RESERVES: BAN, BBL, BCN, CBB, LCP, OCP, PNM, PYR, RRG. **COUNTIES:** BAR, BEL, BOO, CHR, FAY, JEF, MCC, MEA, OLD, OWS, POW, RUS. **DATES:** Mar. 20–Nov. 18. Common in some years.

Genus ***Feltia*** Wlk.

10670 ***F. jaculifera*** Gn.
Dingy Cutworm Moth
PARKS AND RESERVES: ASC, BAN, CAC, CAV, CCR, FKN, GBT, HWS, KLK, LCP, PNM. **COUNTIES:** BAR, BAT, BEL, BUL, CAM, CAR, CRL, FAY, GRV, GRY, HRD, JEF, MEA, MRL, OLD, RUS, SCO. **DATES:** May 4, July 18, Aug. 16–Oct. 27. Common.

10674 ***F. subgothica*** (Haw.)
Subgothic Dart
PARKS AND RESERVES: BBL, BMF, BUC, GBT, KLK, LCP. **COUNTIES:** BOO, BRE, BUL, CAD, CRL, FAY, JEF, LAU, MCC, MRL, OWS, PER, RUS, SHE. **DATES:** Aug. 23–Oct. 31. Common.

10675 ***F. tricosa*** (Lint.)
Christian Co., PYR, Sept. 12; Fulton Co., Willingham Bottoms, 4.0 km (2.5 mi) E of Cayce, Sept. 7; Jefferson Co., Valley Station, Sept 31, Oct. 12; Oldham Co., HWS, Sept. 6. Uncommon.

10676 ***F. herilis*** (Grt.)
Master's Dart
PARKS AND RESERVES: ASC, BLA, CAC, CAV, GBO, GBT, GRY, HWS, KLK, LCP, PYR. **COUNTIES:** BAT, CAD, CAM, CAR, CRL, FAY, FUL, GUP, GRY, JEF, MCC, MRL, OLD, RUS. **DATES:** June 15, Aug. 1, 15–Sept. 26, Oct. 5–20. Common.

Genus ***Trichosilia*** Hamp.

10698.1 ***T. manifesta*** (Morr.)
Russell Co., LCP, in light trap, Apr. 21, 1981, C. Cornett. Rare. Moved to *Trichosilia* from *Agrotis* (Hodges et al. 1983, checklist number 10666) by Lafontaine and Kononenko (1986).

10698.2 ***T. geniculata*** (G. & R.)
Knee-joint Dart
PARKS AND RESERVES: BUC, CAV, KLK, PYR, RRG. **COUNTIES:** BAT, BRE, CAD, JEF, MEN, MRL, OWS, PER. **DATES:** Aug. 28–21. **REMARKS:** Common. Moved from *Feltia* (Hodges et al. 1983, checklist number 10680) by Lafontaine and Kononenko (1986).

Genus ***Euxoa*** Hbn.

10705 ***E. messoria*** (Harr.)
Reaper Dart
Fayette Co., "from hemp, Mar. 24, 1903," "Aug. 1906," Sept. 10 (all UK); Harlan Co., BBM, July 14; Todd Co., larval damage reported on corn, May 22. Very local and uncommon.

10756 ***E. campestris*** (Grt.)
Harlan Co., BBM, July 8, 1977, R. A. Henderson. Rare.

10793 ***E. scholastica*** McD.
Harlan Co., BBM, July 9, 1977, C. V. Covell, Jr., and R. A. Henderson. Very local on our tallest mountain.

10812 ***E. bostoniensis*** (Grt.)
Meade Co., OCP, Oct. 14, 1988, R. A. Henderson. Rare.

Genus ***Richia*** Grt.

10870 ***R. acclivis*** (Morr.)
PARKS AND RESERVES: PYR, RGR, RRG. **COUNTIES:** BOO, BRA, BRK, CHR, GRY, JEF, MEN. **DATES:** Aug. 14–23, Sept. 7, 13. **REMARKS:** Formerly in *Loxagrotis* McD., a junior synonym of *Richia* (Poole 1989).

Genus ***Ochropleura*** Hbn.

10891 ***O. implecta*** Lafontaine
Flame-shouldered Dart
PARKS AND RESERVES: BBL, BUC, HWS, PYR, RRG, SWR. **COUNTIES:** BOO, BRA, BRE, CAD, FAY, HND, JEF, MEN, MRL, OLD, OWS, PER. **DATES:** May 1–15, June 6, 22–29, July 6, 20, Aug. 4–15, Sept.10, 12. **REMARKS:** Common. Formerly known as *O. plecta* Linnaeus, a distinct European species (Lafontaine 1998).

Genus ***Euagrotis*** McD.

10901 ***E. lubricans*** (Gn.)
Slippery Dart
PARKS AND RESERVES: BMF, GBT, LBL, LCP, PNM, PYR. **COUNTIES:** BEL, BUL, CAD, CRL, JEF, RUS, TRI. **DATES:** May 15, July 2, 14, 19, Aug. 19, 24, Sept. 3–17. Common.

10902 ***E. forbesi*** Franc.
Forbes' Dart
Two paratypes from Louisville are in the collection of the NMNH (Franclemont 1952). One was collected on July 6, the other July 21, by D. Nettleroth; both from the Jacob Doll Collection.

10903 ***E. illapsa*** (Wlk.)
Snowy Dart
PARKS AND RESERVES: AJJ, ASC, BAN, BBL, BMF, GBT, GRY, HWS, LBL, PYR. **COUNTIES:** BAR, BOO, BRA, BUL, CAM, CHR, CRL, GAL, GRY, JEF, OLD, TRI. **DATES:** May 7–Oct. 14. Common.

Genus *Anicla* Grt.

10911 ***A. infecta*** (Ochs.)
Green Cutworm Moth
PARKS AND RESERVES: BAN, BBL, CAV, GBT, HWS, IRO, LBK, LCP, OCP, PYR, RRG, WVP. **COUNTIES:** BAR, BAT, BOO, CHR, CRL, FAY, FUL, GRV, HND, HOP, JEF, JES, MCC, MEA, OLD, OWS, POW, RUS, TRI. **DATES:** June 3–Oct. 28, Nov. 15, Dec. 5. Common.

Genus *Peridroma* Hbn.

10915 ***P. saucia*** (Hbn.)
Variegated Cutworm Moth
PARKS AND RESERVES: ASC, BAN, BMF, BOW, BUC, CBB, FKN, GBT, HWS, JCF, JWS, KGC, KLK, LBK, LCP, MCV, NBP, OCP, PNM, PYR, RGR, RRG, WVP. **COUNTIES:** BAR, BEL, BOO, BRE, BRK, BUL, CAD, CRL, EDM, FAY, FLO, GRY, HLN, HND, JEF, MCC, MEA, MRL, OLD, OWS, PER, POW, RUS, SHE, TRI. **DATES:** Mar. 9–Dec. 5. **REMARKS:** A common to abundant pest throughout the state feeding on many trees and crops.

Genus *Diarsia* Hbn.

10917 ***D. rubifera*** (Grt.)
Harlan Co., BBM, July 8–12, Aug. 8, and KGC, July 9. Found only at our highest elevations.

10919 ***D. jucunda*** (Wlk.)
Smaller Pinkish Dart
Harlan Co., BBM, July 11, 14, and KGC, 2 specimens, July 11; all L. D. Gibson. Rare; recorded only at highest elevations.

Genus *Spaelotis* Bdv.

10926 ***S. clandestina*** (Harr.)
Clandestine Dart
Harlan Co., BBM, 2 specimens, June 8, July 12, R. A. Henderson. Rare. The larva is called the w-marked cutworm.

Genus *Xestia* Hbn.

10942.1 ***X. dolosa*** Franc.
Greater Black-letter Dart
PARKS AND RESERVES: BAN, BBL, BBM, BIP, BUC, CBB, GBT, HWS, JWS, KLK, LCP, LMP, NBP, OCP, PNM, RBF, RGR, SWR. **COUNTIES:** BAR, BEL, BOO, BRA, BRE, BRK, CRL, FAY, FLO, GAL, GRY, HEN, HLN, HND, JEF, MEA, MEN, MRL, MUH, OLD, OWS, PER, PIK, POW, RUS, SHE. **DATES:** Apr. 22–Oct.10. Common to abundant.

10943 ***X. normaniana*** (Grt.)
Norman's Dart
PARKS AND RESERVES: BBL, CAC, CAV, GBT, OCP, RGR, RRG. **COUNTIES:** BAT, BOO, BRK, CAR, CRL, GRY, HLN, JEF, LAU, MEA, MEN, OLD, OWS. **DATES:** Aug. 28–Oct. 18. Common.

10944 ***X. smithii*** (Snell.)
Smith's Dart
Carter Co., CAC, Sept. 11; Harlan Co., BBM, July 11, Aug. 8; Laurel Co., DBNF, Craig Creek Rd., Sept. 19; Menifee Co., RRG, Sept. 2, 14; Owsley Co., 4.8 km (3 mi) NE of Booneville, Sept. 2. Uncommon.

10967 ***X. elimata*** (Gn.)
Southern Variable Dart
PARKS AND RESERVES: CAC, CBF, CCR, PNM, RRG. **COUNTIES:** BEL, CAR, JEF, MEN, OWS, POW, ROW, WHI. **DATES:** Sept. 5–12, 26, Oct. 1–23. Locally common.

10969 ***X. dilucida*** (Morr.)
Dull Reddish Dart
Bell Co. PNM, Oct. 15; Menifee Co., RRG, Sept. 22, Oct. 3–12; Owsley Co., 4.8 km (3 mi) NE of Booneville, Sept. 27; Powell Co., Tunnel Ridge, Oct. 6, and NBP, Nov. 22; Rowan Co., DBNF, SR 1274, 3.2 km (2 mi) W of SR 519, Oct. 15. Uncommon.

Genus *Pseudohermanassa* Varga

10950 ***P. bicarnea*** (Gn.)
Pink-spotted Dart
PARKS AND RESERVES: BMF, CAC, CAV, CBF, CCR, GBT, HWS, PNM, RRG. **COUNTIES:** BAT, BEL, BUL, CAR, CRL, GAL, JEF, LAU, MEN, OLD, OWS, WHI. **DATES:** June 15, Aug. 7–Sept. 20, 29–Oct. 16. Common.

Genus *Agnorisma* Lafontaine

10954 ***A. bugrai*** (Kocak)
Carroll Co., GBT, Oct. 11; Jefferson Co., Louisville, Oct. 12. Uncommon. Formerly known as *Xestia collaris* (G. & R.).

10955 ***A. badinodis*** (Grt.)
Pale-banded Dart
PARKS AND RESERVES: BBL, CCR, FKN, GBT, LCP, PNM. **COUNTIES:** BEL, BOO, CRL, FAY, GUP, HRD, JEF, JES, KEN, MCC, MEA,

OLD, OWS, ROW, RUS, SHE, WOO. **DATES:** Sept. 10–Nov. 25. Common.

10956 ***A. bollii*** (Grt.)
PARKS AND RESERVES: LCP, GBT. **COUNTIES:** CAR, CRL, JEF, KEN, OLD, POW, ROW, RUS. **DATES:** Aug. 17, Sept. 22, Oct. 4–30. Locally common.

Genus *Cerastis* Ochs.

10994 ***C. tenebrifera*** (Wlk.)
Reddish Speckled Dart
PARKS AND RESERVES: BBM, BMF, CAC, GBT, HWS, NBP, OCP, WVP. **COUNTIES:** BOO, BRK, BUL, CAR, CRL, FAY, GAL, HLN, JEF, MCC, MEA, OLD, POW. **DATES:** Mar. 15–May 2. Common.

Genus *Choephora* G. & R.

10998 ***C. fungorum*** G. & R.
Bent-line Dart
PARKS AND RESERVES: BBL, BIP, FKN, GBT, LCP, OCP, PNM, RRG. **COUNTIES:** BEL, BOO, CRL, FAY, FUL, HRD, JEF, KEN, MCC, MEA, MEN, NEL, OLD, OWS, PIK, POW, RUS, WAR. **DATES:** June 26, Aug. 30, Sept. 12, 21–Oct. 28, Nov. 12, 22. Common.

Genus *Anaplectoides* McD.

11000 ***A. prasina*** (D. & S.)
Green Arches
Harlan Co., BBM, July 9–14. Very local on Kentucky's highest peak.

11002 ***A. brunneomedia*** McD.
Brown-lined Dart
Harlan Co., BBM, July 9–14. Very local. It flies with green arches.

Genus *Protolampra* McD.

11006 ***P. brunneicollis*** (Grt.)
Brown-collared Dart
PARKS AND RESERVES: ASC, BBL, BLA, BUC, CAV, GBT, HWS, KLK, OCP, PYR, RRG, SWR, WVP. **COUNTIES:** BAT, BOO, BRA, BRE, BUL, CAD, CRL, FAY, FUL, GRY, HND, JEF, MEA, MEN, MRL, OLD, OWS, PER, ROW, WAR, WOO. **DATES:** May 28–Oct. 6. Locally common.

Genus *Eueretagrotis* Sm.

11007 ***E. sigmoides*** (Gn.)
Sigmoid Dart
Harlan Co., BBM, July 9–14. Very local at 1,250 m (4,100 ft) elevation.

11009 ***E. attenta*** (Grt.)
Attentive Dart
Harlan Co., BBM, July 9, 14. Same occurrence as the sigmoid dart.

Genus *Lycophotia* Hbn.

11010 ***L. phyllophora*** (Grt.)
Bell Co., PNM, July 18; Harlan Co., BBM, July 9, and KGC, July 9. Uncommon.

Genus *Abagrotis* Sm.

11029 ***A. alternata*** (Grt.)
Greater Red Dart
PARKS AND RESERVES: ASC, BBL, BIP, BMF, CAC, GBO, KLK, LBL, LCP, OCP, PNM, SWR. **COUNTIES:** BEL, BOO, BUL, CAR, GRY, GUP, HND, JEF, MCC, MEA, MEN, MRL, OWS, PIK, ROW, RUS, TRI, WAR, WHI, WOO. **DATES:** May 7–Oct. 15. **REMARKS:** Common to abundant. The larva is the mottled gray cutworm, which can damage many plants.

11043.1 ***A. magnicupida*** Lafontaine
Bracken Co., 0.8 km (0.5 mi) W of Foster, July 3; Bullitt Co., 11.1 km (6.9 mi) E of Shepherdsville, June 23; Harlan Co., BBM, July 12; Menifee Co., June 21; Woodford Co., June 7. Uncommon.

11045 ***A. anchocelioides*** (Gn.)
PARKS AND RESERVES: BBM, BMF, CCR, OCP. **COUNTIES:** BUL, BRA, HLN, JEF, LAU, MEA. **DATES:** June 9, 11, 19, July 1–16, Aug. 14, Sept. 15. Uncommon.

Subfamily **Heliothentinae**

NOTE: Changes of the subfamily name from "Heliothinae," and synonymies here follow Hardwick (1996).

Genus *Pyrrhia* Hbn.

11063 ***P. adela*** Lafontaine & Mikkola
PARKS AND RESERVES: BMF, BOW, BUC, HWS, OCP, PNM, RGR, WVP. **COUNTIES:** BEL, BOO, BRE, BRK, BUL, FAY, GAL, GRY, GUP, JEF, MCC, MEA, NEL, OLD, OWS, PER, ROW. **DATES:** Apr. 20–June 12, 25–July 25, Aug. 5–Sept. 10, 19, 24, Oct. 6, 24. **REMARKS:** Common. Nomenclature follows Rings et al. (1992). Formerly considered to be *P. umbra* (Hufn.).

11064 ***P. exprimens*** (Wlk.)
Purple-lined Sallow
PARKS AND RESERVES: NBP, RRG. **COUNTIES:** HLN, LAU, MEN, OWS, POW, ROW. **DATES:** May 21, 30, June 28–July 23. Uncommon.

Genus ***Rhodoecia*** Hamp.

11065 ***R. aurantiago*** (Gn.)
Orange Sallow
Bath Co., CAV, FS 918, Aug. 28, Sept. 5, 11; Jefferson Co.,Valley Station ("Jan. 2" and "Feb. 21" labels of S. Scholz are so far from the normal flight season as to suggest that they are in error); Owsley Co., 4.8 km (3 mi) NE of Booneville, Aug. 27. Uncommon.

Genus ***Erythroecia*** Hamp.

11067 ***E. hebardi*** Skin.
Greenup Co., 2.6 km (1.6 mi) S of SR 10 in Portsmouth, and 20.3 km (12.6 mi) SW of South Shore on SR 784, both in light traps, Aug. 26, 27, 1989, J. D. Hooper. Rare. This species was diligently sought for several years before it was located.

Genus ***Helicoverpa*** Hardwick

11068 ***H. zea*** (Boddie)
Corn Earworm Moth
PARKS AND RESERVES: BAN, BIP, BMF, BUC, CAC, FAO, GBT, KLK, LBL, LCP, LMP, OCP, PNM, PYR, RGR, WVP. **COUNTIES:** BAR, BEL, BRA, BRE, BRK, BUL, CAD, CAR, CRL, FAY, FUL, GAL, GRV, GRY, HND, JEF, LAU, MAD, MCC, MEA, MOR, MRL, MUH, OWS, PER, PIK, ROW, RUS, WOO. **DATES:** May 4–Dec. 10. **REMARKS:** A common to abundant pest of corn throughout Kentucky. Formerly placed in *Heliothis* (Poole 1989).

Genus ***Heliothis*** Ochs.

11070 ***H. subflexa*** (Gn.)
Subflexa Straw
PARKS AND RESERVES: BUC. **COUNTIES:** BOO, BRE, FAY, HIC, JEF, MEN, PER, ROW. **DATES:** July 21, 24, Aug. 4, 16, 18, Sept. 3–13. Uncommon.

11071 ***H. virescens*** (F.)
Tobacco Budworm Moth
PARKS AND RESERVES: BMF, LCP, PNM, SWR. **COUNTIES:** BEL, BUL, FAY, FUL, GAL, GRV, HND, HRD, JEF, MCC, OWS, ROW, RUS. **DATES:** May 23, June 18, Aug. 4–14, Sept., Oct. 3–11. Common. A major pest of tobacco.

Genus ***Heliocheilus*** Grt.

11073 ***H. turbatus*** (Wlk.)
PARKS AND RESERVES: HWS, OCP, RGR. **COUNTIES:** BRK, FAY, GRY, JEF, MEA, OLD, WAR. **DATES:** July 2–7, 15, 17, Aug. 10, 12, 25–31. Uncommon.

11074 ***H. paradoxus*** (Grt.)
Fayette Co., Lexington, Aug. 10, 1937, L. H. Townsend. Rare.

Genus ***Schinia*** Hbn.

11105 ***S. bina*** (Gn.)
Bina Flower Moth
McCracken Co., Ray Black farm, 9.7 km (6 mi) SW of Paducah, July, 1975, and Aug. 18, 1976; W. R. Black, Jr. Uncommon and local.

11117 ***S. lynx*** (Gn.)
Lynx Flower Moth
PARKS AND RESERVES: BBL, BMF, HWS, MUR, OCP, PYR. **COUNTIES:** BOO, BRA, BRK, BUL, CHR, FAY, GAL, HIC, JEF, MCC, MEA, OLD, OWS, ROW. **DATES:** May 5–Oct. 11. Common.

11118 ***S. obscurata*** Stkr.
Obscure Flower Moth
Henderson Co., SWR, Sept. 10; Meade Co., OCP, Aug. 15. Rare.

11128 ***S. arcigera*** (Gn.)
Arcigera Flower Moth
PARKS AND RESERVES: ASC, GBO, HWS, LCP, RRG, SWR. **COUNTIES:** BOO, BUL, FAY, FUL, GRY, GUP, HND, JEF, MAD, MCC, MEN, OLD, OWS, ROW, RUS, WAR. **DATES:** June 4–Oct. 16. Common.

11135 ***S. rivulosa*** (Gn.)
Ragweed Flower Moth
PARKS AND RESERVES: BBL, HWS, KGC, LCP, LMP, OCP, PYR, RGR. **COUNTIES:** BOO, BRA, BRK, CHR, FAY, GAL, GRE, GRY, HLN, JEF, MEA, MEN, MRL, MUH, OLD, ROW, RUS, WHI. **DATES:** May 11–Oct. 1, Nov. 22. Common.

11140 ***S. saturata*** (Grt.)
Brown Flower Moth
Meade Co., OCP, Aug. 9, 25, 1983, R. A. Henderson. Rare.

11141 ***S. thoreaui*** (G. & R.)
Thoreau's Flower Moth
PARKS AND RESERVES: BBL, BCN, BOW, CER, RGR. **COUNTIES:** BOO, BRA, BRK, FAY, GRY, JEF, MEN, ROW. **DATES:** Aug. 10–Sept. 1. Uncommon.

11147 ***S. gracilenta*** Hbn.
Bath Co., CAV, Aug. 17; Henderson Co., SWR, Sept. 11; Jefferson Co., Valley Station, Aug. 12; Oldham Co., Goshen, Aug. 20. Uncommon. Formerly known as *bifascia* Hbn.

11148 ***S. oleagina*** Morr.
Christian Co., PYR, Aug. 22; Meade Co., OCP, Aug. 25; both L. D. Gibson. Rare.

11149 ***S. trifascia*** Hbn.
Three-lined Flower Moth
PARKS AND RESERVES: ASC, BBL, CCR, HWS, LCP, OCP, PNM, PYR, RGR, WVP. **COUNTIES:** BEL, BOO, BRK, CAD, CHR, FAY, FUL, GRY, JEF, MEA, MRL, OLD, OWS, RUS, WHI. **DATES:** July 4–Sept. 13, Oct. 5. Locally common.

11164 ***S. florida*** (Gn.)
Primrose Moth
PARKS AND RESERVES: BBM, PNM, WVP. **COUNTIES:** BEL, CAM, HLN, JEF, MEA, OWS. **DATES:** July 9, 19–31, Aug. 11–12. **REMARKS:** Usually found in daytime on its foodplant, evening primrose. Uncommon.

11168 ***S. gaurae*** (J. E. Smith)
Cloudod Crimoon
"Kentucky near Cincinnati, Ohio," Aug. 10–14, A. F. Braun; Bullitt Co., PCP, Aug. 18 (larvae and adults found on foodplant, *Gaura biennis*), adult at light, July 21; Meade Co., OCP, larvae found, reared, and emerged June 15. Probably not as rare as lack of records indicates.

11171 ***S. sanguinea*** (Gey.)
Glorious Flower Moth
Hardin Co., Elizabethtown, Aug.; Jefferson Co., Valley Station, July 4; Meade Co., OCP, in light trap, July 20. Rarely seen. Formerly known as *S. gloriosa* (Stkr.)

11177 ***S. nundina*** (Drury)
Goldenrod Flower Moth
PARKS AND RESERVES: LBL, PCP, PYR. **COUNTIES:** BUL, CHR, HRD, JEF, LYO, MCC. **DATES:** Aug. 13–Sept. 14. Uncommon.

11199 ***S. chrysella*** (Grote)
Caldwell Co., exit 56 of I-24, 20 km S of Princeton, Sept. 4, 1992, Aug. 27, 1994 and Aug. 29, 1995; all found by J. K. Adams at the same gas station lights each of the two years. Rare.

SPECIES ERRONEOUSLY ATTRIBUTED TO KENTUCKY

1970 ***Gnorimoschema albimarginella*** (Cham.) (Gelechiidae). This species was probably described from Colorado, not Kentucky. A syntype in the MCZ is mislabelled as being from Kentucky, according to Miller and Hodges (1990).

4576 ***Neonympha areolata*** (J. E. Smith) (Satyridae). The Georgia satyr is not known to occur in Kentucky, although Pyle (1981) included Kentucky in its range.

LITERATURE CITED

Adamski, D., and R. L. Brown. 1989. Morphology and systematics of North American Blastobasidae (Lepidoptera: Gelechioidea). Mississippi Agriculture and Forestry Experiment Station Technical Bulletin 165.

Annual Summary of Insect Conditions in Kentucky. 1976. Kentucky Agriculture Experiment Station, Lexington, Kentucky, USA.

Barnes, W., and J. H. McDunnough. 1918. Illustrations of the North American species of the genus *Catocala*. Memoirs of the American Museum of Natural History 3 (1).

Beebe, R. 1955. *Zeuzera* and *Adopaea* in Kentucky. Lepidopterists' News 9:142.

———. 1960. Field season summary for 1959 (Kentucky). News of the Lepidopterists' Society Number 3.

Blanchard, A., and E. C. Knudson. 1983. New North American species of Eucosmini (Lepidoptera: Tortricidae). Proceedings of the Entomological Society of Washington 85:845–852.

Borror, D. J., C. A. Triplehorn, and N. F. Johnson. 1989. An introduction to the study of insects. Sixth edition. Harcourt Brace College Publishers, New York, New York, USA.

Braun, A. F. 1908. Revision of the North American species of the genus *Lithocolletis* Hübner. Transactions of the American Entomological Society 34:269–357.

———. 1948. Elachistidae of North America (Microlepidoptera). Memoirs of the American Entomological Society 13.

———. 1963. The genus *Bucculatrix* in America north of Mexico (Microlepidoptera). Memoirs of the American Entomological Society 18.

———. 1972. Tischeriidae of America north of Mexico (Microlepidoptera). Memoirs of the American Entomological Society 28.

Brower, A. E. 1974. A list of the Lepidoptera of Maine. Part 1. The Macrolepidotera. Life Sciences and Agriculture Experiment Station, University of Maine at Orono, Technical Bulletin 66.

———. 1983. A list of the Lepidoptera of Maine. Part 2. The Microlepidoptera. Section 1. Limacodidae through Cossidae. Maine Department of Conservation and Department of Entomology, Maine Agriculture Experiment Station Technical Bulletin 109.

———. 1984. A list of the Lepidoptera of Maine. Part 2. The Microlepidoptera. Section 2. Cosmopterigidae through Hepialidae. Maine Department of Conservation and Department of Entomology, Maine Agriculture Experiment Station Technical Bulletin 114.

Brown, R. L. "1986" [1987]. Resurrection of *Catastega* Clemens and revision of the *Epinotia vertumnana* (Zeller) species-group (Tortricidae: Olethreutinae). Journal of the Lepidopterists' Society 40:327–346.

Burns, J. M. 1985. *Wallengrenia otho* and *W. egeremet* in eastern North America (Lepidoptera: Hesperiidae: Hesperiinae). Smithsonian Contributions to Zoology Number 423.

Burr, B. M., and M. L. Warren, Jr. 1986. A distributional atlas of Kentucky fishes. Kentucky Nature Preserves Commission Scientific and Technical Series 4:1–398.

Butler, L. 1987. A new species of nearctic *Bomolocha* (Noctuidae) from the Appalachian area. Journal of the Lepidopterists' Society 41: 104–107.

Calhoun, J. V. 1989. A visit to Kentucky by Edward Doubleday. Kentucky Lepidopterist 15:1.

Carothers, W. 1981a. Aerial detection survey of forest insect and disease activity, Tennessee Valley Authority, Land Between the Lakes, Kentucky. United States Department of Agriculture, Forest Service, Forest Pest Management Report 81–3–6.

———. 1981b. Aerial detection survey of forest insect and disease activity, federal, state and private lands, Kentucky. United States Department of Agriculture, Forest Service, Forest Pest Management Report 81–3–9.

Cassie, B., J. Glassberg, P. A. Opler, R. K. Robbins, and G. Tudor. 1995. North American Butterfly Association (NABA) checklist and English names of North American butterflies. North American Butterfly Association, Morristown, New Jersey, USA.

Chambers, V. T. 1871a. A new species of *Cemiostoma* (Micro-Lepidoptera. Tineina). Canadian Entomologist 3:23–25.

———. 1871b. Micro-Lepidoptera. Canadian Entomologist 3:54–58, 84–88, 108–112, 127–130, 146–149, 161–166, 182–185, 205–209, 221–224.

———. 1872. Micro-Lepidoptera. Canadian Entomologist 4:7–12, 25–29, 41–44, 65–69, 88–92, 106–108, 126–133, 146–150, 169–175, 191–195, 206–209, 222–226.

———. 1873. Micro-Lepidoptera. Canadian Entomologist 5:12–15, 44–50, 72–75, 85–91, 110–115, 124–128, 147–152, 173–176, 185–190, 229–232.

———. 1874a. Micro-Lepidoptera. Canadian Entomologist 6:8–11, 49–52, 72–77, 96–97, 128–130, 149–153, 166–170, 197–198, 217–220.

———. 1874b. Tineina from Texas. Canadian Entomologist 6:229–249.

———. 1874c. Prof. Frey, of Zürich, and some American Teneina [*sic*]. Cincinnati Quarterly Journal of Science 1:193–211.

———. 1875a. Tineina of the Central United States. Cincinnati Quarterly Journal of Science 2:97–121.

———. 1875b. Teneina [*sic*] of the United States. Cincinnati Quarterly Journal of Science 2: 226–259.

———. 1875c. Teneina [*sic*] of Colorado. Cincinnati Quarterly Journal of Science 2:289–305.

———. 1875d. Tineina from Texas. Canadian Entomologist 7:7–12, 30–35, 51–56, 73–75, 92–95, 105–108.

———. 1876a. Micro-Lepidoptera. Canadian Entomologist 8:18–19.

———. 1876b. Tineina. Canadian Entomologist 8:30–35, 103–106, 135–138, 158–160, 171–173, 217–220.

———. 1877a. Tineina. Canadian Entomologist 9:13–15, 108–110, 123–127, 145–147, 194–196, 206–208.

———. 1877b. Tineina from Texas. Canadian Entomologist 9:22–26, 71–74.

———. 1878a. Tineina. Canadian Entomologist 10:50–54.

———. 1878b. Micro-Lepidoptera. Canadian Entomologist 10:74–78, 109–114, 238–239.

———. 1878c. Descriptions of new Tineina from Texas, and others from more northern localities. Bulletin of the United States Geological and Geographical Survey of the Territories 4:79–106.

———. 1878d. Index to the described Tineina of the United States and Canada. Bulletin of the United States Geological and Geographical Survey of the Territories 4:125–167.

———. 1879. Micro-Lepidoptera. Canadian Entomologist 11:5–10, 72–75, 89–93, 118–119, 125–127.

———. 1880a. Some new species of Tineina from North America. Canadian Entomologist 12: 225–226.

———. 1880b. Descriptions of some new Tineina, with notes on a few old species. Journal of the Cincinnati Society of Natural History 2:179–194.

———. 1881. New species of Tineina. Journal of the Cincinnati Society of Natural History 3: 289–296.

Cook, C. 1948. Early spring collecting in Kentucky. Lepidopterists' News 2:22.

Covell, C. V., Jr. 1968. The University of Louisville Insect Collection. Bulletin of the Entomological Society of America 14:283.

———. 1969. Some pre-flood butterfly records from the Land Between the Lakes. Kentucky Naturalist 23:8–9.

———. 1974. A preliminary checklist of the butterflies of Kentucky. Journal of the Lepidopterists' Society 28:253–256.

———. 1978. A new hindwing aberration of *Catocala micronympha* (Guenée) from Kentucky. Journal of the Lepidopterists' Society 32: 221–223.

———. 1981. The occurrence of *Poanes yehl* (Skinner) (Hesperiidae) in Kentucky. Journal of the Lepidopterists' Society 35:155.

———. 1984. A field guide to the moths of eastern North America. Houghton Mifflin Company, Massachusetts, USA.

———. 1994. Field observations of matings between female *Limenitis archippus* and male *L. arthemis* subspecies (Nymphalidae). Journal of the Lepidopterists' Society 48:199–204.

———. 1999. 1998 field trip review. Kentucky Lepidopterist 25:1–3.

Covell, C. V., Jr., and P. H. Freytag. 1979. The Malaise trap as a means of sampling butterfly populations in Kentucky. Journal of the Lepidopterists' Society 33:153–161.

Covell, C. V., Jr., L. D. Gibson, R. A. Henderson, and M. L. McInnis. 1979. Six new state butterfly records from Kentucky. Journal of the Lepidopterists' Society 33:189–191.

Covell, C. V., Jr., and M. Medley. 1986. *Polychrysia morigera* (Noctuidae) trapped in a southern lady's-slipper labellum in Tennessee. Journal of the Lepidopterists' Society 40:128–129.

Covell, C. V., Jr., and E. H. Metzler. 1992. Two new species of moths (Noctuidae: Acronictinae, Cuculliinae) from midland United States. Journal of the Lepidopterists' Society 46:220–232.

Cox, D. L., and D. A. Potter. 1986. Aerial dispersal behavior of larval bagworms, *Thyridopteryx ephemeraeformis* (Lepidoptera, Psychidae). Canadian Entomologist 118:525–536.

Crumb, S. E. 1956. The larvae of the Phalaenidae. United States Department of Agriculture Technical Bulletin 1135.

Dang, P. T. 1990. Redefinition of Tribe Bactrini Falkovitsch and revised status of genera *Taniva* Heinrich and *Hulda* Heinrich (Tortricidae: Olethreutinae). Journal of the Lepidopterists' Society 44:77–87.

Davis, D. R. 1967. A revision of the moths of the subfamily Prodoxinae (Lepidoptera: Incurvariidae). Bulletin of the United States National Museum 255.

Donahue, J. P. 1967. New records of Acrolophidae (Lepidoptera) from Kentucky. Michigan Entomologist 1:130.

Doubleday, E. 1838. Communications on the Natural History of North America. Entomological Magazine 5:21–34, 199–206, 269–300, 409–417.

Dyar, H. G. "1903" [1904]. A list of North American Lepidoptera and key to the literature of this order of insects. Bulletin of the United States National Museum 52.

———. 1904. Descriptions of new forms of the genus *Illice* Walker. Proceedings of the Entomological Society of Washington 6:197–199.

Eichlin, T. D., and H. B. Cunningham. 1978. The Plusiinae (Lepidoptera: Noctuidae) of America north of Mexico, emphasizing genitalic and larval morphology. United States Department of Agriculture Technical Bulletin 1567.

Ferguson, D. C. 1978. The moths of America north of Mexico. Fascicle 22.2. Noctuoidea: Lymantriidae. E.W. Classey Ltd., and The Wedge Entomological Research Foundation, Washington, District of Columbia, USA.

———. 1985a. The moths of America north of Mexico. Fascicle 18.1. Geometroidea: Geometridae (part): Geometrinae. E.W. Classey Ltd., and The Wedge Entomological Research Foundation, Washington, District of Columbia, USA.

———. 1985b. Contributions toward reclassification of the world genera of the tribe Arctiini, Part 1 - Introduction and a revision of the *Neoarctia-Grammia* group (Lepidoptera: Arctiidae; Arctiinae). Entomography 3:181–275.

Forbes, W. T. M. 1923. The Lepidoptera of New York and neighboring states. Part 1, primitive forms, microlepidoptera, pyraloids, Bombyces. Cornell University Agricultural Experiment Station Memoir 68.

———. 1948. The Lepidoptera of New York and neighboring states. Part 2, Geometridae, Sphingidae, Notodontidae, Lymantriidae. Cornell University Agricultural Experiment Station Memoir 274.

———. 1954. The Lepidoptera of New York and neighboring states. Part 3, Noctuidae. Cornell University Agricultural Experiment Station Memoir 329.

———. 1960. The Lepidoptera of New York and neighboring states. Part 4, Agaristidae through Nymphalidae, including butterflies. Cornell University Agricultural Experiment Station Memoir 371.

Ford, A. 1952. Audubon's butterflies, moths, and other studies. Studio Publications and Thomas Y. Crowell, New York, New York, USA.

Franclemont, J. G. 1952. Four undescribed moths from eastern North America with notes on closely related species (Lepidoptera, Phalaenidae olim Noctuidae). Bulletin of the Brooklyn Entomological Society 47:123–137.

———. 1973. The moths of America north of Mexico. Fascimile 20.1, Mimallonoidea and Bombycoidea. E.W. Classey Ltd., and R.B.D. Publications, Incorporated, London, England.

———. 1985. A new species of *Meganola* Dyar from eastern North America (Lepidoptera: Noctuidae: Nolinae). Proceedings of the Entomological Society of Washington 87:871–874.

———. 1986. New species of tortricid moths from eastern North America (Lepidoptera: Tortricidae). Proceedings of the Entomological Society of Washington 88:56–62.

Freeman, T. N. 1958. The Archipinae of North America (Lepidoptera: Tortricidae). Canadian Entomologist 90, Supplement 7.

Gall, L. F. 1990. Systematics of moths in the Genus *Catocala* (Lepidoptera: Noctuidae). II. Type material at the Museum of Comparative Zoology, Harvard University, with lectotype designations. Psyche 97:121–129.

Garman, H. 1889–1944. Unpublished coded book of insect collection and rearing records. Department of Entomology, University of Kentucky, Lexington, Kentucky, USA.

Ghent, J. H., and C. Dull. 1979. Evaluation of the linden looper and spring cankerworm outbreak on Mammoth Cave National Park, Kentucky, 1979. United States Department of Agriculture Forest Insect and Disease Management Report 79–1–28.

Ghent, J. H., W. Carothers, R. Dorsett, and C. Mitchell. 1980. Evaluation of defoliation on Mammoth Cave National Park, Kentucky – 1980. United States Department of Agriculture Forest Service, Forest Insect and Disease Management report 80–1–19.

Gibson, L. D., and W. E. Miller. 1994. Two new synonymies in Nearctic *Eucosma* (Tortricidae: Olethreutinae). Journal of the Lepidopterists' Society 48:69–71.

Hardwick, D. F. 1996. A monograph to the North American Heliothentinae (Lepidoptera: Noctuidae). Privately published by the author, Ottawa, Canada.

Harvey, D. J. 1991. Higher classification of the Nymphalidae. Pages 255–273 *in* H.F. Nijhout, editor. The development and evolution of butterfly wing patterns. Smithsonian Institution Press, Washington, Disctrict of Columbia, USA.

Hay, C. J. 1974. Survival and mortality of red oak borer larvae on black, scarlet, and northern red oak in eastern Kentucky. Annals of the American Entomological Society 67:981–986.

Heady, S. E., R. G. Lambert, and C. V. Covell, Jr. 1982. Determination of free amino acids in larval insect and gall tissues of the goldenrod *Solidago canadensis* L. Comparative Biochemistry and Physiology 73B:641–644.

Heinrich, C. 1926. Revision of the North American moths of the subfamilies Laspeyresiinae and Olethreutinae. Bulletin of the United States National Museum 132.

———. 1956. American moths of the subfamily Phycitinae. Bulletin of the United States National Museum 207.

Heitzman, J. R. 1974. A new species of *Papilio* from the eastern United States (Papilionidae). Journal of Research on the Lepidoptera 12:1-10.

Henderson, R. A. 1976. *Amblyscirtes belli* (Hesperiidae): A new record for Kentucky. Journal of the Lepidopterists' Society 30:68.

Heppner, J. B. 1985. The sedge moths of North America (Lepidoptera: Glyphipterigidae). Flora and Fauna Publications, Handbook Number 1.

———. 1997. New *Glyphipterix* sedge moths from the southeastern United States (Lepidoptera: Glyphipterigidae). Holarctic Lepidoptera 4:67–69.

Hodges, R. W. 1961. A review of the genus *Walshia* Clemens with descriptions of new species (Lepidoptera: Gelechioidea). Bulletin of the Brooklyn Entomological Society 56:66–80.

———. 1962a. A review of the genus *Periploca* with descriptions of nine new species (Lepidoptera: Gelechioidea). Pan-Pacific Entomologist 38:83–97.

———. 1962b. Revision of the Cosmopterigidae of America north of Mexico. Entomologica Americana 42:1–171.

———. 1974. A new species of *Aroga* (Lepidoptera: Gelechiidae) from North America. Canadian Entomologist 106:987–990.

———. 1978. The moths of America north of Mexico. Fascicle 6.1. Gelechioidea: Cosmopterigidae. E.W. Classey Ltd., and The Wedge

Entomological Research Foundation, Washington, District of Columbia, USA.

———. 1986. The moths of America north of Mexico. Fascicle 7.1. Gelechioidea: Gelechiidae (part). E.W. Classey Ltd., and The Wedge Entomological Research Foundation, Washington, District of Columbia, USA.

Hodges, R. W., T. Dominick, D. R. Davis, D. C. Ferguson, J. G. Franclemont, E. G. Munroe, and J. A. Powell, editors. 1983. Check list of the Lepidoptera of America north of Mexico. E.W. Classey and The Wedge Entomological Research Foundation, London, England.

Holland, W. J. 1903. The moth book. Doubleday, Page and Company, New York, New York, USA.

———. 1931. The butterfly book. Doubleday and Company, Garden City, New York, USA.

Horak, M., and R. L. Brown. 1991. Taxonomy and phylogeny. Pages 23–48 *in* L. P. S. van der Geest and H. H. Evenhuis, editors. World crop pests. Tortricoid pests. Their biology, natural enemies and control. Elsevier, Amsterdam.

Iftner, D. C., J. A. Shuey, and J. V. Calhoun. 1992. Butterflies and skippers of Ohio. Bulletin of the Ohio Biological Survey New Series Volume 9, Number 1.

Kentucky Cooperative Economic Insect Report. 1974. Kentucky Agricultural Experiment Station, Lexington, Kentucky, USA.

———. 1975. Kentucky Agricultural Experiment Station, Lexington, Kentucky, USA.

———. 1976. Kentucky Agricultural Experiment Station, Lexington, Kentucky, USA.

Kentucky Cooperative Plant Pest Report. 1977. Kentucky Agricultural Experiment Station, Lexington, Kentucky, USA.

———. 1978. Kentucky Agricultural Experiment Station, Lexington, Kentucky, USA.

———. 1980. Kentucky Agricultural Experiment Station, Lexington, Kentucky, USA.

[KSNPC] Kentucky State Nature Preserves Commission. 1996. Rare and extirpated plants and animals of Kentucky. Transactions of the Kentucky Academy of Science 57:69–91.

Kimball, C. P. 1965. Arthropods of Florida and neighboring land areas. Volume 1: Lepidoptera of Florida. Division of Plant Industry, Florida Department of Agriculture, Gainesville, FL, USA.

Ladd, H. O. 1952. Letter to the editor. Club Notes, The Moth and Butterfly Club 7:2.

———. 1953. Notes on moths and butterflies. Club Notes, The Moth and Butterfly Club 8:2.

———. 1954. Butterflies of Kentucky and their prevalence. Notes on moths and butterflies. Club Notes, The Moth and Butterfly Club 8:4.

Lafontaine, J. D. 1987. The moths of America north of Mexico. Fascicle 27.2. Noctuoidea: Noctuidae (part). Noctuinae (part - *Euxoa).* The Wedge Entomological Research Foundation, Washington, District of Columbia, USA.

———. 1998. The moths of America north of Mexico. Fascicle 27.3. Noctuoidea: Noctuidae (part). Noctuinae (part - Noctuini). The Wedge Entomological Research Foundation, Washington, District of Columbia, USA.

Lafontaine, J. D., and J. S. Kononenko. 1986. A revision of the genus *Trichosilia* Hampson (Lepidoptera: Noctuidae) with descriptions of four new species. Canadian Entomologist 118: 1079–1113.

Landry, J.-F. 1991. Systematics of nearctic Scythrididae (Lepidoptera: Gelechioidea): phylogeny and classification of supraspecific taxa, with a review of described species. Memoirs of the Entomological Society of Canada 160.

Landry, J.-F., and B. Wright. 1993. Systematics of the nearctic species of metallic-green *Coleophora* (Lepidoptera: Coleophoridae). Canadian Entomologist 125:549–618.

Lauer, M. J., J. J. Krupa, K. E. Haskins, and D. G. Miller, III. 1996. New county records for the West Virginia White (Lepidoptera: Pieridae: *Pieris virginiensis*) in Kentucky. Transactions of the Kentucky Academy of Science 57:128-129.

McCabe, T. L. 1980. A reclassification of the *Polia* complex for North America (Lepidoptera: Noctuidae). New York State Museum Bulletin 432.

McDunnough, J. H. 1937. The *laudabilis* group of *Lacinipolia* (Lepid., Phalaenidae). Canadian Entomologist 69:177–185.

———. 1954. The species of the genus *Hydriomena* occurring in America north of Mexico (Geometridae, Larentiinae). Bulletin of the

American Museum of Natural History 104: 241–358.

Mengel, R. M. 1965. The birds of Kentucky. American Ornithologists' Union Monograph Number 3. The Allen Press, Lawrence, Kansas, USA.

Merritt, J. R. 1948. List of the butterflies of Jefferson County, Kentucky. Annals of Kentucky Natural History 1:27–32.

Metzler, E. H., and J. G. Franclemont. 1991. A review of four species names of *Paectes* from North America (Noctuidae: Euteliinae). Journal of the Lepidopterists' Society 45:34–41.

Miller, S. E., and V. O. Becker. 1989. North American moths described by L. A. G. Bosc D'Antic (Lepidoptera: Noctuidae, Pyralidae). Proceedings of the Entomological Society of Washington 9:22–25.

Miller, S. E., and R. W. Hodges. 1990. Primary types of microlepidoptera in the Museum of Comparative Zoology (with a discussion on V. T. Chambers' work). Bulletin of the Museum of Comparative Zoology 152:45–87.

Miller, W. E. 1979. The genus *Olethreutes*: identity corrections and description of a new species (Lepidoptera: Tortricidae: Olethreutinae). Annals of the American Entomological Society 72: 232–236.

———. 1982. *Grapholita delineana* (Walker), a Eurasian hemp moth, discovered in North America. Annals of the American Entomological Society 75:184–186.

———. 1983a. Genus *Phaneta*: new synonymies and a new species (Lepidoptera: Tortricidae). Annals of the American Entomological Society 76:98–103.

———. 1983b. *Eucosmomorpha albersana* (Hübner), a Palearctic species collected in North America (Tortricidae, Grapholitini). Journal of the Lepidopterists' Society 37:88–89.

———. 1985. Nearctic *Eucosma* (Lepidoptera: Tortricidae): four new species and three new synonymies. Annals of the American Entomological Society 78:240–247.

———. 1987. Guide to the Olethreutine moths of midland North America (Tortricidae). United States Department of Agricultural Handbook Number 660.

Monroe, B. L., Jr. 1952. Butterflies of the Louisville, Kentucky, region. Undergraduate Thesis, University of Louisville, Louisville, Kentucky, USA.

———. 1953. A hybrid *Limenitis*. Lepidopterists' News 17:53.

Munroe, E. G. 1973. The moths of America north of Mexico. Fascicle 13.1C. Pyraloidea, Pyralidae (part): Evergestinae. E. W. Classey Ltd., and R. B. D. Publications, Incorporated, London, England.

———. 1976. The moths of America north of Mexico. Fascicle 13.2a and b. Pyraloidea, Pyralidae (part): Pyraustinae (Pyraustini, part). E. W. Classey Ltd., and The Wedge Entomological Research Foundation, Washington, District of Columbia, USA.

Neary, T. J., and D. L. Dahlman. 1989. Observations of *Chalcoela iphitalis* (Pyralidae) inhabiting *Polistes fuscatus* (Vespidae) nests in Kentucky. Transactions of the Kentucky Academy of Science 50:202–204.

Neunzig, H. H. 1988. A taxonomic study of the Genus *Salebriaria* (Lepidoptera: Pyralidae: Phycitinae) in America north of Mexico. North Carolina Agriculture Research Service Technical Bulletin 287.

———. 1990. The moths of America north of Mexico. Fascicle 15.3. Pyraloidea, Pyralidae (part). The Wedge Entomological Research Foundation, Washington, District of Columbia, USA.

Opler, P. A. 1995. Species richness and trends in western butterflies and moths. Pages 172–174 *in* E. T. LaRoe, G. S. Farris, C. E. Puckett, P. D. Doran, and M. J. Mac, editors. Our living resources: a report to the nation on the distribution, abundance, and health of U.S. plants, animals, and ecosystems. National Biological Service, Washington, District of Columbia, USA.

Opler, P. A., and W. Krizek. 1984. Butterflies east of the Great Plains. Johns Hopkins University Press, Baltimore, Maryland, USA.

Opler, P. A., and V. Malikul. 1992. A field guide to eastern butterflies. Houghton Mifflin Company, Boston, Massachusetts, USA.

Osborn, H. 1937. Fragments of entomological history. Published by the author, Columbus, Ohio, USA.

Packard, A. S. 1876. A monograph of the geometrid moths or Phalaenidae of the United States. Pages 1–607 *in* F.V. Hayden. Report of the Geological Survey of the United States of the Territories, Volume 10.

Pogue, M. G., and D. C. Ferguson. 1998. A revision of the cypress-feeding moths of the genus *Cutina* Walker (Lepidoptera: Noctuidae). Proceedings of the Entomological Society of Washington 100: 331–352.

Pogue, M. G., and T. P. Friedlander. 1987. *Cochylis caulocatax* Razowski (Lepidoptera: Tortricidae: Cochylini): a redescription of the male with new descriptions of the female, larva, and pupa. Journal of the New York Entomological Society 95:320–327.

Poole, R. W. 1989. Lepidopterorum Catalogus. Fascicle 118: Noctuidae, Part 1. E.J. Brill/Flora and Fauna Publishers, Leiden, The Netherlands.

Powell, J. A. 1973. A systematic monograph of new world ethmiid moths (Lepidoptera: Gelechioidea). Smithsonian Contributions to Zoology 120.

———. 1995. Lepidoptera inventories in the continental United States. Pages 168-170 *in* E. T. LaRoe, G. S. Farris, C. E. Puckett, P. D. Doran, and M. J. Mac, editors. Our living resources: a report to the nation on the distribution, abundance, and health of U.S. plants, animals, and ecosystems. National Biological Service, Washington, District of Columbia, USA.

Powell, J. A., and W. E. Miller. 1978. Nearctic pine tip moths of the genus *Rhyacionia*: biosystematic review. United States Forest Service Agricultural Handbook 514.

Pyle, R. M. 1981. The Audubon Society field guide to North American butterflies. Alfred A. Knopf, New York, New York, USA.

Quarterman, E., and R. L. Powell. 1978. Potential ecological/geological natural landmarks on the Interior Low Plateaus. National Park Service, United States Department of the Interior, Washington, District of Columbia, USA.

Rauschenberger, J. L., W. M. Ciesla, and H. L. Lambert. 1966. Forest insect and disease conditions at Mammoth Cave National Park, Kentucky. United States Department of Agriculture Forest Service Report 66–1–23.

Rindge, F. H. 1955. The type material in the J. B. Smith and G. D. Hulst collections of Lepidoptera in the American Museum of Natural History. Bulletin of the American Museum of Natural History 106:91–172.

———. 1971. A revision of the moth Genus *Lytrosis* (Lepidoptera, Geometridae). American Museum Novitates 2474.

———. 1975. A revision of the new world *Bistonini* (Lepidoptera, Geometridae). Bulletin of the American Museum of Natural History 156: 71–155.

Rings, R. W., E. H. Metzler, F. J. Arnold, and D. H. Harris. 1992. The owlet moths of Ohio – Order Lepidoptera, Family Noctuidae. Ohio Biological Survey Bulletin New Series Volume 9, Number 2.

Robinson, G. S. 1986. Fungus moths: a review of the Scardiinae (Lepidoptera: Tineidae). Bulletin of the British Museum (N. H.): Entomology Series 52:37–181.

Robinson, G. S., and E. S. Nielsen. 1983. The microlepidoptera described by Linnaeus and Clerck. Systematic Entomology 8:191–242.

Sargent, T. D. 1976. Legion of night – the underwing moths. University of Massachusetts Press, Amherst, Massachusetts, USA.

Scoble, M. J. 1992. The Lepidoptera: form, function and diversity. Oxford University Press, Oxford, United Kingdom.

Shaffer, J. C. 1968. A revision of the Peoriinae and Anerastiinae (auctorum) of America north of Mexico (Lepidoptera: Pyralidae). United States National Museum Bulletin 280. Smithsonian Institution Press, Washington, District of Columbia, USA.

Shull, E. M. 1987. The butterflies of Indiana. Indiana University Press, Bloomington, Indiana, USA.

Smith, M. E. 1938. A revision of the genus *Apantesis* Walker (Lepidoptera, Arctiidae). Thesis abstract, University of Illinois, Urbana, IL, USA.

Solis, M. A. 1993. A phylogenetic analysis and reclassification of the genera of the *Pococera* complex (Lepidoptera: Pyralidae: Epipaschiinae). Journal of the New York Entomological Society 101:1–83.

———. 1997. Snout moths: unraveling the taxonomic diversity of a speciose group in the neotropics. Pages 231-242 *in* M.L. Reaka-Kudla, D. E. Wilson, and E. O. Wilson, editors. Biodiversity II: understanding and protecting our biological resources. Joseph Henry Press, Washington, District of Columbia, USA.

Wagner, W. H., Jr., and A. H. Showalter. 1976. Ecological notes on *Celastrina ebenina* (Lycaenidae). Journal of the Lepidopterists' Society 30: 310–312.

Wagner, W. H., Jr., and T. L. Mellichamp. 1978. Foodplant, habitat, and range of *Celastrina ebenina* (Lycaenidae). Journal of the Lepidopterists' Society 32:20–36.

Warner, H. H. 1894. Kentucky butterflies. Canadian Entomologist 26:289–291.

———. 1895. A synopsis of the diurnal Lepidoptera of Kentucky. Department of Zoology and Entomology of the State College, Lexington, Kentucky, USA.

Weniger, D. K. 1946. A survey of the Rhopalocera of Jefferson County, Kentucky. Senior Thesis, University of Louisville, Louisville, Kentucky, USA.

Wheat, F. M. "1908" [1909]. The diurnal Lepidoptera of Kentucky with key for their identification. Bulletin of the State University of Kentucky 1.

Wright, D. J., R. L. Brown, and L. D. Gibson. 1997. A new species of *Phaneta* with taxonomic diagnoses and seasonal and geographical data on four related species (Tortricidae). Journal of the Lepidopterists' Society 51:119–127.

Yeargan, K. V. 1988. Ecology of a bolas spider, *Mastophora hutchinsoni*: phenology, hunting tactics, and evidence for aggressive chemical mimicry. Oecologiia (Berlin) 74:524–530.

INDEX TO BUTTERFLY AND MOTH TAXA

Taxa are arranged alphabetically by genus and species as well as species within genera. Taxonomic categories higher than genus are in boldface. Common names are also included for species that have been assigned them. Only the page numbers containing the main text for each taxon have been included.

A

B

C

D

E

F

G

H

I

J

K

L

O

Q

R

S

T

U

V

W

X

Y

Z

INSTRUCTIONS TO AUTHORS

This series is open to papers dealing with all aspects of the natural history of Kentucky. Papers of 30 pages or longer will be considered for this series. Manuscripts should be submitted to the Director, Kentucky State Nature Preserves Commission, 801 Schenkel Lane, Frankfort, Kentucky 40601. Authors should provide phone, fax, and e-mail numbers, and will be notified of receipt of the manuscript. A designated subject-matter editor is responsible for review and acceptance or rejection of the manuscript. Technical correspondence concerning the manuscript may be directed to the subject-matter editor. Turn-around time for manuscripts is usually a minimum of 75 to 90 days after receipt by the Director. Delays in obtaining reviews may prolong this period. If a manuscript is returned for revision, the author should submit a revised manuscript to the subject-matter editor within 6 months. Manuscripts undergoing revision for longer than 6 months may (at the discretion of the subject-matter editor) be considered as new manuscripts and submitted again to review. The author should provide a forwarding address if he moves from the return address shown on the manuscript. Style and format should conform to guidelines adopted by *Ecology*, and recent issues should be consulted. The most recent edition of *Scientific style and format: the CBE manual for authors, editors, and publishers* is also recommended for details on style. Manuscripts that do not conform in most respects to the instructions below will be returned to authors for correction before consideration.

Preparation of Manuscript

Write with precision, clarity, and economy; use the active voice and first person whenever appropriate. Every sentence should convey the exact truth as simply as possible. Type all material (text, tables, quotations, literature cited, etc.) double-spaced in 12-point font, and on only one side of standard size paper approximately 21 by 28 cm (8.5 x 11 inches). Leave one inch margins on all sides, and do not right-hand justify. Send triplicate hard copies of the manuscript, and a file in Microsoft Word format on a 3.5 inch diskette or via e-mail. All hard copies must include illustrations. Retain at least one complete copy of the manuscript in your files. DO NOT submit original India-ink figures due to the risk of loss or damage in the mail. Titles of articles should be informative and short, usually not more than 10 words. Do not underline any words. Scientific names should be italicized. Avoid footnotes; most footnote material can be incorporated in the text to the benefit of readers, editors, and printers. Literature Cited must be double-spaced. Before submitting the manuscript, check each citation in the text against the Literature Cited for agreement. The listing should conform in sequencing and punctuation to that in recent issues of *Ecology*. Titles of journals should be spelled out completely. Authors are to use the International System of Units (SI) for measurements. English measurements may be placed in parentheses. Tables must each be typed and double-spaced on separate sheets. Number tables consecutively and give each a short title. Do not rule tables unless essential to avoid ambiguity. Never repeat the same material in figures and tables; when either is equally clear a figure is preferable. Avoid unnecessary material, especially tables and long lists of little interest save to the author; this material will be edited rigorously. Photographs should be black and white printed for contrasts.

Proofing and reprints

Two sets of proofs and the edited manuscript will be returned to the author. Correct (do not revise) both sets of proofs identically, retain one, and return one promptly, with the edited manuscript, to the subject-matter editor. DO NOT MAKE ANY MARKS on the edited manuscript. If proof is not returned promptly to the subject-matter editor, the manuscript may be published without the benefit of proof corrections. Ten final copies will be provided to the senior author. Further copies may be purchased.

Kentucky State Nature Preserves Commission
Scientific and Technical Series

(All prices include tax and shipping)

No. 1. Ferns and Fern Allies of Kentucky, by Ray Cranfill. 284 pages. 1980. Price = $6.27.

No. 2. The Fishes of the Dix River, Kentucky, by Branley A. Branson and Donald L. Batch. 25 pages. 1981. Price = $3.36.

No. 3. Fishes of the Lower Green River Drainage, Kentucky, by Michael E. Retzer, Brooks M. Burr, and Melvin L. Warren, Jr. 48 pages. 1983. Price = $3.89.

No. 4. A Distributional Atlas of Kentucky Fishes, by Brooks M. Burr and Melvin L. Warren, Jr. 398 pages. 1986. Price = $12.84. **Out of Print.**

No. 5. Aquatic and Wetland Plants of Kentucky, by Earnest O. Beal and John W. Thieret. 314 pages. 1986 reprint. Price = $26.00.

No. 6. The Butterflies and Moths (Lepidoptera) of Kentucky: *An Annotated Checklist*, by Charles van Orden Covell, Jr. 220 pages. 1999. Price = $18.02

Numbers in this series are shipped prepaid. Please make checks payable to "Kentucky State Nature Preserves Commission." Order from:

Kentucky State Nature Preserves Commission
801 Schenkel Lane
Frankfort, Kentucky 40601–1403

KENTUCKY NATURE PRESERVES COMMISSION